The
Plant Viruses

Volume 1
POLYHEDRAL VIRIONS WITH
TRIPARTITE GENOMES

THE VIRUSES

Series Editors
HEINZ FRAENKEL-CONRAT, *University of California*
Berkeley, California

ROBERT R. WAGNER, *University of Virginia School of Medicine*
Charlottesville, Virginia

THE VIRUSES: Catalogue, Characterization, and Classification
Heinz Fraenkel-Conrat

THE ADENOVIRUSES
Edited by Harold S. Ginsberg

THE HERPESVIRUSES
Volumes 1–3 • Edited by Bernard Roizman
Volume 4 • Edited by Bernard Roizman and Carlos Lopez

THE PARVOVIRUSES
Edited by Kenneth I. Berns

THE PLANT VIRUSES
Volume 1 • Edited by R. I. B. Francki

THE REOVIRIDAE
Edited by Wolfgang K. Joklik

The Plant Viruses

Volume 1
POLYHEDRAL VIRIONS WITH TRIPARTITE GENOMES

Edited by
R. I. B. FRANCKI
Waite Agricultural Research Institute
The University of Adelaide
Glen Osmond, Australia

PLENUM PRESS • NEW YORK AND LONDON

Library of Congress Cataloging in Publication Data

Main entry under title:

Polyhedral virions with tripartite genomes.

(The Plant viruses: v. 1) (The Viruses)
Includes bibliographies and index.
1. Viruses, RNA. 2. Plant viruses. I. Francki, R. I. B., 1930–   . II. Title: Tripartite
genomes. III. Series. IV. Series: Viruses.
QR357.P58   1985 vol. 1   576′.6483 s   85-3528
ISBN 0-306-41958-0   [576′.6483]

©1985 Plenum Press, New York
A Division of Plenum Publishing Corporation
233 Spring Street, New York, N.Y. 10013

Printed in the United States of America

Contributors

Patrick Argos, Department of Biological Sciences, Purdue University, West Lafayette, Indiana 47907

J. A. Cooper, Plant Research Institute, Burnley Gardens, Burnley, Victoria 3121, Australia

R. I. B. Francki, Department of Plant Pathology, Waite Agricultural Research Institute, The University of Adelaide, Glen Osmond, South Australia 5064, Australia

R. G. Garrett, Plant Research Institute, Burnley Gardens, Burnley, Victoria 3121, Australia

R. I. Hamilton, Agriculture Canada Research Station, Vancouver, British Columbia V6T 1X2, Canada

R. IIull, John Inncs Institutc, Norwich NR1 7UH, England

John E. Johnson, Department of Biological Sciences, Purdue University, West Lafayette, Indiana 47907

Giovanni P. Martelli, Dipartimento di Patologia vegetale, Università degli Studi di Bari and Centro di studio del CNR sui virus e le virosi delle colture mediterranee, Bari, Italy

A. J. Maule, John Innes Institute, Norwich NR4 7UH, England

Marcello Russo, Dipartimento di Patologia vegetale, Università degli Studi di Bari and Centro di studio del CNR sui virus e le virosi delle colture mediterranee, Bari, Italy

E. P. Rybicki, Microbiology Department, University of Cape Town, Rondebosch 7700, South Africa

P. R. Smith, Plant Research Institute, Burnley Gardens, Burnley, Victoria 3121, Australia

Robert H. Symons, Department of Biochemistry, The University of Adelaide, Adelaide, South Australia 5000, Australia

L. van Vloten-Doting, Department of Biochemistry, University of Leiden, Leiden, The Netherlands

M. B. von Wechmar, Microbiology Department, University of Cape Town, Rondebosch 7700, South Africa

Series Editors' Note

We recognize that the term *plant viruses* has no definitive taxonomic meaning. Many viruses that cause plant disease also replicate in their insect vectors. Many viruses of plants clearly belong to taxonomically defined animal virus families, namely the Reoviridae and Rhabdoviridae, and these have been or will be dealt with in volumes covering these families (e.g., *The Reoviridae*, edited by W. K. Joklik, Plenum Press, New York, 1983). Following the present volume, books are in preparation covering the isometric plant viruses with monopartite genomes (edited by R. Koenig), with isometric particles and bipartite genomes (edited by B. D. Harrison and A. F. Murant), the helical rod-shaped and tubular plant viruses (edited by M. H. V. van Regenmortel and H. Fraenkel-Conrat), the filamentous plant viruses (edited by R. G. Milne), the DNA-containing plant viruses (edited by R. Hull), and the viroids (edited by T. O. Diener).

Heinz Fraenkel-Conrat
Robert R. Wagner

Preface

It has been known for a long time that the majority of plant viruses contain RNA and in the past decade and a half it has been realized that many have genomes consisting of three molecules of single-stranded RNA with positive polarity. Among these are viruses belonging to four groups recognized by the International Committee for Virus Taxonomy: the Bromovirus and Cucumovirus groups whose genomes are encapsidated in small icosahedral particles or the Ilarvirus and alfalfa mosaic virus groups with spheroidal or bacilliform particles. In addition to their tripartite genomes, these viruses share a number of other properties and it has been proposed that they should perhaps be grouped in a single virus family for which the name Tricornaviridae has been suggested, the *tri* indicating the tripartite nature of the genome, the *co* emphasizing the cooperation of the three genome parts required to initiate infection, and the *rna* indicating that the genome is composed of RNA.

Viruses of this "family" are less uniform in their biological properties. A number of them are widespread, causing very destructive plant diseases. Viruses such as those of cucumber mosaic and alfalfa mosaic have very extensive host ranges and are responsible for serious crop losses in many parts of the world. Others such as prunus necrotic ringspot or prune dwarf viruses are more restricted in their host ranges but nevertheless infect important perennial hosts such as stone fruits and reduce productivity considerably. Some of the viruses are transmitted by vectors such as insects or nematodes whereas others use seeds or pollen as vehicles of spread.

In recent years much has been learnt about the Bromoviruses, Cucumoviruses, Ilarviruses, and alfalfa mosaic virus, especially concerning their molecular biology. Many of the data indicate some striking similarities among the viruses whereas others point to some significant differences. In this volume an effort has been made to present a comparative account of the existing knowledge concerning the viruses and hence members of all four groups are discussed in each chapter prepared by an acknowledged authority on the subject. In the first chapter, all the viruses to be discussed later in more detail are listed and their salient features summarized; their current classification is outlined and possible alternatives discussed. The next two chapters present the currently available knowledge of the viral capsid and genome structures, respectively. Chapter 4 concerns the different approaches to studying multiplication of the viruses and summarizes the results of these studies. Genetic studies on the viruses are covered extensively in Chapter 5, and Chapter 6 deals with the effects of infection on their hosts, with special emphasis on the cytopathology. The serology and immunochemistry of the viruses are discussed in Chapter 7. The last two chapters deal with methods by which the viruses spread, their ecology and epidemiology, and the steps that can be taken to protect crops from them. It is hoped that this volume will be of help to all virologists, whether students, teachers, practitioners, or researchers, who wish to acquaint themselves with a general current picture of plant viruses with tripartite genomes and polyhedral particles.

R. I. B. Francki

Adelaide

Contents

Chapter 3

Viral Genome Structure

Robert H. Symons

Chapter 4

Virus Multiplication

R. Hull and A. J. Maule

Chapter 5

Virus Genetics

L. van Vloten-Doting

Chapter 6

Virus–Host Relationships: Symptomatological and Ultrastructural Aspects

Giovanni P. Martelli and Marcello Russo

Chapter 7

Serology and Immunochemistry

E. P. Rybicki and M. B. von Wechmar

Chapter 8

Virus Transmission

R. I. Hamilton

Chapter 9

Virus Epidemiology and Control

R. G. Garrett, J. A. Cooper, and P. R. Smith

CHAPTER 1

The Viruses and Their Taxonomy

R. I. B. FRANCKI

I. INTRODUCTION

In the past decade and a half the International Committee on Taxonomy of Viruses (ICTV) has evolved a virus taxonomy which is accepted and used by the majority of virologists (Wildy, 1971; Fenner, 1976; Matthews, 1979, 1982). This is in sharp contrast to the previous efforts of numerous individual workers who endeavored to establish suitable methods of nomenclature and classification (reviewed by Francki, 1981, and Matthews, 1983) and most of which now appear rather naive and unacceptable by contemporary notions. Currently, the majority of viruses infecting vertebrates are classified into families and genera on the recommendation of the Vertebrate Virus Subcommittee of the ICTV. This approach has been followed for the classification of bacterial and invertebrate viruses by the respective ICTV subcommittees. Contrary to this, the Plant Virus Subcommittee (PVS) has divided viruses into loosely defined groups with the exception of those which show affinities to viruses of vertebrates and arthropods; they are included in the families Reoviridae and Rhabdoviridae. In the case of the Reoviridae, members infecting plants have been subdivided into two genera, *Phytoreovirus* and *Fijivirus* (Matthews, 1982). Thus, at present, plant viruses are divided among two families and 24 groups (Matthews, 1982).

The different approaches to classification taken by plant virologists and their colleagues dealing with viruses infecting other organisms have historical reasons. Plant virologists seem to be far more cautious about the establishment of taxa, probably as a reaction to the overenthusiastic

R. I. B. FRANCKI • Department of Plant Pathology, Waite Agricultural Research Institute, The University of Adelaide, Glen Osmond, South Australia 5064, Australia.

efforts of earlier generations of their colleagues. The early systems of virus taxonomy were based on inadequate data, almost entirely biological, which yielded elaborate systems of nomenclature and classification like that devised by Holmes (1948). This system drew special attention and much criticism, some bordering on ridicule (Andrewes, 1955). It is only relatively recently that there is sufficient information about the intrinsic properties of viruses on which to start building a meaningful taxonomy. However, the plant virologist's fear of creating yet another taxonomic monster has not been completely alleviated.

In its initial pronouncement, the ICTV divided 99 reasonably well-characterized viruses into 16 groups (Wildy, 1971; Harrison et al., 1971). This was done with an essentially Adansonian approach (Gibbs, 1969) in that all available data regarding 49 defined virus characters were used without weighting (Harrison et al., 1971). Although some doubt has since been cast about the taxonomic significance of some of the characters used (Francki, 1983), the general validity of the classification seems vindicated in that all the 16 groups still remain unchanged (Matthews, 1982). Among these groups are three of the four discussed in this volume; the brome mosaic (Bromovirus), cucumber mosaic (Cucumovirus), and alfalfa mosaic virus (AMV) groups. Subsequently, 10 more plant virus groups have been established by the ICTV, including viruses of the family Rhabdoviridae and the two genera of the Reoviridae (Fenner, 1976; Matthews, 1979, 1982). The approach to delimiting these groups was not strictly Adansonian but rather one of recognizing viruses which obviously did not fit into the existing groups but which formed clusters by having a number of common properties. Among these clusters, tobacco streak (TSV) and related viruses are the fourth group considered in this volume for which the name Ilarvirus was approved by the ICTV (Matthews, 1979). The name was derived from the properties of the viruses which have isometric, labile particles and usually produce ringspot symptoms (Fenner, 1976).

At present, the ICTV recognizes the Bromoviruses, Cucumoviruses, Ilarviruses, and AMV as four distinct virus groups (Matthews, 1982). Nevertheless, all four groups share a number of important features (Jaspars, 1974; van Vloten-Doting, 1976; van Vloten-Doting and Jaspars, 1977; Lane, 1979a,b, 1981; Jaspars and Bos, 1980; Kaper and Waterworth, 1981; Fulton, 1981, 1983). All the viruses have genomes consisting of three unique single-stranded (ss), positive-sense RNA molecules which are encapsidated in small polyhedral or bacilliform particles which sets them apart from all other plant virus taxa (Matthews, 1982, 1983). Because of this, it has been proposed by van Vloten-Doting et al. (1981) that the four groups should perhaps be considered as belonging to one family for which the name Tricornaviridae was suggested. This family name summarizes the salient features of its members; trico, referring to the tripartite genome, and rna to the type of genomic nucleic acid (RNA). The ending viridae is that approved by the ICTV for virus family names

TABLE I. Physical Properties of the Bromoviruses

Property	BMV[a]	BBMV[b]	CCMV[c]	MYFV[d]	CYBV[e]
Virus particles					
Diameter (nm)	~26	~26	~25	~25	~26
Structure (icosahedral)	T = 3	T = 3	T = 3	?	?
$s_{20,w}$	87	85	88	88	85
Coat protein					
Subunit M_r ($\times 10^3$)	20.9	21.0	19.4	20.3	20.8
Viral RNA					
M_r RNA 1 ($\times 10^6$)	1.1	1.1	1.2	1.2	1.1
RNA 2 ($\times 10^6$)	1.0	1.0	1.1	1.1	1.0
RNA 3 ($\times 10^6$)	0.72	0.90	0.81	1.0	0.78
RNA 4 ($\times 10^6$)	0.30	0.36	0.28	0.3	0.31

[a] Lane (1977).
[b] Gibbs (1972).
[c] Bancroft (1971).
[d] Hollings and Horváth (1981).
[e] Dale et al. (1984).

(Matthews, 1982). Although these suggestions have not been considered by the ICTV, for convenience, I will refer to the four virus groups here as a "family", the "Tricornaviridae."

The only other group of plant viruses with tripartite genomes are the Hordeiviruses (Matthews, 1982). However, particles of these viruses have helical symmetry and differ in some other respects; they are not discussed in this volume.

II. MEMBERS OF THE TRICORNAVIRIDAE

A. Bromovirus Group

The ICTV lists three members of the group, brome mosaic virus (BMV), broad bean mottle virus (BBMV), and cowpea chlorotic mottle virus (CCMV), and one possible member, melandrium yellow fleck virus (MYFV). Both MYFV (Hollings and Horváth, 1981) and the more recently described cassia yellow blotch virus (CYBV) (Dale et al., 1984) appear to have the characteristics of Bromoviruses (Table I).

BMV, BBMV, and CCMV have all been studied extensively (Lane, 1979b, 1981) and shown to be very distantly related serologically, the coat protein subunits more so than the intact virus particles (Rybicki and von Wechmar, 1981). Neither MYFV nor CYBV appear to be serologically related to BMV, BBMV, and CCMV (Hollings and Horváth, 1981; Dale et al., 1984) although very distant relationships cannot be ruled out on the presently available data. MYFV and CYBV have not been compared serologically.

The physical properties of all the Bromoviruses are remarkably similar (Table I). Except for MYFV which appears to be stable at neutral pH (Hollings and Horváth, 1981), the other Bromoviruses are all in a stable, compact form at pH between 3 and 6 but swell when the pH is raised above 6.5, becoming sensitive to ribonucleases, trypsin, and chymotrypsin (Pfeiffer and Hirth, 1975).

The Bromoviruses can be defined as a group of viruses whose particles are polyhedral, about 26 nm in diameter (Fig. 1A), built from 180 identical polypeptides of M_r about 20×10^3 with $T = 3$ symmetry. The particles encapsidate four species of ssRNA of M_r ranging from 1.1 to 0.3×10^6 packaged in such a way that all particles sediment at about 86 S. These properties distinguish the Bromoviruses from all other groups established by the ICTV (Matthews, 1982).

The biological properties of Bromoviruses are less easily defined but they can all be readily transmitted by mechanical inoculation. Most have been transmitted by beetles and there are reports that some may be transmitted by nematodes, aphids, or through seed (Lane, 1981; Rybicki and von Wechmar, 1982; Murant et al., 1974).

B. Cucumovirus Group

Three members of the group have been approved by the ICTV, cucumber mosaic virus (CMV), tomato aspermy virus (TAV), and peanut stunt virus (PSV). In addition, cowpea ringspot virus (CPRSV) has been included as a possible member (Matthews, 1982) whose affinities with the approved members are not yet clear (Phatak et al., 1976). CMV, TAV, and PSV are all distantly related serologically (Devergne and Cardin, 1975; Devergne et al., 1981) but do not appear to have any significant base sequence homologies between their respective RNAs (Gonda and Symons, 1978; Piazzolla et al., 1979).

The membership of the Cucumovirus group is somewhat complicated by a number of isolates which have been described as distinct viruses in the past but which are clearly strains of the three approved Cucumoviruses. For example, viruses such as those of spinach blight and tomato fern leaf are synonymous with CMV (Francki et al., 1979); those of chrysanthemum mosaic or chrysanthemum aspermy with TAV (Hollings and Stone, 1971); and robinia mosaic (= black locust true mosaic) and clover blotch viruses have been shown to be strains of PSV (Richter et al., 1979). The recent listing of black locust true mosaic and soybean stunt viruses as distinct Cucumoviruses (Boswell and Gibbs, 1983) seems unjustified in view of previously published data on these viruses (Richter et al., 1979; Hanada and Tochihara, 1982).

The physical properties of the Cucumoviruses are very similar (Table II) which is not unexpected as they are serologically interrelated. Moreover, they are very similar to those of the Bromoviruses (Table I) except

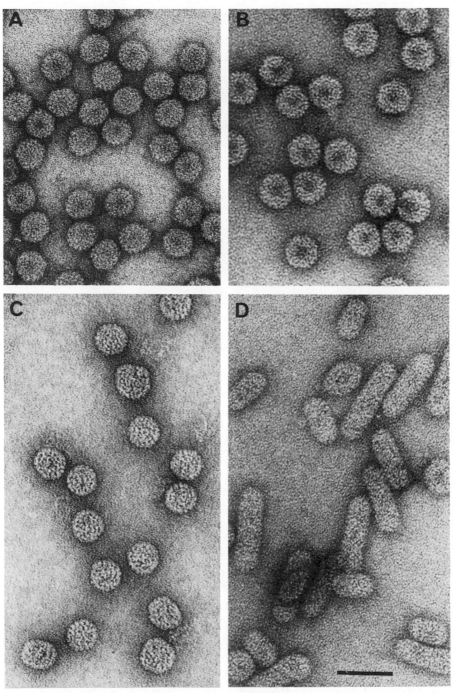

FIGURE 1. Particles of A, broad bean mottle virus (BBMV); B, cucumber mosaic virus (CMV); C, prune dwarf virus (PDV); D, alfalfa mosaic virus (AMV). Preparations of all viruses were negatively stained with uranyl acetate; bar = 50 nm. Micrographs A, B, and D kindly supplied by Dr. T. Hatta and C, by Mr. I. Roberts.

TABLE II. Physical Properties of the Cucumoviruses

Property	CMV[a]	TAV[b]	PSV[c]	CPRSV[d]
Virus particles				
Diameter (nm)	29	29	30	26
Structure (icosahedral)	$T=3$	$T=3$	$T=3$	?
$S_{20,w}$	99	99	98	93
Coat protein				
Subunit M_r ($\times 10^3$)	26.2	26.1	~26	24.0
Viral RNA				
M_r RNA 1 ($\times 10^6$)	1.3	1.3	1.3	1.2
RNA 2 ($\times 10^6$)	1.0	1.1	1.0	?
RNA 3 ($\times 10^6$)	0.75	0.90	0.75	0.86
RNA 4 ($\times 10^6$)	0.35	0.43	0.35	0.38

[a] Habili and Francki (1974) and Francki et al. (1979).
[b] Habili and Francki (1974) and Hollings and Stone (1971).
[c] Mink (1972), Lot and Kaper (1976), and Kaper and Waterworth (1981).
[d] Phatak et al. (1976).

that they seem to have slightly larger particles (Fig. 1A and B) constructed from larger coat protein subunits encapsidating marginally larger RNA molecules (Tables I and II). However, Cucumovirus particles differ from those of the Bromoviruses in that they are not prone to swelling at neutral pH.

The Cucumoviruses can be defined as a small group of serologically interrelated viruses each of which includes numerous biological variants. They have particles about 29 nm in diameter (Fig. 1B) built from 180 identical polypeptides of M_r about 26×10^3 with $T=3$ symmetry. The particles encapsidate four species of ssRNA of M_r ranging from 1.3 to 0.3 $\times 10^6$ packaged so that all particles sediment at about 99 S. These properties distinguish the Cucumoviruses from all other groups established by the ICTV (Matthews, 1982).

All Cucumoviruses are readily transmitted by mechanical inoculation to a very wide range of plant hosts belonging to numerous families and are seed-borne in some of them. With the exception of CPRSV which has not been studied extensively, all the other Cucumoviruses are transmitted by aphids in a non-persistent manner. Their wide host ranges and efficient vector transmission account for their world-wide importance as plant disease agents (Kaper and Waterworth, 1981).

C. Ilarvirus Group

The ICTV presently recognizes 11 distinct Ilarviruses (Matthews, 1982) but citrus crinkly leaf virus (CiCLV) and asparagus virus II (AVII) probably also qualify for inclusion in the group (Francki et al., 1985). Recently, a new virus isolated from hydrangea has been isolated and

TABLE III. Viruses of the Ilarvirus Group, Their
Synonyms and Suggested Subgroupings[a]

Subgroup 1[b]
  Tobacco streak virus (TSV)
    Black raspberry latent virus
    Asparagus stunt virus
Subgroup 2[b]
  Tulare apple mosaic virus (TAMV)
  Citrus leaf rugose virus (CiLRV)
  Citrus variegation virus (CVV)
  Citrus crinkly leaf virus (CiCLV)
  Elm mottle virus (EMoV)
  Asparagus virus II (AVII)
Subgroup 3[b]
  Prunus necrotic ringspot virus (PNRSV)
    Cherry rugose mosaic virus
    Rose mosaic virus (some isolates)
    Hop virus C
  Apple mosaic virus (ApMV)
    Danish plum line pattern virus
    Rose mosaic virus (some isolates)
    Hop virus A
Subgroup 4[b]
  Prune dwarf virus (PDV)
Subgroup 5[b]
  American plum line pattern virus (APLPV)
Subgroup 6[c]
  Spinach latent virus (SPLV)
Subgroup 7[c]
  Lilac ring mottle virus (LRMV)
Subgroup 8[c]
  Hydrangea mosaic virus (HydMV)

[a] Modified after Francki *et al.* (1985).
[b] Subgroups based on serological relationships (indented names refer to synonyms or virus strains).
[c] Viruses which at present have not been shown to have serological affinities with any other group members.

named hydrangea mosaic virus (HydMV); its properties indicate that it should belong to the Ilarvirus group (Thomas *et al.*, 1983). Some authors have suggested ways in which the group may be subdivided (Uyeda and Mink, 1983; Fulton, 1983; Francki *et al.*, 1985) and this has also been done here in Table III. In addition, two other viruses of uncertain affinities, those causing pelargonium zonate spot and raspberry bushy dwarf have been considered as possible members of the Ilarvirus group (Francki *et al.*, 1985).

The physical properties of those Ilarviruses which have been investigated in any detail are summarized in Table IV. The data in some cases are probably not very precise because of the inherent difficulties of working with Ilarviruses. Most of the viruses are unstable and many occur in

TABLE IV. Physical Properties of Ilarviruses[a]

Property	TSV	TAMV	CiLRV	CVV	EMoV	AVII	PNRSV	ApMV	PDV	APLPV	SPLV	LRMV	HydMV
Virus particles													
Diameter (nm)	35 30 27	31 30 28[b]	32 31 26 25	33 31 28	25–30	32 28 26	~23	29 25[b]	23 20[b]	33 31 28 26	~27[b]	~27[b]	30 × 32–38 30 × 30 28 × 30
$s_{20,w}$	113 98 90	114 108 93	105 98 89 79	110 93 83 79	101 88 83	104 95 90	95 90 72	117 ?[c] 88	113 99 85 81 75	126 114 100 95	108 98 87	98 83	105 97 86
Coat protein													
Subunit M_r (× 10^3)	30	19	26	26	25	—[d]	25	25	24	—	28	28	26
Viral RNA													
M_r RNA 1 (× 10^6)	1.0	1.0	1.1	1.0	1.3	—	1.3	1.2	1.3	—	1.3	1.2	1.3
RNA 2 (× 10^6)	0.98	0.92	1.0	1.0	1.2	—	0.89	1.0	0.95	—	1.2	1.1	1.1
RNA 3 (× 10^6)	0.78	0.74	0.7	0.7	0.82	—	0.69	0.69	0.76	—	0.91	0.90	0.83
RNA 4 (× 10^6)	?		0.3	0.3	0.39 + 0.30	—	0.31	0.31	—	—	0.35 + 0.27	0.37	0.36 + 0.27

[a] Data from Fulton (1983), Thomas et al. (1983), and Francki et al. (1985).
[b] Virus preparations also contain some bacilliform particles.
[c] Indicates that component is present but was not quantified.
[d] Indicates that the property has not been determined.

TABLE V. Physical Properties of Alfalfa Mosaic Virus[a]

Virus particles[b]	B	M	T_b	T_a
Dimensions (nm)	56×18	43×18	35×18	30×18
$s_{20,w}$	94	82	73	66
RNA M_r ($\times 10^6$)	1.04	0.73	0.62	0.27
	(RNA 1)	(RNA 2)	(RNA 3)	(RNA 4, 2 copies)

[a] Data from Jaspars and Bos (1980).
[b] All particles built from one species of polypeptide of M_r about 24×10^3.

low concentrations in plants, making it difficult to obtain sufficient highly purified virus for critical studies (Fulton, 1981).

The Ilarviruses can be defined as a fairly large group of viruses which have quasi-isometric particles varying in size but around 30 nm in diameter (Fig. 1C) which sediment as three or four components between 72 and 126 S. Some of the viruses also form a small proportion of bacilliform particles. Their capsomeric structure is unknown but all particles are built from one species of polypeptide of M_r between 25 and 28×10^3 which encapsidate four species of ssRNA. These properties distinguish the Ilarviruses from all other virus groups established by the ICTV (Matthews, 1982).

Ilarviruses can all be transmitted by mechanical inoculation but individual members differ considerably in their biological properties. Most will infect a wide range of plants and many of the hosts are woody plants. Seed transmission is common among the Ilarviruses and some are transmitted through pollen to pollinated plants. Ilarviruses are not usually spread by vectors although individual viruses have been reported to be transmitted by a mite, a nematode, and thrips (Fulton, 1983; see also Chapter 8). However, transmission by these vectors needs to be confirmed. In the field, the spread of most Ilarviruses can be accounted for by seed or pollen transmission or by agricultural practices involving vegetative propagation (see Chapters 8 and 9).

D. Alfalfa Mosaic Virus Group

Although the ICTV has designated this as a group, alfalfa mosaic virus (AMV) is the sole member (Matthews, 1982). The physical properties of AMV are summarized in Table V. The salient features of the virus are its typical particles (Fig. 1D) built from a single species of protein subunit of M_r about 24×10^3. The particles are of at least four different types sedimenting as distinct boundaries between 68 and 94 S. The smallest particle is spheroidal (about 30×18 nm) and the remainder are a series of bacilliform shapes of the same width but differing lengths (Fig. 1D). These particles encapsidate four species of ssRNA of M_r ranging from about 0.3 to 1.0×10^6 (Jaspars and Bos, 1980).

AMV is an important plant disease agent with a world wide distribution (Hull, 1969; Jaspars and Bos, 1980). The virus occurs as numerous biological variants which, however, have very similar physical properties. AMV is readily transmitted by mechanical inoculation and also nonpersistently by aphids to a very wide range of plants in numerous dicotyledonous families. It is also transmitted through seed in some species (Hull, 1969; Jaspars and Bos, 1980). In many respects, especially its physical properties, AMV resembles the Ilarviruses (van Vloten-Doting and Jaspars, 1977; van Vloten-Doting et al., 1981) but biologically, it is more like the Cucumoviruses in having a wide range of similar hosts and aphid vectors.

III. TAXONOMIC CONSIDERATIONS

A. Features Common to All Tricornaviridae

In addition to being tripartite, the genomes of all the Tricornaviridae have a number of other common features. At present the genomes of relatively few of the viruses have been studied extensively but a considerable amount of information has accumulated about at least the type members of each of the four groups (Table VI; see also Chapter 3) and it is probably reasonable to assume that other members are essentially similar.

The genome sizes of all the Tricornaviridae are similar as is the arrangement of the genes on their three RNA segments; RNAs 1 and 2 are monocistronic mRNAs whereas RNA 3 is dicistronic. Near the 5' end of each RNA 3 there is always a cistron which is readily translated in vitro to yield a nonstructural protein (3a protein) about 300 amino acids long and a silent cistron near the 3' end. Furthermore, sequences of the 3'-terminal region of RNA 3 including the coat protein cistron, are present on a subgenomic RNA 4; a monocistronic mRNA which is readily translated in vitro.

The 5' ends of BMV, CMV, and AMV are all capped with 7-methylguanosine and it will be interesting to see if members of the Ilarvirus group also share this property. The 3' ends of all Tricornaviridae studied so far are not polyadenylated.

Tricornaviridae of all the four groups have similar coat protein subunit size (Tables, I, II, IV, and V) and their capsids readily dissociate in strong salt solutions, indicating that they are stabilized by electrovalent bonds (see Chapter 2). All these properties shared by the Bromovirus, Cucumovirus, Ilarvirus, and AMV groups, set them well apart from all other plant viruses and argue for their classification in one taxon which could be given the status of a family.

TABLE VI. Comparison of the Properties of the Type Members of the Four Groups of Tricornaviridae[a]

	BMV (Bromovirus)	CMV (Cucumovirus)	TSV (Ilarvirus)	AMV
Viral RNA (number of nucleotides)[b]				
RNA 1	3234	~3410	~2940	3644
RNA 2	2865	3035	~2770	2593
RNA 3	2111–2117	2193	2205	2037
RNA 4	876	1027	~850	881
5'-terminus	m^7 Gppp	m^7 Gppp	?	m^7 Gppp
3'-termini	Aminoacylated by tyrosine	Aminoacylated by tyrosine	Not aminoacylated	Not aminoacylated
Viral coat protein (number of amino acids)	189	236	237	222
3a protein (number of amino acids)	303	333	290	301
Virus particles				
Morphology	Icosahedral (T=3)	Icosahedral (T=3)	Spheroidal (some bacilliform)	Bacilliform (some spheroidal)
Dimensions (nm)	~26	~29	~35	~56 × 18
			~30	~43 × 18
			~27	~35 × 18
				~30 × 18
$s_{20,w}$	85	99	113	94
			98	82
			90	73
				66
Biological properties				
Host range	Narrow	Very wide	Wide	Very wide
Vectors	Beetles	Aphids	Seed and pollen	Aphids

[a] For further details on data summarized here, see the appropriate chapters in this volume.
[b] Data from complete nucleotide sequences of RNAs except where indicated as approximate (~) in which case numbers are estimates from relative electrophoretic mobility in polyacrylamide gels.

B. Comparison of the Bromoviruses with the Cucumoviruses

The physical properties of the Bromoviruses and Cucumoviruses are remarkably similar (Tables I, II, and VI). It can be said that the Bromoviruses are simply slightly smaller editions of Cucumoviruses (Figs. 1A and B), with shorter RNAs and smaller particles built from smaller protein subunits. Furthermore, there is marked similarity in RNA sequences of the three Bromoviruses (BMV, BBMV, and CCMV) and CMV, especially at their 3'-terminal 190-nucleotide residues. These regions are tRNA-like in that they can be aminoacylated, all with tyrosine (Kohl and Hall, 1974; Ahlquist et al., 1981a; Joshi et al., 1983).

The only really significant difference in the particles of the Bromoviruses and Cucumoviruses is that the Bromovirus particles swell in response to changes of pH and ionic environment (Pfeiffer *et al.*, 1976) whereas those of the Cucumoviruses do not (see also Chapter 2).

The most significant differences between the viruses of the two groups are in their biological properties. Whereas the Bromoviruses infect relatively few plant species, the Cucumoviruses have very wide host ranges. Also, the Bromoviruses are transmitted primarily by beetles whereas the Cucumoviruses are vectored nonpersistently by aphids (Lane, 1974, 1981; Kaper and Waterworth, 1981; see also Chaper 8). The cytopathic effects induced by viruses of the two groups also differ significantly (Francki *et al.*, 1985; see also Chapter 6).

When the Bromoviruses and Cucumoviruses were initially described as two distinct taxonomic groups (Harrison *et al.*, 1971), much more was known about their biological than their molecular properties. Hence, in spite of the Adansonian approach to their classification, differences between them appeared to outweigh their similarities. However, with our present knowledge of the viruses, persuasive arguments can be advanced in support of their inclusion in a single taxon.

C. Comparison of AMV with the Ilarviruses

Like the Bromovirus and Cucumovirus groups, AMV and the Ilarviruses were assigned to separate groups primarily by virtue of their biological properties (Harrison *et al.*, 1971; Shepherd *et al.*, 1976). At the time of their establishment, the molecular characterization of AMV was fairly well advanced but much less was known about the physical properties of any of the Ilarviruses.

Biologically, AMV resembles the Cucumoviruses in having a wide host range and aphid vectors whereas Ilarviruses are not transmissible by arthropods but usually spread through seeds and by pollen (see Chapter 8). It has also been known for some time that the particles of AMV differ from those of the Ilarviruses in their gross morphology (Figs. 1C and D). With this knowledge, it is not surprising that AMV and the Ilarviruses were not classified in the same group.

Data which have emerged since the mid 1970s reveal a number of similarities between AMV and TSV, the most extensively studied of the Ilarviruses. Both viruses require coat protein for initiating infection, which sets them apart from the Bromoviruses and Cucumoviruses. Moreover, although serologically unrelated and having quite distinct peptide maps, the coat proteins of AMV and TSV are interchangeable in activating each other's genomes. Such activation can also be achieved with protein of other Ilarviruses but not with protein from viruses of other groups including the Bromoviruses and Cucumoviruses (Bol *et al.*, 1971; van Vloten-Doting, 1975; Gonsalves and Garnsey, 1975; Gonsalves and Fulton, 1977; van Volten-Doting and Jaspars, 1977). There is some evidence which suggests that the coat proteins bind to the viral RNAs at the 3'-

terminal regions to be recognized by the viral replicase (Houwing and Jaspars, 1978). Whereas there is virtually no similarity in the primary structure of the AMV RNA 3'-terminal regions with those of TSV RNA, they do have somewhat similar hairpin structures flanked by the tetranucleotide AUGC (Barker *et al.*, 1983; Cornelissen *et al.*, 1984). These structures may be responsible for the mutual recognition of their heterologous coat proteins (Koper-Zwarthoff and Bol, 1980). Unlike the Bromoviruses and Cucumoviruses, AMV and Ilarvirus RNAs do not accept amino acids (van Vloten-Doting and Jaspars, 1977).

Although the molecular structure of AMV capsids is reasonably well understood, little is known about that of the Ilarviruses (see Chapter 2). The particles look quite different in their gross morphology (Figs. 1C and D) but there are indications that these differences may reflect only minor changes in their coat protein subunit interactions. For example, AMV-like bacilliform particles have been seen in preparations of some Ilarviruses (Basit and Francki, 1970; Halk and Fulton, 1978) and an AMV spontaneous mutant has been isolated which forms Ilarvirus-like particles (Roosien and van Vloten-Doting, 1983). It would seem that much of the comparative molecular biological data on AMV and the Ilarviruses which has emerged in the past decade can be used to argue that AMV could be included in the Ilarvirus group.

D. Significance of Genome Primary Structure to Taxonomy

In recent years considerable effort has gone into the determination of the entire nucleotide sequences of the genomic RNAs of several Tricornaviruses (Table VI, see also Chapter 3). The primary structure of all four encapsidated BMV and AMV RNAs are already known whereas RNAs 2, 3, and 4 of CMV and RNA 3 of TSV have also been determined.

Knowledge of the complete sequences of RNAs 3 of BMV, CMV, TSV and AMV (Ahlquist *et al.*, 1981b; Gould and Symons, 1982; Barket *et al.*, 1983; Cornelissen *et al.*, 1984) has allowed detailed computer-aided comparisons of the various RNA regions (Murthy, 1983; Savithri and Murthy, 1983; A. J. Gibbs, personal communication). It was revealed that whereas there is little if any homology between the coat protein cistrons of any of the four viral RNAs, some interesting homologies among the 3a protein cistrons have been found. The homology between the RNAs 3 of BMV and CMV were shown to be significantly greater than those betwen either BMV or CMV and AMV or TSV. Similarly the homology between AMV and TSV was very much greater than that between AMV or TSV and BMV or CMV (A. J. Gibbs, personal communication). However, no serological relationships could be detected between the *in vitro* synthesized 3a proteins of BMV, CMV, AMV, and TSV using an antiserum raised against *in vitro* synthesized AMV 3a protein (van Tol and van Vloten-Doting, 1981). The similarities of the noncoding 3'-terminal regions of BMV and

CMV on the one hand and those of AMV and TSV on the other, have already been mentioned.

Computer analysis of the amino acid sequences of nonstructural proteins encoded by BMV, CMV, and AMV RNAs 1 and 2 revealed that they all have significant homologies and that there are also homologies with the proteins encoded by RNAs of other plant and animal viruses such as those of tobacco mosaic and Sindbis (Haseloff *et al.*, 1984; Cornelissen and Bol, 1984; Rezaian *et al.*, 1984). However, the relationship of CMV to BMV RNAs 2 was closer than that of CMV to AMV or any of these to the RNA of tobacco mosaic virus (Rezaian *et al.*, 1984). As more data concerning the sequences of these and other Tricornaviruses become available, valuable information relevant to taxonomy should emerge. However, even the existing data go some way to support the view that the Tricornaviridae are a related group of viruses and that the Bromoviruses are more closely related to the Cucumoviruses than to either AMV or the Ilarviruses and that AMV is more closely related to the Ilarviruses than to either the Bromoviruses or the Cucumoviruses.

IV. CONCLUSIONS

From the preceding discussion it seems clear that the Tricornaviridae include viruses that are very similar in many respects and hence constitute a coherent cluster which can be looked upon as a virus family. It is also clear that within the Tricornaviridae, in molecular terms, there are close affinities between the Bromoviruses and the Cucumoviruses on the one hand and AMV and the Ilarviruses on the other. However, on the basis of their biological properties, the Cucumoviruses and AMV are similar but differ from both the Bromoviruses and the Ilarviruses. Thus, there is merit in retaining the current classification approved by the ICTV. The argument that AMV should be included in the Ilarvirus group (van Vloten-Doting *et al.*, 1981) will hinge largely on whether the biological properties of viruses are considered as important taxonomic criteria. Views on this may vary considerably but applied biologists including many plant pathologists, may well support the view that the biological properties should be considered as important. On the other hand, others may argue that the biological properties of viruses are unstable and hence should not loom high in taxonomic considerations. Nevertheless, it seems that for the present at least, the current ICTV classification satisfies the needs of most virologists and any changes could cause unnecessary confusion.

The reluctance of plant virologists to recognize taxa other than groups has been discussed at length elsewhere (Francki, 1983; Matthews, 1983). It has been argued by many, especially plant pathologists, that other taxa are at present unnecessary. In the case of the viruses discussed in this volume, it can be argued that the four groups approved by the ICTV correspond to genera. The inclusion of these four groups or genera within a family also seems justified. If these were approved by the ICTV together with the suggested family name, this volume could have been entitled simply, the Tricornaviridae.

ACKNOWLEDGMENTS.   I thank Dr. L. van Vloten-Doting for critically reviewing the manuscript; Drs. B. J. C. Cornelissen, J. Haseloff, and M. A. Rezaian and their colleagues for access to manuscripts prior to publication; and Drs. T. Hatta and I. Roberts for electron micrographs. Research in my laboratory was supported by generous grants from the Australian Research Grants Scheme.

REFERENCES

Ahlquist, P., Dasgupta, R., and Kaesberg, P., 1981a, Near identity of 3' RNA secondary structure in Bromoviruses and cucumber mosaic virus, Cell **23**:183.

Ahlquist, P., Luckow, V., and Kaesberg, P., 1981b, Complete nucleotide sequence of brome mosaic virus RNA 3, J. Mol. Biol. **153**:23.

Andrewes, C. H., 1955, The classification of viruses, J. Gen. Microbiol. **12**:358.

Bancroft, J. B., 1971, Cowpea chlorotic mottle virus, CMI/AAB Descriptions of Plant Viruses No. 49.

Barker, R. F., Jarvis, N. P., Thompson, D. V., Loesch-Fries, L. S., and Hall, T. C., 1983, Complete nucleotide sequence of alfalfa mosaic virus RNA 3, Nucleic Acids Res. **11**:2881.

Basit, A. A., and Francki, R. I. B., 1970, Some properties of rose mosaic virus from South Australia, Aust. J. Biol. Sci. **23**:1197.

Bol, J. F., van Vloten-Doting, L., and Jaspars, E. M. J., 1971, A functional equivalence of top component a RNA and coat protein in the initiation of infection by alfalfa mosaic virus, Virology **46**:73.

Boswell, K. F., and Gibbs, A. J., 1983, Viruses of Legumes 1983: Descriptions and Keys from VIDE, Canberra Publishing and Printing Co., Canberra.

Cornelissen, B. J. C., and Bol, J. F., 1985, Homology between the proteins encoded by tobacco mosaic virus and two Tricornaviridae, Plant Mol. Biol. (in press).

Cornelissen, B. J. C., Janssen, H., Zuidema, D., and Bol, J. F., 1984, Complete nucleotide sequence of tobacco streak virus RNA 3, Nucleic Acids Res. **12**:2427.

Dale, J. L., Gibbs, A. J., and Behncken, G. M., 1984, Cassia yellow blotch virus: A new Bromovirus from an Australian native legume, Cassia pleurocarpa, J. Gen. Virol. **65**:281.

Devergne, J. C., and Cardin, L. 1975, Relations serologiques entre cucumovirus (CMV, TAV, PSV), Ann. Phytopathol. **7**:255.

Devergne, J. C., Cardin, L., Burckard, J., and van Regenmortel, M. H. V., 1981, Comparison of direct and indirect ELISA for detecting antigenically related Cucumoviruses, J. Virol. Methods **3**:193.

Fenner, F., 1976, Classification and nomenclature of viruses: Second report of the International Committee on Taxonomy of Viruses, Intervirology **7**:1.

Francki, R. I. B, 1981, Plant virus taxonomy, in: Handbook of Plant Virus Infections: Comparative Diagnosis (E. Kurstak, ed.), p. 3, Elsevier/North-Holland, Amsterdam.

Francki, R. I. B., 1983, Current problems in plant virus taxonomy, in: A Critical Appraisal of Viral Taxonomy (R. E. F. Matthews, ed.), p. 63, CRC Press, Boca Raton, Florida.

Francki, R. I. B., Mossop, D. W., and Hatta, T., 1979, Cucumber mosaic virus, CMI/AAB Descriptions of Plant Viruses No. 213.

Francki, R. I. B., Milne, R. G., and Hatta, T., 1985, An Atlas of Plant Viruses Volume II, CRC Press, Boca Raton, Florida (in press).

Fulton, R. W., 1981, Ilarviruses, in: *Handbook of Plant Virus Infections: Comparative Diagnois* (E. Kurstak, ed.), p. 377, Elsevier/North-Holland, Amsterdam.

Fulton, R. W., 1983, Ilarvirus group, *CMI/AAB Descriptions of Plant Viruses* No. 275.

Gibbs, A., 1969, Plant virus classification, *Adv. Virus Res.* **14:**263.

Gibbs, A. J., 1972, Broad bean mottle virus, *CMI/AAB Descriptions of Plant Viruses* No. 101.

Gonda, T. J., and Symons, R. H., 1978, The use of hybridization analysis with complementary DNA to determine the RNA sequence homology between strains of plant viruses: Its application to several strains of Cucumoviruses, *Virology* **88:**361.

Gonsalves, D., and Fulton, R. W., 1977, Activation of prunus necrotic ringspot virus and rose mosaic virus by RNA 4 components of some Ilarviruses, *Virology* **81:**398.

Gonsalves, D., and Garnsey, S. M., 1975, Infectivity of heterologous RNA–protein mixtures from alfalfa mosaic, citrus leaf rugose, citrus variegation and tobacco streak viruses, *Virology* **67:**319.

Gould, A. R., and Symons, R. H., 1982, Cucumber mosaic virus RNA 3: Determination of the nucleotide sequence provides the amino acid sequences of protein 3A and viral coat protein, *Eur. J. Biochem.* **126:**217.

Habili, N., and Francki, R. I. B., 1974, Comparative studies on tomato aspermy and cucumber mosaic viruses. I. Physical and chemical properties, *Virology* **57:**392.

Halk, E. L., and Fulton, R. W., 1978, Stabilization and particle morphology of prune dwarf virus, *Virology* **91:**434.

Hanada, K., and Tochihara, H., 1982, Some properties of an isolate of the soybean stunt strain of cucumber mosaic virus, *Phytopathology* **72:**761.

Harrison, B. D., Finch, J. T., Gibbs, A. J., Hollings, M., Shepherd, R. J., Valenta, V., and Wetter, C., 1971, Sixteen groups of plant viruses. *Virology* **45:**356.

Haseloff, J., Goelet, P., Zimmern, D., Ahlquist, P., Dasgupta, R., and Kaesberg, P., 1984, Striking similarities in amino acid sequences among nonstructural proteins encoded by RNA viruses that have dissimilar genomic organization, *Proc. Natl. Acad. Sci. USA* **81:**4358.

Hollings, M., and Horváth, J., 1981, Melandrium yellow fleck virus, *CMI/AAB Descriptions of Plant Viruses* No. 236.

Hollings, M., and Stone, O. M., 1971, Tomato aspermy virus, *CMI/AAB Descriptions of Plant Viruses* No. 79.

Holmes, F. O., 1948, Order Virales: The filtrable viruses, in: *Bergey's Manual of Determinative Bacteriology*, 6th ed. (R. S. Breed, E. G. D. Murray, and A. P. Hitchens, eds.), p. 126, Ballière, Tindall & Cox, London.

Houwing, C. J., and Jaspars, E. M. J., 1978, Coat protein binds to the 3′-terminal part of RNA 4 of alfalfa mosaic virus, *Biochemistry* **17:**2927.

Hull, R., 1969, Alfalfa mosaic virus, *Adv. Virus Res.* **15:**365.

Jaspars, E. M. J., 1974, Plant viruses with a multipartite genome, *Adv. Virus Res.* **19:**37.

Jaspars, E. M. J., and Bos, L., 1980, Alfalfa mosaic virus, *CMI/AAB Descriptions of Plant Viruses* No. 229.

Joshi, R. L., Joshi, S., Chapeville, F., and Haenni, A.-L., 1983, tRNA-like structures of plant viral RNAs: Conformational requirements for adenylation and aminoacylation, *EMBO J.* **2:**1123.

Kaper, J. M., and Waterworth, H. E., 1981, Cucumoviruses, in: *Handbook of Plant Virus Infections: Comparative Diagnosis* (E. Kurstak, ed.), p. 257, Elsevier/North-Holland, Amsterdam.

Kohl, R. J., and Hall, T. C., 1974, Aminoacylation of RNA from several viruses: Amino acid specificity and differential activity of plant, yeast and bacterial synthetases, *J. Gen. Virol.* **25:**257.

Koper-Zwarthoff, E. C., and Bol, J. F., 1980, Nucleotide sequence of the putative recognition site for coat protein in the RNAs of alfalfa mosaic virus and tobacco streak virus, *Nucleic Acids Res.* **8:**3307.

Lane, L. C., 1974, The Bromoviruses, *Adv. Virus Res.* **19:**151.

Lane, L. C., 1977, Brome mosaic virus, *CMI/AAB Descriptions of Plant Viruses* No. *180.*

Lane, L. C., 1979a, The nucleic acids of multipartite, defective, and satellite plant viruses, in: *Nucleic Acids in Plants* (T. C. Hall and J. W. Davies, eds.), Volume 2, p. 65, CRC Press, Boca Raton, Florida.

Lane, L. C., 1979b, Bromovirus Group, *CMI/AAB Descriptions of Plant Viruses* No. *215.*

Lane, L. C., 1981, Bromoviruses, in: *Handbook of Plant Virus Infections: Comparative Diagnosis* (E. Kurstak, ed.), p. 333, Elsevier/North-Holland, Amsterdam.

Lot, H., and Kaper, J. M., 1976, Physical and chemical differentiation of three strains of cucumber mosaic virus and peanut stunt virus, *Virology* **74:**209.

Matthews, R. E. F., 1979, Classification and nomenclature of viruses: Third report of the International Committee on Taxonomy of Viruses, *Intervirology* **12:**129.

Matthews, R. E. F., 1982, Classification and nomenclature of viruses: Fourth report of the International Committee on Taxonomy of Viruses, *Intervirology* **17:**1.

Matthews, R. E. F., 1983, *A Critical Appraisal of Viral Taxonomy,* CRC Press, Boca Raton, Florida.

Mink, G. I., 1972, Peanut stunt virus, *CMI/AAB Descriptions of Plant Viruses* No. *92.*

Murant, A. F., Abu-Salih, H. S., and Goold, R. A., 1974, Viruses from broad bean in the Sudan, *Rep. Scott. Host Res. Inst. for 1973,* p. 67.

Murthy, M. R. N., 1983, Comparison of the nucleotide sequences of cucumber mosaic virus and brome mosaic virus, *J. Mol. Biol.* **168:**469.

Pfeiffer, P., and Hirth, L., 1975, The effect of conformational changes in brome mosaic virus upon its sensitivity to trypsin, chymotrypsin and ribonuclease, *FEBS Lett.* **56:**144.

Pfeiffer, P., Herzog, M., and Hirth, L., 1976, Stabilization of brome mosaic virus, *Philos. Trans. R. Soc. London Ser. B.* **276:**99.

Phatak, H. C., Diaz-Ruiz, J. R., and Hull, R., 1976, Cowpea ringspot virus: A seed transmitted Cucumovirus, *Phytopathol. Z.* **87:**132.

Piazzolla, P., Diaz-Ruitz, J. R., and Kaper, J. M., 1979, Nucleic acid homologies of eighteen cucumber mosaic virus isolates determined by competition hybridization, *J. Gen. Virol.* **45:**361.

Rezaian, M. A., Williams, R. H. V., Gordon, K. H. J., Gould, A. R., and Symons, R. H., 1984, Nucleotide sequence of cucumber mosaic virus RNA 2 reveals a translation product significantly homologous to corresponding proteins of other viruses, *Eur. J. Biochem.* **143:**277.

Richter, J., Proll, E., and Musil, M., 1979, Serological relationships between robinia mosaic, clover blotch and peanut stunt viruses, *Acta Virol.* **23:**489.

Roosien, J., and van Vloten-Doting, L., 1983, A mutant of alfalfa mosaic virus with an unusual structure, *Virology* **126:**155.

Rybicki, E. P., and von Wechmar, M. B., 1981, The serology of the Bromoviruses. I. Serological interrelationships of the Bromoviruses, *Virology* **109:**391.

Rybicki, E. P., and von Wechmar, M. B., 1982, Characterization of an aphid-transmitted virus disease of small grains: Isolation and partial characterisation of three viruses, *Phytopathol. Z.* **103:**306.

Savithri, H. S., and Murthy, M. R. N., 1983, Evolutionary relationship of alfalfa mosaic virus with cucumber mosaic virus and brome mosaic virus, *J. Biosci.* **5:**183.

Shepherd, R. J., Francki, R. I. B., Hirth, L., Hollings, M., Inouye, T., MacLeod, R., Purcifull, D. E., Sinha, R. C., Tremaine, J. H., Valenta, V., and Wetter, C., 1976. New groups of plant viruses approved by the International Committe on Taxonomy of Viruses, September 1975, *Intervirology* **6:**181.

Thomas, B. J., Barton, R. J., and Tuszynski, A., 1983, Hydrangea mosaic virus, a new Ilarvirus from *Hydrangea macrophylla* (Saxifragaceae), *Ann. Appl. Biol.* **103:**261.

Uyeda, I., and Mink, G.I., 1983, Relationships among some Ilarviruses: Proposed revision of subgroup A, *Phytopathology* **73:**47.

van Tol, R. G. L., and van Vloten-Doting, L., 1981, Lack of serological relationship between the 35K nonstructural protein of alfalfa mosaic virus and the corresponding proteins of three other plant viruses with the tripartite genome, *Virology* **109:**444.

van Vloten-Doting, L., 1975, Coat protein is required for infectivity of tobacco streak virus: Biological equivalence of the coat proteins of tobacco streak and alfalfa mosaic viruses, *Virology* **65**:215.

van Vloten-Doting, L., 1976, Similarities and differences between viruses with a tripartite genome. *Ann. Microbiol. (Inst. Pasteur)* **127A**:119.

van Vloten-Doting, L., and Jaspars, E. M. J., 1977, Plant covirus systems: Three component systems, in: *Comprehensive Virology* (H. Fraenkel-Conrat and R. R. Wagner, eds.), Volume 11, p. 1, Plenum Press, New York.

van Vloten-Doting, L., Francki, R. I. B., Fulton, R. W., Kaper, J. M., and Lane, L. C., 1981, Tricornaviridae—A proposed family of plant viruses with tripartite, single-stranded RNA genomes, *Intervirology* **15**:198.

Wildy, P., 1971, Classification and nomenclature of viruses: First report of the International Committee on Nomenclature of Viruses, *Monogr. Virol.* **5**:1.

CHAPTER 2

Virus Particle Stability and Structure

JOHN E. JOHNSON AND PATRICK ARGOS

I. INTRODUCTION

The Bromo-, Cucumo-, Ilar-, and alfalfa mosaic viruses are grouped together in this volume as they contain four distinct sizes of RNA molecules which can be separated by polyacrylamide gel electrophoresis. Abbreviations used for the various viruses discussed in the text are listed
below.* This chapter will concentrate on the virion particle structures
and is divided into two major sections. The first reviews current knowledge of the atomic resolution structures for the capsid proteins of three
spherical, icosahedral plant viruses, namely, TBSV, SBMV, and STNV.
The second section presents detailed empirical results and related references for each of the four viral classes. Table VI of Chapter 1 lists the
gross structural parameters for the Bromo, Cucumo, alfalfa mosaaic, and
Ilarvirus groups.

*The abbreviations used for viruses referred to in this chapter are: SBMV (southern bean
mosaic virus), STNV (satellite tobacco necrosis virus), TBSV (tomato bushy stunt virus),
BMV (brome mosaic virus), CCMV (cowpea chlorotic mottle virus), BBMV (broad bean mottle virus), CMV (cucumber mosaic virus), TAV (tomato aspermy virus), PSV (peanut stunt
virus), AMV (alfalfa mosaic virus), CLRV (citrus leaf rugose virus, CVV (citrus variegation
virus), TSV (tobacco streak virus).

JOHN E. JOHNSON AND PATRICK ARGOS • Department of Biological Sciences, Purdue
University, West Lafayette, Indiana 47907.

II. ATOMIC RESOLUTION STRUCTURES OF SPHERICAL VIRUSES

A. Icosahedral Symmetry of Protein Capsids

The protein shells of simple, spherical plant viruses generally encapsidate RNA, utilize (numerous) copies of an identical polypeptide in the 20- to 40-kilodalton range with the number of subunits being discrete multiples of 60, and display icosahedral symmetry. Crick and Watson (1956, 1957) were the first to propose the regular polyhedral design for viral capsids and emphasized the nucleic acid economy in multiple use of an identical protein in small, infectious particles. Platonic solids and their surfaces would allow similar environments for the quaternary association of the protein monomers. Of the available cubic point group symmetries, viruses have chosen the symmetry of the icosahedron, probably because the largest number of subunits, and therefore the least molecular weight requiring minimal coding information, can symmetrically surround the RNA. Furthermore, the quasi-equivalent requirements (*vide infra*) are least stringent for the highly symmetric icosahedron, demanding only a fivefold pentamer to become a quasi-equivalent sixfold hexamer as juxtaposed, for example, to an octahedron necessitating the transformation of a tetramer to a quasi-equivalent hexamer.

An icosahedral surface lattice consists of 20 equilateral triangles with 5 arranged about two pentameric vertices at the zenith and nadir of the polyhedron and 10 banding about the center (Fig. 1). The polygon possesses 10 threefold, 6 fivefold, and 15 twofold symmetry axes, all of which pass through its center. The icosahedral asymmetric unit, from which the entire polyhedron can be generated by application of the symmetry axes, is one-third of a triangular face, bounded by a threefold and two adjacent fivefold symmetry axes. Other asymmetric units can be chosen (e.g., the volume bounded by a fivefold and two adjacent threefold symmetry axes). Consecutive application on the asymmetric unit of adjacent three-, two-, and fivefold axes on a great circle and a twofold axis orthogonal to the first applied twofold will yield the entire icosahedron (3 × 2 × 5 × 2 = 60).

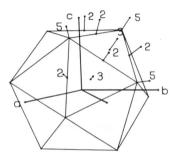

FIGURE 1. Arrangement of equilateral triangles on the surface lattice of an icosahedron. The polyhedron displays five-, two-, and threefold symmetry, most readily observed by the repeating triangular faces. The axes *a*, *b*, and *c* are shown for spatial reference; *a* is perpendicular to *b* and *c* and the angle between *b* and *c* is 120°.

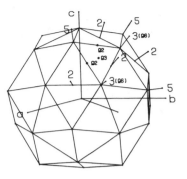

FIGURE 2. Arrangement of symmetry elements on an icosahedron for a T = 3 particle utilizing 180 identical protein subunits. The quasi-symmetry axes are indicated with a "Q" and the angle between *b* and *c* is 120°. The icosahedral symmetry axes are in the same orientation as in Fig. 1 which illustrates the T = 1 surface lattice.

If three protein subunits are laid on each of the triangular faces, displaying threefold symmetry, 60 protein monomers would be utilized. However, the majority of viral capsids are composed of 180, 240, 420, . . . subunits corresponding to 3, 4, 7, . . . multiples of 60. These facts prompted Caspar and Klug (1962) to introduce the concepts of quasi-equivalence and triangulation number (T = 1, 3, 4, 7, . . .). Subunits arranged about the icosahedral symmetry axes exhibit the same quaternary contracts (T = 1 capsids) while higher T values display a breakdown of some of the equivalences and call for quasi-symmetry between some subunits. For example, in T = 3 viral capsids (Fig. 2), three protein chains (referred to as A, B, and C), related by a quasi-threefold and located on one-third of the triangular surface, would allow a total of 180 subunits where each quasi-threefold cluster would obey the exact icosahedral symmetry. The quasi-threefolds which are parallel to neighboring icosahedral two-folds in turn imply the existence of quasi-twofolds and sixfolds (Fig. 3). Caspar and Klug (1962) have demonstrated, through the construction of hollow figures made with scissor cuts in a hexagonal net, that only

FIGURE 3. Diagrammatic representation of the spatial distribution of the subunits (A, B, and C) in T = 3 virus protein capsids. The open threadlike structures indicate the ordered β_A-strand and intertwining β-annulus structures observed only for the C subunits. The icosahedral symmetry axes are indicated by the appropriate symbols while the quasi-threefold relates protein monomers A, B, and C. Figure 2 illustrates the orientation of the icosahedral (2, 3, 5) and quasi-symmetry axes (Q2, Q3, Q6). The quasi-sixfold is exemplified by the B_2C relationship and the quasi-twofold by AB_5. The contact surfaces as observed in SBMV (Rossmann *et al.*, 1983a) are indicated by various patterns which, if identical, refer to similar subunit contacts.

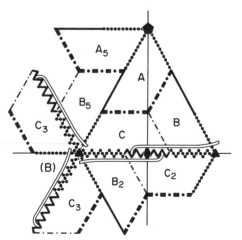

certain triangulation numbers (e.g., T = 2 not permitted) allow equivalent and quasi-equivalent symmetry, itself required by use of identical protein subunits. Of course, the larger triangulations permit an increasing diameter of the encapsidating shell, given subunits of similar size. The elastic deformation or distortion in the local protein structure, necessary to effect quasi-symmetry as postulated by Caspar and Klug (1962), would involve about 0.5-Å movements in chemically and spatially equivalent atoms and no more than 5° in bond direction alterations. Since Caspar and Klug (1962) considered only two-dimensional nets, the local translational distortions in quasi-equivalent related units could be as large as 15 Å, given the 40-Å thickness of the protein shells, Some spherical viruses utilize more than one type of protein in the coat; nonetheless, the icosahedral asymmetric unit will consist of all the protein types and repeat itself according to the polyhedral symmetry.

B. Structures of TBSV, SBMV, and STNV

1. Introduction

The tertiary architecture of three spherical plant virus capsids have been determined to better than 3.5-Å resolution; they include TBSV (Harrison et al., 1978), SBMV (Abad-Zapatero et al., 1980), and STNV (Liljas et al., 1982). The former two have T = 3 capsids and the latter, T = 1. The structure of a T = 1 assembled aggregate of AMV capsid protein has also been elucidated at 4.5-Å resolution and is discussed in a later section (Fukuyama et al., 1983). The capsids of each of these viruses are formed from multiple use of a single polypeptide chain.

The structural determinations were the result of an X-ray crystallographic analysis which involves crystallization of the viruses, collection of X-ray diffraction intensities from native and heavy atom-substituted crystals, solution of the X-ray phases, and calculation of the particle electron density through Fourier combination of the structure factor amplitudes and their relative phases. The density maps allow tracing of the capsid protein peptide backbone and placement of the residue side chains (Blundell and Johnson, 1976). Disappointingly, the interior RNA structure could not be elucidated, apparently due either to dissimilar orientation of the possibly symmetric nucleic acid in the particle crystalline arrays or to the lack of the RNA icosahedral symmetry as expressed by the coat protein whose surface controls crystal packing. However, the tertiary architecture of the capsid protomers has allowed several structural suggestions for protein–RNA interaction

2. Topology of Capsid Protein and Possible Protein–RNA Interactions

Coat proteins of all three viruses possess an N-terminal region that is likely to interact with the RNA. In STNV the first 10 residues are disordered and not observed in the electron density and the next 15 amino acids exist in an α-helical conformation with one side hydrophobic and in contact with two other helices contributed by threefold related subunits. The other helical side is largely polar and basic and is exposed within the RNA interior of the particle. Though the RNA structure cannot be visualized, supposedly the basic side chains prefer interaction with the nucleic acid phosphates. In TBSV and SBMV coat proteins, the N-terminal 40 amino acids, which are composed of many basic residues for RNA association, appear disordered though the most N-terminally ordered amino acid is within the RNA-containing particle interior. Rossmann et al. (1983c) have suggested a further mode of RNA binding, arising from basic residues on the backside of the folded protein monomer which faces the capsid interior. A model allows the placement of several Lys and Arg side chains in the proximity of nuleic acid phosphates arranged in the double-stranded A-RNA conformation (Arnott et al., 1976).

The next 25 residues of TBSV and SBMV coat proteins form a β-annulus structure followed by a $β_A$-strand configuration (Figures 4A–C). The annulus is a coiled-coil of intertwining turn structures contributed by three adjacent C subunits related by an icosahedral threefold. This observation has led Harrison (1980) to propose an assembly mechanism involving the initial formation of a T = 1, C-subunit open scaffold to which A and B subunits are subsequently added. The $β_A$-strand is an individual part of the three C-subunit folds. In the A and B monomers this structure is disordered and in STNV it is absent.

In all three viruses an eight-stranded antiparallel β-barrel follows. It consists of two back-to-back four-stranded β-sheets and traces along the so-called "jelly roll" topological pattern (Richardson, 1981) where pairs of consecutive antiparallel strands are wound into a helix (Fig. 4). This barrel collection of secondary structural elements, consisting of about 200 residues, has been named the S (or surface) domain as it provides the protective shell for the nucleic acid. Though the S domain structural topology is similar for all three viruses, the amino acid homology for spatially equivalent $C_α$ positions (Rossmann et al., 1983b) is not recognizable in SBMV and STNV. The major structural differences in the barrels are insertions and deletions in the turn regions connecting the β-strands. In SBMV, the most notable insertions are the $α_C$, $α_D$, and $α_E$ helices (Fig. 4A). Quite strikingly, the short, tight-turn spans which face the icosahedral fivefold axes show little or no insertions, apparently demanded by the wedge-shaped barrel whose pointed edge must cluster about the fivefold.

SBMV goes on to form a five-residue "vestigial" hinge region at its C-terminus while TBSV possesses an actual hinge that connects to the C-terminal P (or protruding) domain which is another, yet smaller, β-barrel structure. The P domain consists of about 110 amino acids, provides strong contacts for the TBSV icosahedral dimer (Golden and Harrison, 1982), and is a likely result of gene duplication (Argos et al., 1980).

SBMV

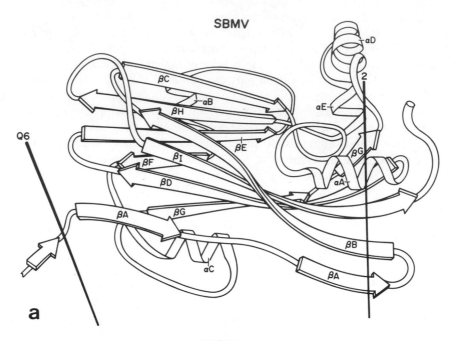

a

TBSV

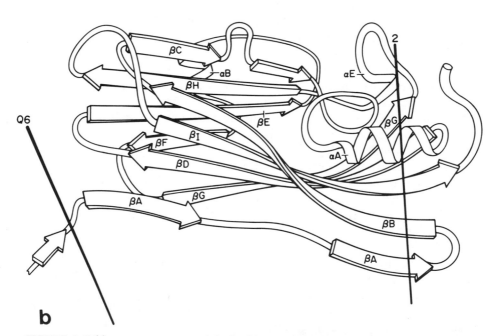

b

FIGURE 4. Ribbon representation of the backbone folding of (A) SBMV, (B), TBSV, and (C) STNV. The C$_\alpha$ peptide tracings are in comparable orientations.

STNV

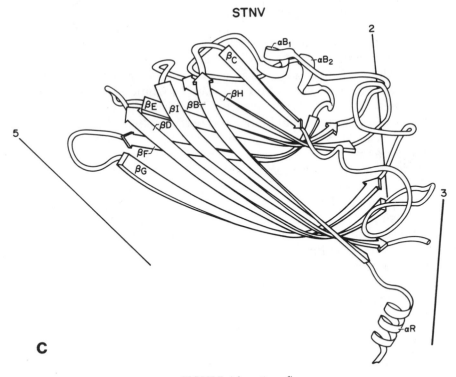

C

FIGURE 4 *(continued)*

STNV does not contain the P domain. Of further interest is the protruding turn region inserted within the β_C conformation of TBSV (Fig. 4B); it has been called the "doorstopper," propping the P domain into its dimeric association. The importance of the P domains and their change in structural orientation during the TBSV swelling process has been discussed by Robinson and Harrison (1982).

3. Subunit Contacts

Rossman *et al.* (1983a; Rossmann, 1984) have performed a detailed analysis of the subunit associations in SBMV. The monomeric contacts along the icosahedral fivefold (AA₅, Fig. 3) and threefold (CC₃) axes are respectively similar as are those about the quasi-threefolds (AB, BC, and CA). Contacts about the quasi-sixfold (B₅C and CB₂) differ considerably with B₅C similar to those of AA₅. The quasi-twofold related monomers (AB₅) also differ from those of the icosahedrally related dimers (CC₂). The C$_\alpha$ positions of the twofold and quasi-sixfold related subunits achieve the best spatial equivalence only after a rotation of more than 35°, which is

in contrast to the local deformation anticipated from the concepts of Caspar and Klug (1962). The two quantal states of the dimer formation or quasi-sixfold monomeric relationships have led Rossmann (1984) to propose an assembly process involving cooperative steps of dimer assimilations. The capsid maturation process would appropriately utilize the two available dimeric states from solution or would induce the required associations. It might also be possible that different trimeric or pentameric states of contact could build the particle shell. Caspar (1980) and Fuller and King (1980) have earlier proposed such self-controlled switching mechanisms in subunit association states for other viruses. Caspar (1980) describes the capsid self-assembly as "autostery" such that a relaxed subunit association is induced to a less sociable relationship during capsid formation. The "switch" assembly pathway proposal is contrary to the C-subunit scaffolding suggested by Harrison (1980).

The subunit interactions are largely hydrophobic in nature, notably exemplified by a hydrophobic pocket created by SBMV Trp *107* and Trp *99* from respective quasi-twofold related protomers. Contacts between carboxyl groups of Glu's and Asp's are also visible and are generally mediated by metal ions.

4. Metal Ion Binding Sites

Calcium ion binding sites have been observed in the electron density maps of all three viruses. They are located in similar positions between adjacent subunits. In SBMV a further cation is found on the quasi-three-fold axis (Abdel-Meguid et al., 1981). The calcium ions are coordinated by the carboxyls of Glu and Asp and the carbonyls of Asn. The swelling in all three particles is controlled by pH and cation removal, suggesting that carboxylate electrostatic repulsion in the absence of calcium provides the driving force for subunit separation during swelling and possibly *in vivo* disassembly. The cations and pK_a of the nearby carboxyls may also be responsible for the different oligomeric association states induced during capsid maturation (Savithri and Erickson, 1983).

5. Spatial Comparison of the Capsid Architectures

Rossmann and Argos (1981) have devised an automated technique to select spatially superimposable backbone C_α sites which allows recognition of appropriate insertions and deletions in the peptide topologies of two proteins. The degree of C_α equivalence is generally assessed by three criteria: the mean percentage of possible equivalences in the two molecules, the root-mean-square deviation (Å) of the superposed C_α's, and the mean minimum base change per codon (MBC/C) of amino acids associated with paired C_α's. In comparing the S domains of SBMV and TBSV (Rossmann et al., 1983b), approximately 85% of the peptide C_α

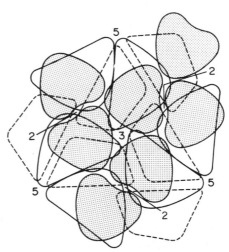

FIGURE 5. The relationship of the icosahedral asymmetric units in T = 3 SBMV and TBSV (dashed), T = 1 STNV (continuous), and T = 1 AMV (dotted). The T = 3 units contain three quasi-equivalent capsid monomers while the T = 1 units consist of only one capsid protein polypeptide.

atoms can be *equivalenced* with an rms separation of 2.2 Å. Similar calculations for the TBSV–STNV and SBMV–STNV relationships are 65%, 3.4 Å and 58%, 3.7 Å with the latter comparison yielding a mean MBC/C of 1.34, close to the random value of about 1.45. The primary structure of TBSV is yet to be published [see Hermodson *et al.* (1982) and Ysebaert *et al.* (1980) for the respective SBMV and STNV amino acid sequences]. The β-strands of the respective barrel domains constitute the bulk of the superimposable atoms while structural insertions and deletions, least in the SBMV–TBSV equivalencing, are in the backbone segments connecting the strands. However, the short reverse-turn regions which form the wedge-shaped barrel edge near the fivefold axes are strikingly conserved in all three viruses, apparently necessitated by spatial constraints. The quaternary associations (i.e., the barrel orientation relative to the icosahedral axes) in the T = 3 TBSV and SBMV are quite similar while the STNV barrel is rotated about 30° relative to the other viruses (Fig. 5), altering the position of the subunit contact areas relative to the barrel topology of SBMV and TBSV. Further conservations extracted from the STNV–SBMV equivalences include Thr's and Ser's at the hexagonal–pentagonal wedge-shaped end of the barrel, Asp's near cation binding sites, hydrophobic residues in the barrel cavity (MBC/C = 1.00), ionic interactions at fivefold contacts, and Lys's and Arg's at the protein–RNA interface. The capsid equivalence calculations, when compared to those of other proteins most likely related by divergent evolution, suggest a similar fate for the three viruses (Rossmann *et al.*, 1983b). Secondary structural predictions elicited from the primary sequences of other plant viruses suggest the barrel topology as even more widespread (Argos, 1981).

III. VIRUS DESCRIPTIONS AND STRUCTURE

A. Alfalfa Mosaic Virus

1. Introduction

AMV has been the subject of intense biological and physical study for more than a decade. Reviews include those by Hull (1969), Jaspars (1974), and van Vloten-Doting and Jaspars (1977).

AMV and the Ilarviruses appear to be unique in terms of structure, as they are the only simple RNA plant viruses that display nonspherical particles (Hull, 1969) of various sizes proportioned to the amount and type of RNA encapsidated. Both AMV and members of the Ilarviruses are properly referred to as heterocapsidic (van Vloten-Doting and Jaspars, 1977). Furthermore, AMV and Ilarviruses require either coat protein or RNA 4 in addition to RNA 1, RNA 2, and RNA 3 for infection.

The discussion of the AMV particle structure will follow closely the methods of investigation. The protein quaternary structure has been investigated by high-resolution electron microscopy and image-processing techniques. Solution, neutron, and X-ray scattering studies reveal the interactive disposition of RNA and protein within the particles and provide refined dimensions in the fully hydrated state. The recently reported single-crystal X-ray diffraction results are examined and the protein tertiary and secondary structure discussed. Nuclear magnetic resonance studies of AMV provide a detailed model of protein–RNA interactions and suggest how the virus behaves in a dynamic manner under various conditions. The chemical stability of AMV is approached through a discussion of disassembly–assembly experiments.

Four different ribonucleoprotein particles can be distinguished in preparations of AMV (see Hull, 1969). The particles are bacilliform in shape with diameter of 180 Å but with lengths of 570, 430, 350, and 300 Å (Fig. 6). The three largest particles each encapsidate a single RNA molecule with the particle length proportional to the mass of the RNA encapsidated; however, the smallest particle encapsidates two identical RNA molecules. The nucleic acid species encapsidated are designated RNA 1–4 with the genetic information necessary for infection contained in the three largest molecules. RNA 4 is the mRNA for the coat protein and is a copy of the 3'-terminal portion of RNA 3 (see Chapter 3 for details of RNA structures). The particle components are designated, in decreasing size, as B(bottom), M(middle), and T_b(top$_b$) and T_a(top$_a$) (Fig. 6). The T_a band is actually composed of two structurally distinct particles, T_a-b and T_a-t as shown in Fig. 6. The latter two components can only be separated by their differing solubilities in $MgCl_2$ (T_a-b precipitates and T_a-t is soluble). Infection occurs when susceptible plants are inoculated with B, M, and T_b. If isolated RNA molecules are used to inoculate plants,

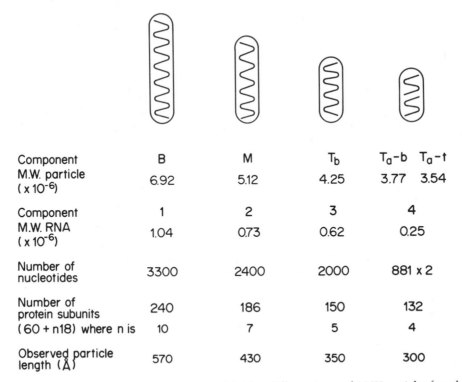

Component	B	M	T_b	T_a-b	T_a-t
M.W. particle ($\times 10^{-6}$)	6.92	5.12	4.25	3.77	3.54
Component	1	2	3	4	
M.W. RNA ($\times 10^{-6}$)	1.04	0.73	0.62	0.25	
Number of nucleotides	3300	2400	2000	881 x 2	
Number of protein subunits	240	186	150	132	
(60 + n18) where n is	10	7	5	4	
Observed particle length (Å)	570	430	350	300	

FIGURE 6. Diagrammatic representation of the four different types of AMV particles found in infected plant tissue. Tabulated from Heijtink *et al.* (1977) and Brederode *et al.* (1980). From Fukuyama *et al.* (1983) with permission.

all four RNAs are required despite the redundant coding information in RNA 4 (Bol *et al.*, 1971).

The protein subunits of three strains of AMV have been sequenced: strain 425 (M_r 24,250; Van Beynum *et al.*, 1977), strain VRU (M_r 24,056; Castel *et al.*, 1979), and strain S (M_r 23,655; Collot *et al.*, 1976). The alignment of the homologous sequences is shown in Fig. 7.

2. Particle Structure

Electron microscopic examination of negatively stained preparations of AMV strain 425 suggests that the particles are likely to be composed of hexamers of protein distributed on a cylindrical P6 lattice (Hull *et al.*, 1969) such that the tube portion of the particles consists of six interlocking rows of hexagons. Particle elongation results from an addition of three hexamers, suggesting that the difference in the number of subunits in the various particles should be a multiple of 18. It is further proposed that tube ends are closed by half of an icosahedral shell with its threefold axis coincident with the cylinder axis. The total number of subunits per

```
         1         10        20        30        40        50        60
AMV-425  SSSQKKAGGKAGKPTKRSQNYAALRKAQLPKPPALKVPVVKPTNTILPQTGCVWQSLGTP
AMV-S    ------------------------------------A-------------AV-----
AMV-VRU  ------------------------------------A------------L-------


                   70        80        90       100       110       120
AMV-425  LSLSSFNGLGARFLYSFLKDFVGPRILEEDLIYRMVFSITPSHAGTFCLTDDVTTEDGRA
AMV-S    ------D---V-----------T------------------------------------
AMV-VRU  -----S---------H------AA---------F-----------S-------------


                  130       140       150       160       170       180
AMV-425  VAHGNPMQEFPHGAFHANEKFGFELVFTAPTHAGMQNQNFKHSYAVALCLDFDAQPGGSK
AMV-S    ----D-----------------------------------------------E-E---
AMV-VRU  ------------------R-----------------------------------L----A


                  190       200       210       220
AMV-425  NPSFRFNEVWVERKAFPRAGPLRSLITVGLFDEADDLDRH
AMV-S    D------------------------------L------
AMV-VRU  ---T-------------------------D----Q-
```

FIGURE 7. The coat protein amino acid sequences of strains 425, S, and VRU of alfalfa mosaic virus. If an amino acid is not given, they are identical to those of strain 425.

particle would then be 60 + ($n \times 18$), given the 12 pentamers of the two cylindrical caps and the $3n$ hexamers of the tube surface. The model agrees well with the number of subunits observed in the different AMV components (Fig. 6) and suggests approximately 14 ribonucleotides per protein subunit. High-resolution electron microscopy, employing image-processing techniques, shows that the tube hexagonal lattice has a structural repeat of 84 Å and that the majority of the stain-excluding matter resides at the twofold axes of the lattice (Mellema and van den Berg, 1974; Mellema, 1975). The VRU strain of AMV was found to be consistent with the same hexameric structure, but to be in a helical arrangement rather than a stacked lattice (Cremers et al., 1981). Figure 8 shows the overall quaternary structure derived for the AMV particles. The VRU strain differs only slightly in amino acid sequence from that of AMV 425 (Fig. 7), yet the angular freedom in bond formation between subunits is altered (Castel et al., 1979). Nevertheless, the VRU strain is also capable of forming the stacked lattice, reflecting the polymorphism of the assembled protein. The proposed hexagonal surface lattice possesses twofold symmetry axes andd quasi-three- and quasi-sixfold axes. Two types of dimer interactions are in evidence, one producing a flat face (a dihedral angle of 180° between trimers) while the other utilizes a dihedral angel of about 140°, similar to that observed between trimers of an icosahedron. Figure 9 illustrates the two dimeric intersubunit contacts and their locations in the cylindrical particle.

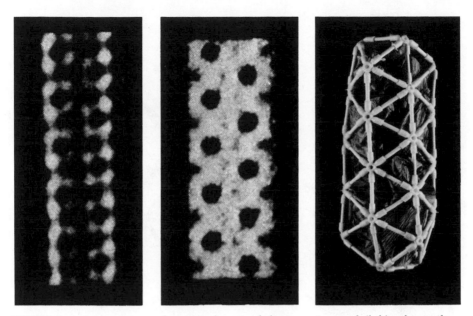

FIGURE 8. An image-processed, negatively stained electron micrograph (left) and a similary oriented model projection of AMV (center). This is compared with a geostix model of a T_a particle (right). From Fukuyama et al. (1983) with permission.

Low-angle neutron scattering studies have been performed on the bottom component of AMV 425 in solution (Cusack et al., 1981). Fitting of the diffraction data to a cylindrically shaped particle resulted in an outside radius of 94 Å with protein extending from an inner radius somewhat less than 65 Å. The RNA, which is uniformly packed within the 65-Å radial limit, slightly penetrates the protein shell and occupies approximately 20% of the interior volume available. Neutron scattering from the 30 S icosahedral particle assembled from AMV protein (Driedonks et al., 1977) was effectively modeled using 60 spherical subunits off 19-Å radius arranged symmetrically about the 30 twofold axes of an icosahedron. Attempts to fit the data with a uniform protein shell failed (Cusack et al., 1981). Neutron diffraction analyses of the elongated AMV VRU particles (Oostergetel et al., 1983) yielded a radial extension of 100 Å, which is in good agreement with the previous studies (Cusack et al., 1981). Oostergetel et al. (1983) suggest that roughly 30 residues of the VRU protein interact directly with, or penetrate into the RNA (Fig. 10). Chemical studies (vide infra) point to the N-terminal basic amino acids of the coat protein being the RNA interaction sites. It is suggested that radii 15 to 30 form an α-helix with one side of the helical surface providing hydrophobic side chains for interaction with the capsid protein and with the positively charged helical side in contact with phosphate groups of the RNA. Residues 1 to 14 point toward the center of the particle and

A. Dimer with disordered arm

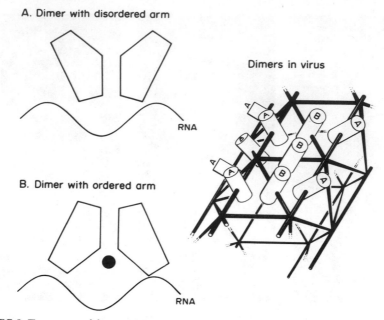

Dimers in virus

B. Dimer with ordered arm

RNA

RNA

FIGURE 9. Two types of dimer environments must occur in the cylindrical portion of AMV. The A-type dimers have a dihedral angle of roughly 140° and have less space in the interior than the B-type dimers, assuming a rigid subunit. The B-type dimers have greater space and may therefore have ordered N-terminal arms in the subunit interface. The icosahedral virus ends would contain only A-type dimers. The figure at the right shows the locations of the two types of dimers in the cylindrical capsid. From Fukuyama *et al.* (1983) with permission.

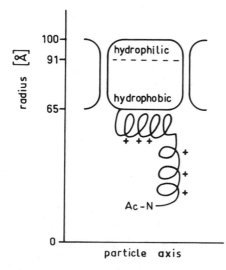

FIGURE 10. Model for the protein in the capsid of AMV strain VRU and conformation of the N-terminus derived from neutron scattering. From Oostergetel (1983) with permission.

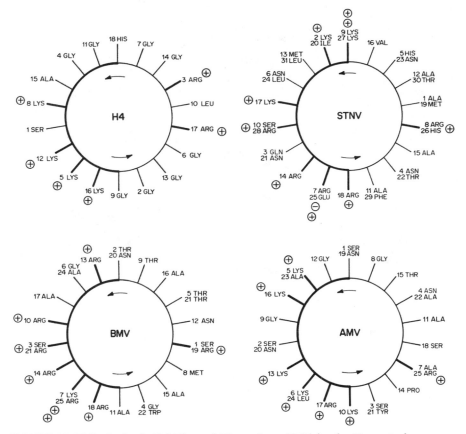

FIGURE 11. Helical wheels (Schiffer and Edmundson, 1967) for the N-terminal segments of calf histone H₄ (Sung and Dixon, 1970), STNV, AMV, and BMV. The views are projections down the helical axis with successive C_α peptide positions rotated by 100° (3.6 residues per turn of a α-helix) along the projected peptide backbones. Side chains with basic charges are indicated by thick-lined spokes; the polar side of the helix is also illustrated. Positively charged side chains are proposed to bind nucleotide phosphates. From Argos (1981) with permission.

also form an α-helix with one side in contact with the RNA phosphate groups and the other side providing polar residues which can hydrogen bond to bases of the RNA (Fig. 11), a model first proposed by Argos (1981) for plant viral N-terminal regions. Small-angle X-ray scattering studies and photon correlation spectroscopy were also performed on the AMV B and AMV T_a-t in the pH range 5.5 to 7.8 at ionic concentrations from 0.01 to 0.1 M. Oostergetel et al. (1981) concluded that AMV does not swell like the Bromoviruses or that native AMV virions are in a swollen state and do not contract. The latter is more likely considering the chemical behavior of the virus (vide infra).

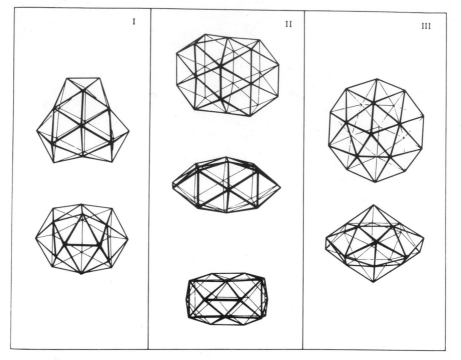

FIGURE 12. Models of nonicosahedral structures visualized as projections along symmetry axes. (I) A model for a structure with 3,2 point group symmetry containing 108 subunits. (II) Model for a 120-subunit structure with 2,2,2 point group symmetry. (III) Model for a structure with 5,2 point group symmetry and containing 120 subunits. From Oostergetel (1983) with permission.

Recent electron microscopy and low-angle scattering investigations (Cusack *et al.*, 1983) indicate that the subunit arrangement of AMV T_a-t particles may not conform with the icosahedral geometry or quasi-equivalent symmetry displayed by many other plant viruses. Electron micrographs show the T_a-t particle to be an oblate ellipsoid with external dimensions $284 \times 284 \times 216$ Å and containing 120 subunits on a deltahedron with 5,2 point group symmetry (Fig. 12).

Crystals that diffract X-rays to 4.5-Å resolution (Fig. 13) have been obtained for reassembled T = 1 AMV protein capsids (Fukuyama *et al.*, 1983) from AMV subunits lacking the 25 N-terminal residues cleaved by mild trypsin treatment (Bol *et al.*, 1974). Particle packing in the crystal yields a diameter of 190 Å for the spherical empty capsids. The X-ray diffraction data indicate a protein shell inner radius of 58 Å and an outer radius of 95 Å. Normally, an electron density map at 4.5-Å resolution does not permit the determination of backbone topology; however, data available from the structures of other spherical viruses (*vide supra*) did permit a possible interpretation of the electron density map (Fukuyama *et al.*, 1983). SBMV and TBSV display holes of roughly 22-Å diameter

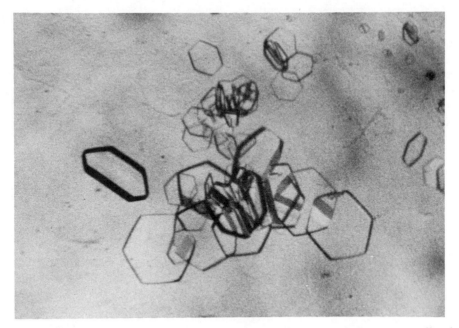

FIGURE 13. Crystalline AMV reassembled capsid protein. The T = 1 aggregate is crystallized in space group P6$_3$, a = 199, c = 314.5 Å containing two molecules in the unit cell. From Fukuyama *et al.* (1981) with permission.

about the five- and sixfold symmetry axes near the surface of the particle; the feature reduces in diameter to virtually nothing at the protein–RNA interface. In contrast, the T = 1 AMV electron density map displays a 38-Å hole at the fivefold axis which penetrates to the interior of the particle and results in a porous shell, probably accounting for nuclease senitivity of encapsidated RNA. The large holes observed in electron micrographs of elongated particles (Driedonks *et al.*, 1977) (Fig. 8) suggest that the icosahedral fivefold axes in the T = 1 structure are quasi-equivalent to the bacilliform particle sixfold axes in the cylindrical region.

Initial analysis of the map suggested a β-barrel (Fig. 14), however, at 4.5-Å resolution the barrel did not close in a manner consistent with that seen in TBSV, SBMV, and STNV. At 3.5-Å resolution (Y. Hatta, K. Fukuyama, I. Fita, and M. G. Rossmann, personal communication) it is clear that AMV has a structure significantly different from the β-barrel observed in the other viruses studied. While it is still not possible to trace the polypeptide chain, the current results indicate that AMV protein has a significant amount of α-helix (probably greater than 30%) and less β-sheet than the original interpretation indicated. Figure 15 shows a schematic diagram of the T = 1 AMV particle organization as proposed by Driedonks *et al.* (1977) which agrees well with the structure derived from X-ray data.

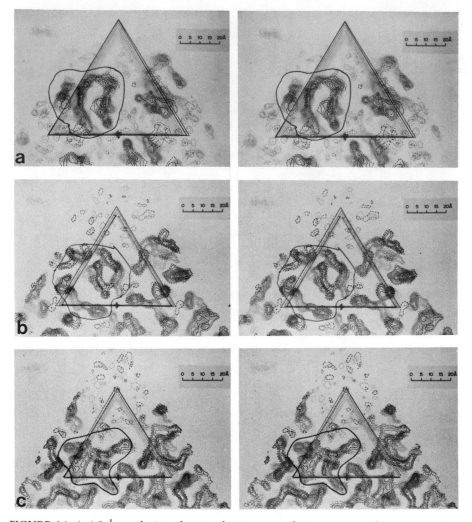

FIGURE 14. A 4.5-Å resolution electron density map of AMV seen as three consecutive stereo views. (a) Sections at R = 92, 90, . . ., 80 Å. (b) Sections at R = 78, 77, . . ., 70 Å. (c) Sections at R = 70, 68, . . ., 60 Å. The icosahedral twofold axis is perpendicular to the page at the center of the base of the triangles. A fivefold axis forms the top vertex and threefold axes form the side vertices of the triangle. Contour levels were draw at 20, 35, 50. . . . The 20 contour is drawn as a broken line. The highest contour is 160. The subunit boundary is outlined. From Fukuyama *et al.* (1983) with permission.

Proton NMR studies (Andree *et al.*, 1981; Kan *et al.*, 1982) on AMV particles with and without the 36 tryspin-removeable N-terminal amino acids indicated segmental mobility of the N-terminal region. Positive identification of signals from Tyr 21, Thr 15, and the N-terminal acetyl group suggests their involvement in protein–RNA interactions. The

FIGURE 15. A model of the T = 1 AMV aggregate proposed by Driedonks *et al.* (1977) based on assembly studies and electron microscopy. These results are consistent with the electron density map in the 4.5-Å studies. From Fukuyama *et al.* (1983) with permission.

study also indicated that the capsid protein undergoes a structural alteration during assembly because of a change in signal from His 220.

3. Particle Disassembly and Assembly

The AMV particle behaves as a complex stabilized by ionic interactions (Bol and Kruseman, 1969). The capsid protein can be readily isolated from the RNA with 1 M $MgCl_2$, thus precipitating the RNA while the protein subunits remain in solution (Kruseman *et al.*, 1971). Though the virus is primarily stabilized by protein–RNA interactions, the predominant soluble species are dimers of the protein subunits after disassembly (Driedonks *et al.*, 1977).

The intact virus is susceptible to proteolytic cleavage by trypsin functioning under suboptimal conditions. The first 25 amino acids are removed from the N-terminal region of the polypeptide, 7 of which are basic. Upon cleavage, the bacilliform particles spontaneously convert to spherical structures which contain approximately 8% RNA by weight (Bol *et al.*, 1974). It is likely that removal of the charged residues which interact with the RNA allows the protein–protein interactions to dominate, yielding icosahedrally symmetric particles. Concentrated acetic acid (not high salt) is necessary to disrupt the empty shells (Bol *et al.*, 1974). If AMV particles are treated with ribonuclease (Bol and Veldstra,

1969), a similar transformation occurs. The spherical particles produced sediment at 36 S and display a diameter of roughly 190 Å, typical of T = 1 structures.

Early studies on AMV assembly from the protein dimer and AMV RNA in solution produced a variety of particle types (Bol and Kruseman, 1969; Hull, 1970; Lebeurier *et al.*, 1971). Driedonks *et al.* (1977, 1978) characterized reassembly of AMV protein with native and a variety of natural and synthetic nucleic acids. The conditions for optimal reassembly for all the nucleic acid types used were found to be pH 6.0, an ionic strength of approximately 0.1 M, and room temperature. Optical diffraction analysis of reassembled particles revealed tubular structures identical to the native particles regardless of the type of nucleic acid used. All evidence favored a cooperative process where initiation was followed by elongation, though heterologous nucleic acids did not produce a specific initiation complex as did the AMV - RNAs. Further investigations have identified the specific, RNA sequence which interacts with the AMV dimer (van Vloten-Doting and Jaspars, 1972; Koper-Zwarthoff and Bol, 1980). Though the same 3'-terminal site is found in all the RNA species, only RNA 1 and RNA 4 have been shown to date to bind protein in the same way (Zuidema *et al.*, 1983a). Two stages in the association have been recognized: a dimer first interacts at the specific RNA site and then two additional dimers bind, yielding a possible initiation hexamer (Houwing and Jaspars, 1982) (Fig. 16). It is clear that the N-terminal region of the protein forms the vital interaction in initiation as its removal prevents complex formation and allows only nonspecific interactions with the RNA molecules (Bol *et al.*, 1974; Zuidema *et al.*, 1983b). The RNA is structurally altered by dimer binding as evidenced by circular dichroism studies with ethidium bromide (Srinivasan and Jaspars, 1982). Analysis of the tube structures formed from AMV coat protein reveals the presence of two types of dimers, labeled A and B in Fig. 9 (*vide supra*). The A-type dimers, are observed in the T = 1 particle while the B-type dimers are a likely result of the ordered portion of the protein N-terminus being placed between two subunits in analogy with the β-annulus structure of the C subunits in SBMV and TBSV. It is possible that the A–B switch is controlled by RNA binding.

B. Ilarviruses

1 Introduction

The Ilarvirus group, whose type member is TSV, consists of about a dozen distinct viruses (see Chapter 1, Table III). With the exception of particle shape, the Ilarvirus and AMV groups have similar properties and several authors have suggested that they be placed in a common taxon (Lister and Saksena, 1976; Gonsalves and Fulton, 1977; Halk and Fulton,

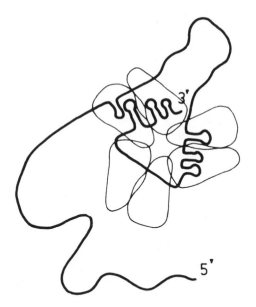

FIGURE 16. A tentative representation of the nucleoprotein complex of AMV coat protein and the 3'-terminal region of the RNA. Three coat protein dimers protect the secondary structural elements indicated in the RNA. From Houwing and Jaspars (1982) with permission.

1978; van Vloten-Doting *et al.*, 1981). On the other hand, the structural differences between AMV (bacilliform shape) and the Ilarviruses (spheroidal shape) have led to proposals for two distinct groups (Matthews, 1979). The occasional appearance of bacilliform particles in preparations of Ilarviruses (Basit and Francki, 1970; Halk and Fulton, 1978) and the recent report of an AMV mutant that forms spheroidal particles (Roosian and van Vloten-Doting, 1983) suggest that subtle changes in subunit interactions lead to the two different particle shapes.

2. Particle Types and Composition

Table IV of Chapter 1 lists the characteristic properties of the Ilarvirus group. The particles in a given virus preparation have nearly identical densities and contain similar RNA/protein ratios, averaging 15 ribonucleotides per protein subunit, which is similar to that reported for AMV. The protein subunits range in M_r from 19,000 to 30,000 with most being near 25,000. The spheroidal particles fall into three size classes with diameters ranging from 250 to 270 Å, 300 to 310 Å, and 320 to 330 Å though some of the viruses show a fourth, smaller particle (approximately 230 Å) containing one or two copies of subgenomic RNA 4. Ghabrial and Lister (1974) have shown that the partial specific volume, M_r and RNA content of the three TSV particles lead to estimates of 149, 179, and 225 subunits per particle. Though the latter two can perhaps be classified as $T=3$ (180 subunits) and $T=4$ (240 subunits) particles, respectively, the smallest does not appear to fit within the scheme of quasiequivalence (Caspar and Klug, 1962). The three TSV genomic RNA molecules have M_r in the ranges $1.1–1.3 \times 10^6$, $0.9–1.1 \times 10^6$, and $0.7–0.9 \times 10^6$.

Electron microscopy of negatively stained particles has not revealed any details of the capsid organization. There is no indication of morphological clustering. The lack of regular structure precludes the use of image processing to analyze the electron micrographs, although the bacilliform particles of PDV (Halk and Fulton, 1978), which display uniform shape and size, may be amenable.

3. Particle Stability

Ilarviruses are apparently stabilized in a similar way to those of AMV. TSV, CLRV, and CVV are disassembled into protein and RNA in high salt (van Vloten-Doting, 1975; Gonsalves and Garnsey, 1975a,b). The particles of TSV are susceptible to protease treatment (van Vloten-Doting, 1975) and the RNA is readily degraded *in situ* by nucleases (Clark and Lister, 1971). All the viruses tested are readily disassembled by low concentrations of SDS (van Vloten-Doting, 1975; Lister and Saksena, 1976). This behavior typifies ionic interactions between protein and RNA (Kaper, 1973).

The Ilarviruses generally lose infectivity within minutes to hours when kept in sap extracted from infected plants (Fulton, 1981) though the gross particle integrity as observed in electron micrographs is not affected. The mechanism of inactivation may involve the action of quinones formed by oxidation (Hampton and Fulton, 1961) or degradation of the nucleic acid.

Ilarvirus capsids are probably more stable than their AMV counterparts. van Vloten-Doting (1975) demonstrated that neither AMV nor TSV RNA would remove protein from nucleoprotein particles of TSV. In contrast, both TSV and AMV RNA taken individually caused protein to redistribute from particles of AMV to the naked RNA. In addition, harsher procedures are required to remove coat protein from TSV than AMV capsids.

In analogy with AMV, it is likely that the N-terminal portion of Ilarvirus protein interacts with the nucleic acid. van Vloten-Doting (1975) has shown that the coat protein is specifically cleaved when treated with 100 times the amount of trypsin required for AMV. The TSV protein lost between 60 and 90 residues, compared with 26 for AMV. The cleaved fragment is unlikely to form part of the "tight" tertiary structure responsible for capsid integrity and could well be an N-terminal region as found for AMV, albeit longer. Furthermore, since the N-terminus of the AMV protein is essential for activating infection and since the capsid protein of TSV, CVV, and CLRV will activate AMV RNA, it seems likely that Ilarvirus proteins possess a predominately basic region within the 90 N-terminal residues. AMV protein will activate infection when plants are inoculated with RNA from the three Ilarviruses (Gonsalves and Garnsey, 1975c). However, there is no serological relationship between AMV and TSV or CVV or CLRV and the tryptic peptide maps of TSV and AMV protein show no similarity (van Vloten-Doting, 1975). Capsid proteins from the Bromo- and Cucumovirus groups will not activate Ilarvirus RNA.

Recently, Roosien and van Vloten-Doting (1983) reported a coat protein mutant of AMV that formed a range of spheroidal particles virtually indistinguishable from those of TSV. The size of the particles was correlated with the amount of RNA encapsidated. Exposure of these nucleoprotein particles to naked AMV RNA caused no redistribution of protein as observed with wild-type AMV. The authors suggested that the mutant behavior may result from an increased tendency of the subunits to form hexamers.

C. Bromoviruses

1. Introduction

The Bromoviruses (Harrison et al., 1971; Lane, 1979) consist of three members that have been studied extensively: BMV, CCMV, and BBMV. The most recent review of Bromovirus research is that of Lane (1981).

These viruses with divided genomes use three identical protein shells to encapsidate four RNA species with M_r of 1.20, 1.07, 0.81, and 0.25 × 10^6 (Bancroft and Flack, 1972; Reijenders et al., 1974). The two largest RNA molecules are enveloped separately in a heavy (RNA 1) and a light (RNA 2) nucleoprotein particle, respectively, while each particle of intermediate density encapsidates a molecule of RNA 3 and RNA 4 (Bancroft and Flack, 1972). RNA 1, 2, and 3 comprise the viral genome necessary for infection while RNA 4 is the subgenomic coat protein messenger (Shih et al., 1972; Bastin et al., 1976). The viral capsid consists of 180 identical polypeptides (t = 3) of M_r near 20,000 (Fig. 17) arranged on a icosahedrally symmetric surface lattice (Caspar and Klug, 1962). Electron microscopic image reconstruction of the CCMV particle shows 12 pentamers at vertices and 20 hexamers on faces of an icosahedron (Horne et al., 1975; Steven et al., 1978).

2. In Vitro Disassociation and Association

Bromoviruses are stable around pH 5.0 and sediment as a homogeneous population of particles of 88 S. At pH 7.5 and high salt concentration (ionic strength > 0.5 M), the particles disassociate into RNA–protein

```
         1         10        20        30        40        50        60
CCMV  STVGTGKLTRAQRRAAARKNKRNT–RVVQPVIVEPIASGQGKAIKAWTGYSVSKWTASCAAAEAKVTSA
BMV      S     M          R R–W A –        L A        IA    I    E  SD IT  A N


         70        80        90       100       110       120       130
CCMV  ITISLPNELSSERNKQLKVGRVLLWLGLLPSVSGTVKSCVTETQTTAAASFQVALAVADNSKDVVAAMY
BMV   MS T  H     K  E                    A RI A   A K AQ E A          S  E


        140       150       160       170       180
CCMV  PEAFKGITLEQLAADLTIYLYSSAALTEGDVIVHLEVEHVRPTFDDSFTPVY–
BMV    TD  R A  GDLL–N Q    A E VPAKA V              F    R
```

FIGURE 17. The amino acid sequences of BMV and CCMV coat proteins as determined by direct RNA 4 and cDNA sequencing without cloning (Dasgupta and Kaesberg, 1982). The single-letter code designates the amino acids; a blank for BMV indicates the same residue as in CCMV while a - refers to a deletion. The numbering scheme is that of CCMV.

complexes and protein subunits (Bancroft and Hiebert, 1967; Hiebert, 1969). The isolated coat protein can reassociate under various conditions of ionic strength, pH, and temperature into a variety of quaternary structures which include sheets, tubes, empty protein shells, double shells, and rosettes (cf. Bancroft et al., 1969; Adolph and Butler, 1974; Pfeiffer and Hirth, 1974). At pH 7.5 and 1 M NaCl the particles disassociate into RNA and protein dimers (Bancroft and Hiebert, 1967; Verduin, 1974; Adolf and Butler, 1977). Dialysis of the dimers to pH 5.0 with 1 M NaCl allows the formation of in vitro empty protein shells (pseudotop) (Verduin, 1974) which display an architecture similar to the protein shells of virus particles isolated from infected plants (Finch and Bancroft, 1968) and yet possess a different surface charge (Bancroft et al., 1968). Cleavage of the 25 N-terminal residues from CCMV coat protein does not prevent formation of the pseudotop component (Bancroft et al., 1968; Childlow and Tremaine, 1971). In the absence of RNA, BMV protein degraded with trypsin which cleaves the N-terminal 63 residues, will form T = 1 particles at pH 7.6 (Cuillel et al., 1981). BBMV differs from CCMV and BMV in that empty protein shells will not form from disassociation products which are a heterogeneous population of protein subunit oligomers dominated by a monomer–trimer equilibrium (Bancroft, 1970).

CCMV protein dimers show a different conformation ("loose") in solution that the "compact" folding of the subunit in the T = 3 empty shells. The transition from compact to loose is governed by deprotonation which is strongly time dependent and reversible (Johnson et al., 1973; Verduin, 1974; Jacrot, 1975). Hydrophobic interactions are largely responsible for the dimer formation as suggested by their stability under high-salt conditions and various temperatures (Verduin, 1978). Circular dichroic ellipticities suggest little helical content and mostly β-secondary structures for the coat protein in the native state (Verduin, 1978).

The aggregation states of BMV protein during assembly at various pH and ionic strengths have recently been examined through the concerted use of small-angle neutron scattering, quasi-elastic light scattering, analytical ultracentrifugation, electron microscopy, and X-ray scattering kinetics (Cuillel *et al.*, 1983a,b). Not only are dimers observed at pH values above neutrality, but also monomers and aggregates of higher molecular weight (possibly including pentamers and hexamers). At lower pH, empty T=3 capsids form spontaneously; however, their radius is larger by 20 to 30 Å than that of the native virus particles and is dependent on ionic strength and the environmental mode of capsid formation (dialysis or pH-jump). These results point to a degree of elasticity in the particle structure, possible polymorphism of the subunit conformation, and a model of quickly formed (< 1 sec) incomplete shells serving as a scaffold for capsid maturation.

3. Swelling of Bromoviruses

At low ionic strength and with an increase in pH to 7.5, the sedimentation coefficient of the Bromoviruses is lowered from 88 S to 78 S (cf. Bancroft *et al.*, 1968; Incardona and Kaesberg, 1964) along with an increased sensitivity to nucleolytic and proteolytic enzymes (Adolph, 1975; Pfeiffer, 1980). Low-angle neutron scattering (Chauvin *et al.*, 1978) and intensity fluctuation spectroscopy (Zulauf, 1977) show a distinct increase in the hydrodynamic radius of BMV particles resulting from a radial extension of the protein and RNA ("swelling"). The BMV particle radius increased from 134 Å at pH 5.0 to 155 Å at pH 7.5 in 0.2 M KCl and 10 mM EDTA (Chauvin *et al.*, 1978). Reversibility of the swelling process cannot be accomplished upon back-titration to pH 5.0; an 84 S particle results with radius at 137 Å. In the presence of 10 mM $MgCl_2$, the process appears reversible; however, the swelling does not achieve its full extent at pH 7.5 (Chauvin *et al.*, 1978).

A distinct polymorphism in the extent of RNA and protein radial swelling and conformation is observed for various environments of pH, ionic strength, temperature, and the presence of divalent ions and spermine (Incardona and Kaesberg, 1964; Incardona *et al.*, 1973; Zulauf, 1977). The neutron investigations (Chauvin *et al.*, 1978) on BMV are particularly revealing in this regard. At pH 5.5 and 1.5 M KCl, the RNA compacts without a change in capsid size. The addition of divalent ions reduces the extent of particle swelling with the greatest reduction for the RNA. Spermine condenses the RNA even more efficiently. Swelling does not significantly alter the thickness of the protein shell. In the case of the BMV pseudotop component, the neutron studies show a radius of 147 Å and a scattering curve very different from that of the normal nucleoprotein particle. Fluorescence studies of CCMV modified with pyridoxal 5'-phosphate also point to structural alterations of the pseudotop component (Kruse *et al.*, 1980). Pfeiffer (1980) has proposed a tentative model to explain the Bromovirus polymorphism. The model is based on some exposure of the RNA upon swelling, allowing nucleolytic cleavage which is negated upon cation or polyamine addition which prevents RNA expansion and protrusion from the protein surface.

Coat protein Ca^{2+} binding is also observed in the tertiary structures of SBMV (Abad-Zapatero et al., 1980), TBSV (Harrison et al., 1978), and STNV (Liljas et al., 1982). The presence of Ca^{2+} in the SBMV particle at alkaline pH prevents swelling (Hsu et al., 1976; Durham, 1977). It has been speculated that the reduced Ca^{2+} concentration in the host cell cytoplasm provides the free energy for particle penetration of the cell membrane and disassembly (Durham, 1977; Durham et al., 1977). Verduin (1978) suggests a protein subunit carboxyl cage of Glu's and Asp's that bind Ca^{2+}; upon cation release, the carboxyl repulsion acts as a spring inducing protein conformational changes, swelling and RNA release. Similar proposals have been made for tobacco mosaic virus (Stubbs et al., 1977) and SBMV for which the pH and divalent cation effects, under various conditions for capsid protein and RNA assembly, suggest the importance of ionizable carboxylate side chains at quasi-threefold interfaces (Savithri and Erickson, 1983). Titration of CCMV and BMV causes the release of about seven protons per subunit; back-titration also results in the absorption of the same number of protons though the pH return displays hysteresis (Jacrot, 1975; Pfeiffer and Durham, 1977).

4. Protein–RNA Interactions

The mixing of Bromovirus coat protein and isolated nucleic acid followed by dialysis at pH 7.4, low ionic strength, and the presence of Mg^{2+} results in nucleoprotein sedimenting at around 80 S and minor fractions greater than 110 S (Bancroft and Hiebert, 1967; Adolph and Butler, 1977; Verduin, 1978). SDS polyacrylamide gel electrophoresis revealed that RNA 3 was primarily encapsidated in the 80 S particles while the > 110 S species contained mainly RNA 3 and 4 (Verduin, 1978; Herzog and Hirth, 1978). The protein alone will not associate under these conditions, suggesting the significance of the RNA in achieving proper protein interaction for particle assembly under neutral conditions. The BMV nucleoproteins will also assemble in the presence of yeast tRNA, AMV 12 S RNA, and oligo(U) (Bancroft et al., 1969; Adolph and Butler, 1977; Cuillel et al., 1981); however, competition experiments show that the favored nucleic acid for envelopment is that of BMV (Cuillel et al., 1981) suggesting preferred base sequences in recognition.

Verduin (1978) has observed through rate zonal centrifugation and fractionation of disassociation products that, at pH 7.5 and 1 M NaCl, the four intact CCMV RNAs each had four to eight bound coat protein subunits; he also found protein dimers in solution. Verduin (1978) thus proposed an assembly model involving protein dimer association with the nucleic acid which directs the protein conformational change from

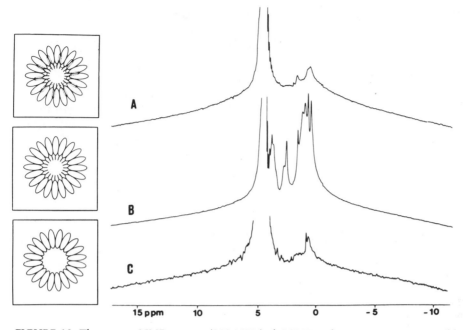

FIGURE 18. The proton NMR spectra (250 MHz) of CCMV and its coat protein assembly products: (A) native virus, (B) empty protein capsids, and (C) empty capsids with protein subunits missing 25 N-terminal residues. Inserts diagram the particle states where dotted lines indicated RNA; solid lines, the N-terminal tail; and ellipses, the remaining coat protein. It is clear that the sharpest NMR peaks exist for case (B) and can be explained by segmental mobility for the N-terminal span. From Vriend *et al.* (1982a) with permission.

loose to compact, followed by further coat protein association into an icosahedral capsid. Verduin (1978) also observed in circular dichroic studies that CCMV RNA *in situ* has less base stacking than the isolated nucleic acid; apparently the protein–RNA interaction results in some melting of the nucleic acid structure. Different conformational states of the CCMV RNA components have also been noted in the presence of polyanions (Dickerson and Trim, 1978).

Recent proton NMR investigations of native CCMV and pseudotop component preparations confirm the importance of the 25 N-terminal residues for interaction with RNA as the observed mobility of the N-terminal region was considerably lessened in the intact particle (Fig. 18). Furthermore, side chain assignments in the empty capsid signal correlated well with the CCMV N-terminal amino acid composition which consists of nine basic residues (Vriend *et al.*, 1981, 1982a). Further proton NMR work revealed that the CCMV protein–RNA interactions are not altered during the pH-dependent swelling of the virus and that only a few CH_2 and CH_3 groups of amino acid side chains become mobile, suggesting some changes in subunit contacts during swelling (Vriend *et al.*, 1982b). Argos (1981) has proposed through helical wheels and secondary structure prediction algorithms, an α-helical structure for the N-terminal region with a hydrophobic side for possible interaction in the grooves of double-stranded RNA and a basic hydrophilic side for phosphate group recognition (Fig. 11). In STNV a helical configuration has been observed in the N-terminal arm structure (Liljas et al., 1982).

Six spherical viruses including BMV have been shown to exhibit magnetically induced birefringence in solution (Torbet, 1983), a phenomenon which can be accounted for by the structural anisotropy of the viral RNA interiors. The anisotropy may be a result of nucleic acid packaging constraints requiring stem and loop secondary structures and probably explains the observed lack of RNA electron density in X-ray diffraction patterns of crystals whose particle packing is controlled by the symmetry of the capsid.

Crystals of CCMV have recently been obtained (Heuss et al., 1981). The tertiary structure determination of the virus by X-ray diffraction techniques will hopefully elucidate the nature of the coat protein conformation and its interaction with nucleic acid.

D. Cucumoviruses

1. Introduction

Cucumoviruses utilize spherical protein capsids, made up of identical protein subunits, which encapsidate RNA (Diener et al., 1964; Kaper and Waterworth, 1981; Lane, 1981). Three native particles are required for infection, each consisting of the same protein shell and yet containing different RNA species. Two of the particles contain respectively genomic RNA 1 (M_r 1.35 × 10⁶) and RNA 2 (M_r 1.16 × 10⁶) while the third particle uses genomic RNA 3 (M_r 0.85 × 10⁶) along with the subgenomic piece RNA 4 (M_r 0.35 × 10⁶) (Kaper and West, 1972; Peden and Symons, 1973; Lot et al., 1974; Lot and Kaper, 1976). If only RNA is used for infection, species 1, 2, and 3 are sufficient (Peden and Symons, 1973) as the dicistronic RNA 3 codes for the coat protein as does RNA 4 which acts as a coat protein messenger. The coat protein monomer M_r is about 26,000 (Gould and Symons, 1982). The capsid of diameter about 300 Å (Scott, 1963) consists of 180 identical subunits, placed on a T=3 icosahedrally symmetric surface lattice. The particle M_r is about 5.3 × 10⁶ (Kaper and Re, 1974) with an RNA content of 19% by mass (Kaper et al., 1965; Francki et al., 1966).

Molecular biology has introduced particle structural considerations, stabilizing interactions, and molecular organization as criteria for classifying plant viruses. Earlier biological distinctions were founded on serological relatedness, base ratio identity, host and vector specificity, and transmission modes. Among the Cucumoviruses, three distinct viruses are generally recognized (Harrison et al., 1971): CMV (Kaper et al., 1965),

PSV (Mink *et al.*, 1969; Boatman *et al.*, 1973), and TAV (Stace-Smith and Tremaine, 1973; Habili and Francki, 1974a,b,c). Though CMV is by far the most thoroughly investigated virus of the group, comparative studies on the physical and chemical properties of CMV and PSV (Lot and Kaper, 1976) and CMV and TAV (Habili and Francki, 1974a,b) have been performed. Various strains of CMV (for a review, see Tolin, 1977) have been recognized by host range and symptomatology (Marrou *et al.*, 1975), serological relationships (Devergnes and Cardin, 1973), and physical and chemical properties (Lot and Kaper, 1976; Piazzolla *et al.*, 1979). Piazzolla *et al.*, (1979), through the use of RNA competition hybridization experiments, were able to distinguish two CMV strain subclasses represented by isolates S and WT, respectively. For the Cucumoviruses, the various differentiation studies based on molecular properties generally correlated with those resulting from serological properties.

2. Stability of Capsids and Protein–RNA Interactions

CMV is unstable in high concentrations of chloride salts; for example, the Y strain will disassociate into protein monomers and RNA in 1.5 M KCl (Kaper *et al.*, 1965; Kaper and Halperin, 1965). Other strains exhibit somewhat different properties; for example, the Q strain requires 2 M LiCl or 1 M $CaCl_2$ for degradation (Francki *et al.*, 1966). Later studies (Kaper and Geelen, 1971) on CMV with 1.3 M LiCl showed that complete disassociation could be achieved between pH 5 and 10, in contrast to BMV and CCMV which displayed salt resistance below pH 6. More extensive investigations correlated the CMV disassembly with the lyotropic or Hofmeister salt series (Kaper, 1975). Furthermore, the virus in the native or formaldehyde-stabilized forms displayed ribonuclease sensitivity between pH 4.0 and 7.2 (Francki, 1968; Kaper and Geelen, 1971; Francki and Habili, 1972). This result once again contrasted with BMV and CCMV which were resistant at pH 6 and susceptible at pH 7 (Incardona and Kaesberg, 1964). Very low concentrations of SDS will also disrupt CMV; in fact, the virus is among the least stable in SDS of all the plant viruses examined (Boatman and Kaper, 1976). CMV can be reassembled from coat protein and RNA merely by lowering the salt concentration, either by dialysis or by dilution (Kaper and Geelen, 1971); however, the formation of empty capsids was never observed, in contrast to BMV or CCMV. CMV nucleoprotein with the same sedimentation rate as native particles were formed from TYMV RNA fragments and CMV protein (Kaper, 1975), suggesting the significance of the polycationic protein and the polyanionic nucleic acid in particle assembly. This strongly suggests that protein–RNA interactions are the dominant factors in CMV stabilization and that the virus is probably in the permanently swollen state, especially given its pH-independent nuclease sensitivity. The DS^- ions are likely to bind basic amino acids in the CMV protein (Kaper, 1976) neutralizing the residue charge and its ability to interact with negatively charged nucleic acid phosphates as well as providing electrostatic repulsion of the phosphates, all leading to the disruption of the CMV particle. Kaper (1976) has suggested that Lys is the likely amino acid to interact with the RNA, probably through uridine and guanosine phosphate residues. This is supported by the facts that the pK for viral degradation is around 10, that the lysyl-specific reagent trinitrobenzenesulfonic acid causes nonspecific protein–RNA reassociation products around pH 10.3, and that the pK of lysyl and uridine and guanosine phosphate residues are 10 and 9.5, respectively. However, it must be noted that within the 21 N-terminal residues of the CMV Q coat protein (Gould and Symons, 1982), there are six Arg's and only one Lys.

Gera *et al.* (1979) have pointed to the importance of the coat protein in determining differences in the transmissibility of the CMV strains by aphids; differences of viral concentration in the plant or the presence of a helper or inhibitory factor had little effect. The coat protein may be important in facilitating adherence of the virions to the aphid foregut.

3. Similarities between Bromoviruses and Cucumoviruses

Cucumoviruses share many physical and chemical properties with the Bromoviruses (Kaper, 1975). These include icosahedral capsids of about the same size, coat proteins with similar molecular weights, SDS and nuclease sensitivity, high-salt disassembly, a strong dependence on protein–RNA interactions for stability, basic N-terminal regions in capsid protein sequences, four RNA molecular species of similar size and particle distribution, infectivity requiring three genomic RNAs and similar 3' terminal nucleotides for the four RNA species. However, there are differences such as those of serology, host symptomatology, ability to swell, dependence on cations, and CMV's inability to form empty capsids. Can these viruses be related by divergent evolution; that is, are they derived from a common ancestor?

Murthy (1983) has recently compared the nucleotide sequences of the two cistrons of CMV RNA3 (Gould and Symons, 1982) with those of BMV RNA3 (Dasgupta and Kaesberg, 1982). He observed that the respective primary structures of the 3a proteins align well (Fig. 19). Only seven gaps were observed with five in BMV and two in CMV; BMV has 5 more residues at the N-terminus while CMV has 44 at its C-terminus. The homology level is 37% in the first 230 residues aligned, with Leu, Gly, Ala, and Pro most conserved and with many of the substituted residues showing physical characteristic conservation. However, the coat proteins do not display any significant homology except the basic N-terminal regions of about 25 residues in length. The noncoding regions show no relationship except for the 114 3'-terminal nucleotides, 52 of which are conserved and largely confined to the single-stranded loop regions in the proposed tRNA-like structural model for this region (Ahlquist

```
              1        10          20         30         40         50         60
CMV       MAFPGPSRTLTQQSSAA---LPDDLQKILFSPDAIKKMATECDLGRHHYMRADNAISVRPLVPQVTSN
BMV   MSNIVSPFSGSSRTTSDVGKQAGGTSDEKLIESLFSEKAVKEIAAECKLGCYNYLKSNEPRNYIDLVPKSHVS

              70         80         90        100        110        120        130
CMV   NLLSFFKSGYDAGELRSKGYMSVPQVLCVVTRTGSTDAEGSLKIYLADLGDKE----LSPIDGNCVTLHNHEL
BMV   AWLSWATSKYDKGELPSRGFMNVPRIVCFLVRTTDSAESGSITVSLCDSGKAAARAGLEAIDNWEATIWLSAL

             140        150        160        170        180        190        200
CMV   PALISFQPTYDGPMELVGNRH---RCFAVVVERHGYIGYGGTTASVCSNWQAQFSSKNNNYTHAAAGKTLVLP
BMV   PALIALTPSYDCPMEVIGGNSGRNRCFGIATQLSGVVGTTGSVAVTHAYWQANFKAKPNNYKLHGPATIMVMP

             210        220        230        240        250        260        270
CMV   YNRLAEHSYPSAVARLLKSQLNNVSSSRYLL--PNVALNQNASGHESEILKESPPIAIGSLSASRNNSFRSQV
BMV   FDRLRQLDKKSLKNYIRGISNQSVDHG-YLLGRPLQSVDQVAQEDLLVEESESP-SALGR-GVKDSKSVSASS

             280        290        300        310        320        333
CMV   NGLCALLRCTYVYIYYVYLPYVNTCESRVPCELSTVDICDDALEDFDHGVSKYTSQF
BMV   AGLPVSSPTLRIK
```

FIGURE 19. Alignment of the 3a protein sequences from BMV and CMV as proposed by Murthy (1983). A - indicates a deletion; the N- and C-terminal deletions are given as blanks. The residues are numbered according to CMV and exactly conserved residues as underlined.

et al., 1981; Gould and Symons, 1982). Murthy (1983) suggests that one of these loops may be involved in RNA replicase binding.

The 3a protein homologies may well point to divergent evolution of the Cucumo- and Bromoviruses from a common ancestor. The lack of significant amino acid homology in the coat proteins of BMV and CMV does not mitigate this suggestion; the relationship could be one of peptide backbone topology as exemplified by STNV and SBMV (Rossmann *et al.*, 1983b). The conservation of 3a can be relegated to a contrained function of viral replication while the coat protein may have experienced rapid evolution for purposes of host specificity. It may be that all minimal plant viruses are derived from a common ancestor with the STNV–SBMV–TBSV–AMV and CMV–BMV relationships merely the tip of the evolutionary iceberg.

ACKNOWLEDGMENT. The authors are deeply grateful to Ruth Rafferty for diligent help in the preparation of the manuscript.

REFERENCES

Abad-Zapatero, C., Abdel-Meguid, S. S., Johnson, J. E., Leslie, A. G. W., Rayment, I., Rossmann, M. G., Suck, D., and Tsukihara, T., 1980, Structure of southern bean mosaic virus at 2.8-Å resolution, *Nature (London)* **286**:33–39.

Abdel-Meguid, S. S., Yamane, T., Fukuyama, K., and Rossmann, M. G., 1981, The location of calcium ions in southern bean mosaic virus, *Virology* **114**:81–85.

Adolph, K. W., 1975, The conformation of the RNA in cowpea chlorotic mottle virus: Dye-binding studies, *Eur. J. Biochem.* **53**:449–455.

Adolph, K. W., and Butler, P. J. G., 1974, Studies on the assembly of a spherical plant virus, *J. Mol. Biol.* **88:**327–341.

Adolph, K. W., and Butler, P. J. G., 1977, Studies on the assembly of a spherical plant virus, *J. Mol. Biol.* **109:**345–357.

Ahlquist, P., Dasgupta, R., and Kaesberg, P., 1981, Near identity of 3' RNA secondary structure in Bromoviruses and cucumber mosaic virus, *Cell* **23:**183–189.

Andree, P. J., Kan, J. H., and Mellema, J. E., 1981, Evidence for internal mobility in alfalfa mosaic virus, *FEBS Lett.* **130:**265–268.

Argos, P., 1981, Secondary structure prediction of plant virus coat proteins, *Virology* **110:**55–62.

Argos, P., Tsukihara, T., and Rossman, M. G., 1980, A structural comparison of concanavalin A and tomato bushy stunt virus protein, *J. Mol. Evol.* **15:**169–179.

Arnott, S., Campbell-Smith, P., and Chandrasekaran, R. 1976, Atomic coordinates and molecular conformations for DNA-DNA, RNA-RNA, and DNA-RNA helices, in *Handbook of Biochemistry and Molecular Biology: Nucleic Acids Section,* Volume II (G. Fasman, ed.), pp. 411–422, CRC Press, Boca Raton, Fla.

Bancroft, J. B., 1970, The self-assembly of spherical plant viruses, *Adv. Virus Res.* **16:**99–134.

Bancroft, J. B., and Flack, I. H., 1972, The behavior of cowpea chlorotic mottle virus in CSCl, *J. Gen. Virol.* **15:**247–251.

Bancroft, J. B., and Hiebert, E., 1967, Formation of an infectious nucleoprotein from protein and nucleic acid isolated from a small spherical virus, *Virology* **32:**354–356.

Bancroft, J. B., Hiebert, E., Rees, M. W., and Markham, R., 1968, Properties of cowpea chlorotic mottle virus, its protein and nucleic acid, *Virology* **34:**224–239.

Bancroft, J. B., Hiebert, E., and Bracker, C. E., 1969, The effects of various polyanions on shell formation of some spherical viruses, *Virology* **39:**924–930.

Basit, A. A., and Francki, R. I. B., 1970, Some properties of rose mosaic virus from South Australia, *Aust. J. Biol. Sci.* **23:**1197–1206.

Bastin, M., Dasgupta, R., Hall, T. C., and Kaesberg, P., 1976, Similarity in structure and function of the 3'-terminal region of the four brome mosaic viral RNAs, *J. Mol. Biol.* **103:**737–745.

Blundell, T. L., and Johnson, L. N., 1976, *Protein Crystallography,* Academic Press, New York.

Boatman, S., and Kaper, J. M., 1976, Molecular organization and stabilizing forces of simple RNA viruses. IV. Selective interference with protein–RNA interactions by use of sodium dodecyl sulfate, *Virology* **70:**1–16.

Boatman, S., Kaper, J. M., and Tolin, S. A., 1973, A comparison of properties of peanut stunt virus and cucumber mosaic virus, *Phytopathology* **63:**801.

Bol, J. F., and Kruseman, J., 1969, The reversible disassociation of alfala mosaic virus, *Virology* **37:**485–488.

Bol, J. F., and Veldstra, H., 1969, Degradation of alfalfa mosaic virus by pancreatic ribonuclease, *Virology* **37:**74–85.

Bol, J. F., van Vloten-Doting, L., and Jaspars, E. M. J., 1971, A functional equivalence of top component *a* RNA and coat protein in the initiation of infection by alfalfa mosaic virus, *Virology* **46:**73–85.

Bol, J. F., Kraal, B., and Brederode, F. J., 1974, Limited proteolysis of alfalfa mosaic virus: Influence on the structural and biological function of the protein, *Virology* **58:**101–110.

Brederode, P. T., Kaper-Zwarthoff, E. C., and Bol, J. F., 1980, Complete nucleotide sequence of alfalfa mosaic virus RNA4, *Nucleic Acids Res.* **8:**2213–2223.

Caspar, D. L. D., 1980, Movement and self-control in protein assemblies, *Biophys. J.* **32:**103–138.

Caspar, D. L. D., and Klug, A., 1962, Physical principles in the construction of regular viruses, *Cold Spring Harbor Symp. Quant. Biol.* **27:**1–24.

Castel, A., Kraal, B. DeGraff, J. M., and Bosch, L., 1979, The primary structure of the coat protein of alfalfa mosaic virus strain VRU: A hypothesis on the occurrence of two conformations in the assembly of the protein shell, *Eur. J. Biochem.* **102:**125–138.

Chauvin, C., Pfeiffer, P., Witz, J., and Jacrot, B., 1978, Structural polymorphism of Bromegrass mosaic virus: A neutron small angle scattering investigation, *Virology* **88:**138–148.

Childlow, J., and Tremaine, J. H., 1971, Limited hydrolysis of cowpea chlorotic mottle virus by trypsin and chymotrypsin, *Virology* **43:**267–278.

Clark, M. F., and Lister, R. M., 1971, Preparation and some properties of nucleic acid of tobacco streak virus, *Virology* **45:**61–74.

Collot, D., Peter, R., Das, B., Wolff, B., and Durantan, H., 1976, Primary structure of alfalfa mosaic virus coat protein (strain S), *Virology* **74:**236–238.

Cremers, A. F. M., Oostergetel, G. T., Schilstra, M. J., and Mellema, J. E., 1981, An electron microscope investigation of the structure of alfalfa mosaic virus, *J. Mol. Biol.* **145:**545–561.

Crick, F. H. C., and Watson, J. D., 1956, Structure of small viruses, *Nature (London)* **177:**473–475.

Crick, F. H. C., and Watson, J. D., 1957, Virus structure: General principles, in: *Ciba Foundation Symposium on the Nature of Viruses* (G. E. W. Wolstenholme and E. C. P. Millar, eds.), pp. 5–13, Little, Brown, Boston.

Cuillel, M., Jacrot, B., and Zulauf, M., 1981, A T = 1 capsid formed by protein of brome mosaic virus in the presence of trypsin, *Virology* **110:**63–72.

Cuillel, M., Zulauf, M., and Jacrot, B., 1983a, Self-assembly of brome mosaic virus protein into capsids, *J. Mol. Biol.* **164:**589–603.

Cuillel, M., Berthet-Coloninas, C., Krop, B., Tardieu, A. Vachette, P., and Jacrot, B., 1983b, Self-assembly of brome mosaic virus capsids: Kinetic study using neutron and X-ray solution scattering, *J. Mol. Biol.* **164:**645–650.

Cusack, S., Miller, A., Krijgsman, P. C. J., and Mellema, J. E., 1981, An investigation of the structure of alfalfa mosaic virus by small angle neutron scattering, *J. Mol. Biol.* **145:**525–543.

Cusack, S., Oostergetel, G. T., Krijgsman, P. C. J., and Mellema, J. E., 1983, The structure of top A-T component of alfalfa mosaic virus: A non icosahedral virion, *J. Mol. Biol.* **171:**139–155.

Dasgupta, R., and Kaesberg, P., 1982, Complete nucleotide sequences of the coat protein messenger RNAs of brome mosaic virus and cowpea chlorotic mottle virus, *Nucleic Acids Res.* **10:**703–713.

Devergnes, J. C., and Cardin, L., 1973, Contribution a l'etude due virus de la mosaique duc concombre (CMV). IV. Essai de classification de plusieurs isolats sur la base de leur structure antigenique, *Ann. Phytopahthol.* **5:**409–430.

Dickerson, P. E., and Trim, A. R., 1978, Conformational states of cowpea chlorotic mottle virus ribonucleic acid components, *Nucleic Acids Res.* **5:**987–998.

Diener, T. O., Scott, H. A., and Kaper, J. M., 1964, Highly infectious nucleic acid from crude and purified preparations of cucumber mosaic virus (Y strain), *Virology* **22:**131–141.

Driedonks, R. A., Krijgsman, P. C. J., and Mellema, J. E., 1977, Alfalfa mosaic virus protein polymerization, *J. Mol. Biol.* **113:**123–140.

Driedonks, R. A., Krijgsman, P. C. J., and Mellema, J. E., 1978, A characterization of alfalfa mosaic virus protein polymerization in the presence of nucleic acid, *Eur. J. Biochem.* **82:**405–417.

Durham, A. C. H., 1977, Do viruses use calcium ions to shut off host cell functions?, *Nature (London)* **267:**375–376.

Durham, A. C. H., Hendry, D. A., and von Wechmar, M. B., 1977, Does Calcium ion binding control plant virus disassembly?, *Virology* **77:**524–533.

Finch, J. T., and Bancroft, J. B., 1968, Structure of the reaggregated protein shells of two spherical viruses, *Nature (London)* **220:**815–816.

Francki, R. I. B., 1968, Inactivation of cucumber mosaic virus (Q strain) nucleoprotein by pancreatic ribo-nuclease, *Virology* **34**:694–700.

Francki, R. I. B., and Habili, N., 1972, Stabilization of capsid structure and enhancement of immunogenicity of cucumber mosaic virus (Q strain) by formaldehyde, *Virology* **48**:309–315.

Francki, R. I. B., Randles, J. W., Chambers, T. C., and Wilson, S. B., 1966, Some properties of purified cucumber mosaic virus (Q strain), *Virology* **28**:729–741.

Fukuyama, K., Abdel-Meguid, S. S., Johnson, J. E., and Rossmann, M. G., 1983, Structure of a T = 1 aggregate of alfalfa mosaic virus coat protein seen at 4.5 Å-resolution, *J. Mol. Biol.* **167**:873–894.

Fuller, M. T., and King, J. A., 1980, Regulation of coat protein polymerization by the scaffolding protein of basteriophage P22, *Biophys. J.* **32**:381–401.

Fulton, R. W., 1981, Ilarviruses, in: *Handbook of Plant Virus Infections and Comparative Diagnosis* (E. Kurstak, ed.) p. 378, Elsevier/North-Holland, Amsterdam.

Gera, A., Loebenstein, G., and Pascah, B., 1979, Protein coats of two strains of cucumber mosaic virus affect transmission by *Aphis gossypii, Phytopathology* **69**:396–399.

Ghabrial, S. A., and Lister, R. M., 1974, Chemical and physicochemical properties of two strains of tobacco streak virus, *Virology* **57**:1–10.

Golden, J. S., and Harrison, S. C., 1982, Proteolytic dissection of turnip crinkle virus subunit in solution, *Biochemistry* **21**:3862–3866.

Gonsalves, D., and Fulton, R. W., 1977, Activation of prunus necrotic ringspot virus and rose mosaic virus by RNA 4 components of some ilar viruses, *Virology* **81**:398–407.

Gonsalves, D., and Garnsey, S. M., 1975a, Functional equivalence of an RNA component and coat protein for infectivity of citrus leaf rugose virus, *Virology* **64**:23–31.

Gonsalves, D., and Garnsey, S. M., 1975b, Nucleic acid components of citrus variegation virus and their activation by coat protein, *Virology* **67**:311–318.

Gonsalves, D., and Garnsey, S. M., 1975c, Infectivity of heterologous RNA–protein mixtures from alfalfa mosaic, citrus leaf rugose, citrus variegation and tobacco streak virus, *Virology* **67**:319–326.

Gould, A. R., and Symons, R. H., 1982, Cucumber mosaic virus RNA 3: Determination of the nucleotide sequence provides the amino acid sequences of protein 3A and viral coat protein, *Eur. J. Biochem.* **126**:217–226.

Habili, N., and Francki, R. I. B., 1974a, Comparative studies on tomato aspermy and cucumber mosaic virus. I. Physical and chemical properties, *Virology* **57**:392–401.

Habili, N., and Francki, R. I. B., 1974b, Comparative studies on tomato aspermy and cucumber mosaic virus. II. Virus stability, *Virology* **60**:29–36.

Habili, N., and Francki, R. I. B., 1974c, Comparative studies on tomato aspermy and cucumber mosaic viruses. III. Further studies on relationship and construction of a virus from parts of the two viral genomes, *Virology* **61**:443–449.

Halk, E. L., and Fulton, R. W., 1978, Stabilization and particle morphology of prune dwarf virus, *Virology* **91**:434–443.

Hampton, R. E., and Fulton, R. W., 1961, The relation of polyphenol oxidase to instability *in vitro* of prune dward and sour cherry necrotic ring spot viruses, *Virology* **13**:44–52.

Harrison, B. D., Finch, J. J., Gibbs, A. J., Hollings, M., Shepherd, R. J., Valenta, V., and Wetter, C., 1971, Sixteen groups of plant viruses, *Virology* **45**:356–363.

Harrison, S. C., 1980, Protein interfaces and intersubunit bonding, *Biophys. J.* **32**:139–153.

Harrison, S. C., Olson, A. J., Schutt, C. E., Winkler, F. K., and Bricogne, C., 1978, Tomato bushy stunt virus at 2.9-Å resolution, *Nature (London)* **276**:368–373.

Heijtink, R. A., Houwing, C. J., and Jaspars, E. M. J., 1977, Molecular weights of particles and RNAs of alfalfa mosaic virus, *Biochemistry* **16**:4684–4693.

Hermodson, M. A., Abad-Zapatero, C., Abdel-Meguid, S. S., Pundak, S., Rossmann, M. G., and Tremaine, J. H., 1982, Amino acid sequence of southern bean mosaic virus coat protein and its relation to the three-dimensional structure of the virus, *Virology* **119**:133–149.

Herzog, M., and Hirth, L., 1978, *In vitro* encapsidation of the four RNA species of brome mosaic virus, *Virology* **86**:4856.

Heuss, K. L., MohanaRao, J. K., and Argos, P., 1981, Crystallization of cowpea chlortic mottle virus, *J. Mol. Biol.* **146**:635–640.

Hiebert, E., 1969, Ph.D. thesis, Purdue University.

Horne, R. W., Hobart, J. M., and Pasquali-Ronchetti, I., 1975, A negative staining-carbon film technique for studying viruses in the electron microscope, *J. Ultrastruct. Res.* **53**:319–330.

Houwing, C. J., and Jaspars, E. M. J., 1982, Protein binding sites in nucleation complexes of alfalfa mosaic virus RNA 4, *Biochemistry* **21**:3408–3414.

Hsu, C. H., Sehgal, O. P., and Pickett, E. E., 1976, Stabilizing effect of divalent metal ions on virions of southern bean mosaic virus, *Virology* **69**:587–595.

Hull, R., 1969, Alfalfa mosaic virus, *Adv. Virus Res.* **15**:365–370.

Hull, R., 1970, Studies on alfalfa mosaic virus. III. Reversible dissociation and reconstitution studies, *Virology* **40**:34–47.

Hull, R., Markham, R., and Hills, J. G., 1969, Studies on alfalfa mosaic virus. II. The structure of the virus components, *Virology* **37**:416–428.

Incardona, N. L., and Kaesberg, P., 1964, A pH-induced structural change in bromegrass mosaic virus, *Biophys. J.* **4**:11–21.

Incardona, N. L., McKee, S., and Flanegan, J. B., 1973, Noncovalent interactions in viruses: Characterization of their role in the pH and thermally induced conformational changes in bromegrass mosaic virus, *Virology* **53**:204–214.

Jacrot, B., 1975, Studies on the assembly of a spherical plant virus, *J. Mol. Biol.* **95**:433–446.

Jaspars, E. M. J., 1974, Plant viruses with a multipartite genome, *Adv. Virus Res.* **19**:37–140.

Johnson, M. W., Wagner, G. W., and Bancroft, J. B., 1973, A titrimetric and electrophoretic study of cowpea chlorotic mottle virus and its protein, *J. Gen. Virol.* **19**:263–273.

Kan, J. H., Andree, P. J., Koreijzer, L. C., and Mellema, J. E., 1982, Proton magnetic resonance studies on the coat protein of alfalfa mosaic virus, *Eur. J. Biochem* **126**:29–33.

Kaper, J., 1973, Arrangement and identification of simple isometric viruses according to their dominating stabilizing forces. *Virology* **55**:299–304.

Kaper, J. M., (ed.), 1975, in: *The Chemical Basis of Virus Structure Dissociation and Reassembly*, pp. 321–352, Elsevier/North-Holland, Amsterdam.

Kaper, J. M., 1976, Molecular organization and stabilizing forces of simple RNA viruses. V. The role of lysyl residues in the stabilization of cucumber mosaic virus strain S, *Virology* **71**:185–198.

Kaper, J. M., and Geelen, J. L. M. C., 1971, Studies on the stabilizing forces of simple RNA viruses. II. Stability, dissociation and reassembly of cucumber mosaic virus, *J. Mol. Biol.* **56**:277–294.

Kaper, J. M., and Halperin, J. E., 1965, Alkaline degradation of turnip yellow mosaic virus. II. *In situ* breakage of the ribonucleic acid, *Biochemistry* **4**:2434–2440.

Kaper, J. M., and Re, G. G., 1974, Redetermination of the RNA content and the limiting RNA size of three strains of cucumber mosaic virus, *Virology* **60**:308–311.

Kaper, J. M., and Waterworth, H. E., 1981, Cucumoviruses, in: *Handbook of Plant Virus Infections and Comparative Diagnosis* (E. Kurstak, ed.), pp. 258–290, Elsevier/North-Holland, Amsterdam.

Kaper, J. M., and West, C. K., 1972, Polyacrylamide gel separation and molecular weight determination of the components of cucumber mosaic virus RNA, *Prep. Biochem.* **2**:251–263.

Kaper, J. M., Diener, T. O., and Scott, H. A., 1965, Some physical and chemical properties of cucumber mosaic virus (strain Y) and of its isolated ribonucleic acid, *Virology* **27**:54–72.

Koper-Zwarthoff, E. C., and Bol, J. F., 1980, Nucleotide sequence of the putative recognition site for coat protein in the RNAs of alfalfa mosaic virus and tobacco streak virus, *Nuleic Acids Res.* **8**:3307–3318.

Kruse, J., Verduin, B. J. M., and Visser, A. J. W. G., 1980, Fluorescence of cowpea-chlorotic-mottle virus modified with pyridoxal 5′-phosphate, *Eur. J. Biochem.* **105**:395–401.

Kruseman, J., Kraal, B., Jaspars, E. M. J., Bol, J. F., Brederode, F. J., and Veldstra, H., 1971, Molecular weight of the coat protein of alfalfa mosaic virus, *Biochemistry* **10**:447–455.

Lane, L. C., 1979, Bromovirus group, in: *Descriptions of Plant Viruses* No. 215, Commonwealth Mycological Institue and Association of Applied Biologists, Surrey.

Lane, L. C., 1981, Bromoviruses, in: *Handbook of Plant Virus Infections and Comparative Diagnosis* (E. Kurstak, ed.) pp. 333–376, Elsevier/North-Holland, Amsterdam.

Lebeurier, G., Fraenkel-Conrat, H., Wurtz, M., and Hirth, L., 1971, Self-assembly of protein subunits from alfalfa mosaic virus, *Virology* **43**:51–61.

Liljas, L., Unge, T., Jones, T. A., Fridborg, K. Lovgren, S., Skoglund, V., and Strandberg, B., 1982, Structure of satellite tobacco necrosis virus at 3.0-Å resolution, *J. Mol. Biol.* **159**:93–108.

Lister, R. M., and Saksena, K. N., 1976, Some properties of tulare apple mosaic and ilar viruses suggesting grouping with tobacco streak virus, *Compr. Virol.* **10**:440–445.

Lot, H., and Kaper, J. M., 1976, Physical and chemical differentiation of three strains of cucumber mosaic virus and peanut stunt virus, *Virology* **74**:209–222.

Lot, H., Marchoux, G., Marrou, J., Kaper, J. M., West, C. K., van Vloten-Doting, L., and Hull, R., 1974, Evidence for three functional RNA species in several strains of cucumber mosaic virus, *J. Gen. Virol.* **22**:81–93.

Marrou, J., Quiott, J. B., and Marchoux, G., 1975, Caracterisation de Douze souches der VMG par leurs apitudes pathogenes: Tentative de classification, Meded. *Fac. Landbouwwet. Rijksuniv. Gent.* **40**:107–122.

Matthews, R. E. F., 1979, Classification and nomenclature of viruses: Third report of the International Committee on Taxonomy of Viruses, *Intervirology* **12**:3–12.

Mellema, J. E., 1975, Model for the capsid structure of alfalfa mosaic virus, *J. Mol. Biol.* **94**:643–648.

Mellema, J. E., and van den Berg, H. J. N., 1974, The quaternary structure of alfalfa mosaic virus, *J. Supramol. Struct.* **2**:17–31.

Mink, G. I., Silbernagel, M. J., and Saksena, K. N., 1969, Host range, purification, and properties of the western strain of peanut stunt virus, *Phytopathology* **59**:1625–1631.

Murthy, M. R. N., 1983, Comparison of the nucleotide sequences of cucumber mosaic virus and brome mosaic virus, *J. Mol. Biol.* **168**:469–476.

Oostergetel, G. T., 1983, Ph.D. thesis, Rijksuniversiteit te Leiden, The Netherlands.

Oostergetel, G. T., Krijgsman, P. C. J., Mellema, J. E., Cusack, S., and Miller, A., 1981, Evidence for the absence of swelling of alfalfa mosaic virions, *Virology* **109**:206–210.

Oostergetel, G. T., Mellema, J. E., and Cusack, S., 1983, A solution scattering study on the structure of alfalfa mosaic virus strain VRU, *J. Mol. Biol.* **171**:157–173.

Peden, K. W. C., and Symons, R. H., 1973, Cucumber mosaic virus contains a functionally divided genome, *Virology* **53**:487–492.

Pfeiffer, P., 1980, Changes in the organization of bromegrass mosaic virus in response to cation binding as probed by changes in susceptibility to degradative enzymes, *Virology* **102**:54–61.

Pfeiffer, P., and Durham, A. C. H., 1977, The cation binding associated with structural transition in bromegrass mosaic virus, *Virology* **81**:419–432.

Pfeiffer, P., and Hirth, L., 1974, Aggregation states of brome mosaic virus protein, *Virology* **61**:160–167.

Piazzolla, P., Diaz-Ruiz, J. R., and Kaper, J. M., 1979, Nucleic acid homologies of eighteen cucumber mosaic virus isolates determined by competition hybridization, *J. Gen. Virol.* **45**:361–369.

Reijenders, L., Aalbers, A. M. J., Van Kammen, A., and Tunring, R. W. J., 1974, Molecular weights of plant viral RNAs determined by gel electrophoresis under denaturing conditions, *Virology* **60**:515–521.

Richardson, J. S., 1981, The anatomy and taxonomy of protein structure, *Adv. Prot. Chem.* **34**:167–339.

Robinson, I. K., and Harrison, S. C., 1982, Structure of the expanded state of tomato bushy stunt virus, *Nature (London)* **297:**563–568.

Roosien, J., and van Vloten-Doting, L., 1983, A mutant of alfalfa mosaic virus with an unusual structure, *Virology* **126:**155–167.

Rossmann, M. G., and Argos, P., 1981, Protein folding, *Annu. Rev. Biochem.* **50:**497–532.

Rossmann, M. G., 1984, Constraints on the assembly of spherical virus particles, *Virology* **134:**1–11.

Rossmann, M. G., Abad-Zapatero, C., Hermodson, M. A., and Erickson, J. W., 1983a, Subunit interactions in southern bean mosaic virus, *J. Mol. Biol.* **166:**37–83.

Rossman, M. G., Abad-Zapatero, C., Murthy, M. R. N., Liljash, L., Jones, T. A., and Strandberg, B., 1983b, Structural comparisons of some small spherical plant viruses, *J. Mol. Biol.* **165:**711–736.

Rossmann, M. G., Chandrasekaran, R., Abad-Zapatero, C., Erickson, J. W., and Arnott, S., 1983c, RNA–protein binding in southern bean mosaic virus, *J. Mol. Biol.* **166:**73–80.

Savithri, H. S., and Erickson, J. W., 1983, The self-assembly of the cowpea strain of southern bean mosaic virus: Formation of T = 1 and T = 3 nucleoprotein particles, *Virology* **126:**328–335.

Schiffer, M., and Edmundson, A. B., 1967, Uses of helical wheels to represent the structures of proteins and to identify segments with helical potential, *Biophys. J.* **7:**121–135.

Scott, H. A., 1963, Purification of cucumber mosaic virus, *Virology* **20:**103–106.

Shih, D. S., Lane, L. C., and Kaesberg, P., 1972, Origin of the small component of brome mosaic virus RNA, *J. Mol. Biol.* **64:**353–362.

Srinivasan, S., and Jaspars, E. M. J., 1982, Alterations of the conformation of the RNAs of alfalfa mosaic virus upon binding of a few coat protein molecules, *Biochim. Biophys. Acta* **696:**260–266.

Stace-Smith, R., and Tremaine, J. H., 1973, Biophysical and biochemical properties of tomato aspermy virus, *Virology* **51:**401–408.

Steven, A. C., Smith, P. R., and Horne, R. W., 1978, Capsid fine structure of cowpea chlorotic mottle virus: From a computer analysis of negatively stained virus arrays, *J. Ultrastruct. Res.* **64:**63–73.

Stubbs, G., Warren, S., and Holmes, K., 1977, Structure of RNA and RNA binding site in tobacco mosaic virus from 4-Å map calculated from X-ray fibre diagrams, *Nature (London)* **267:**216–221.

Sung, M. T., and Dixon, G. H., 1970, Modification of histones during spermiogenesis in trout; A molecular mechanism for altering histone binding to DNA, *Proc. Natl. Acad. Sci. USA* **67:**1616–1623.

Tolin, S. A., 1977, Cucumovirus group, in: *The Atlas of Insects and Plant Viruses Including Mycoplasma Viruses and Viroids*, pp. 303–309, Academic Press, New York.

Torbet, J., 1983, Internal structural anisotropy of spherical viruses studied with magnetic birefringence, *EMBO J.* **2:**63–66.

Van Beynum, G. M. A., DeGraff, J. M., Castel, A., Kraal, B., and Bosch, L., 1977, Structural studies on the coat protein of alfalfa mosaic virus: The complete primary structure, *Eur. J. Biochem.* **72:**63–78.

van Vloten-Doting L., 1975, Coat protein is required for infectivity of tobacco streak virus: Biological equivalence of the coat proteins of tobacco streak and alfalfa mosaic virus, *Virology* **65:**215–225.

van Vloten-Doting, L., and Jaspars, E. M. J., 1972, The uncoating of alfalfa mosaic virus by its own RNA, *Virology* **48:**699–708.

van Vloten-Doting, L., and Jaspars, E. M. J., 1977, Plant co-virus systems: Three component systems, *Compr. Virol.* **11:**1–10.

van Vloten-Doting, L., Francki, R. I. B., Fulton, R. W., Kaper, J. M., and Lane, L. C., 1981, Tricornaviridae proposed family of plant viruses with tripartite, single stranded, RNA genomes, *Intervirology* **15:**198–209.

Verduin, B. J. M., 1974, The preparation of CCMV-protein in connection with its association into a spherical particle, *FEBS Lett.* **45:**50–54.

Verduin, B. J. M., 1978, Ph.D. thesis, Laboratory for Virology, Wageningen.
Vriend, G., Hemminga, M. A., Verduin, B. J. M., deWit, J. L. T., and Schaafsma, T. J., 1981, Segmental mobility involved in protein–RNA interaction in cowpea chlorotic mottle virus, *FEBS Lett.* **134:**167–171.
Vriend, G., Verduin, B. J. M., Hemminga, M. A., and Schaffsma, T. J., 1982a, Mobility involved in protein–RNA interaction in spherical plant viruses, studied by nuclear magnetic resonance spectroscopy, *FEBS Lett.* **145:**49–52.
Vriend, G., Hemminga, M. A., Verduin, B. J. M., and Schaafsma, T. J., 1982b, Swelling of cowpea chlorotic mottle virus studied by proton nuclear magnetic resonance, *FEBS Lett.* **146:**319–321.
Ysebaert, M., van Emmels, J., and Fiero W., 1980, Total nucleotide sequence of a nearly full-size DNA copy of satellite tobacco necrosis virus RNA, *J. Mol. Biol.* **143:**273–287.
Zuidema, D., Bierhuizen, M. F. A., Cornelissen, B. J. C., Bol, J. F., and Jaspars, E. M. J., 1983a, Coat protein binding sites on RNA1 of alfalfa mosaic virus, *Virology* **125:**361–369.
Zuidema, D., Bierhuizen, M. F. A., and Jaspars, E. M. J., 1983b, Removal of the N-terminal part of Alfalfa Mosaic Virus coat protein interferes with the specific binding to RNA 1 and genome activation, *Virology* **129:**255–260.
Zulauf, M., 1977, Swelling of brome mosaic virus as studied by intensity fluctuation spectroscopy, *J. Mol. Biol.* **114:**259–266.

CHAPTER 3

Viral Genome Structure

Robert H. Symons

I. INTRODUCTION

All characterized members of the Bromoviruses, Cucumoviruses, Ilarviruses, and also alfalfa mosaic virus (AMV) contain a tripartite genome in that the three largest of the four major encapsidated RNAs are required for infectivity. The Ilarviruses and AMV are distinguished from the two other virus groups by the additional requirement for infection of either a small amount of coat protein or its subgenomic mRNA, RNA 4 (reviewed in van Vloten-Doting and Jaspars, 1977). A further distinguishing feature is the inability of the RNAs of Ilarviruses and AMV to be aminoacylated in the presence of plant aminoacyl tRNA synthetases, whereas the RNAs of the Bromoviruses and Cucumoviruses that have been tested accept tyrosine (Hall, 1979; Koper-Zwarthoff and Bol, 1980). This division of the viruses into two groups on the basis of the ability or inability of their RNAs to be aminoacylated indicates significant differences in the biological and structural properties of the RNAs.

The increasing availability of sequence data for RNAs of these viruses allows a much more rigorous comparison of RNA structure and function than has been possible in the past. Already the complete sequence of the genomes of AMV and BMV have been determined (see below) and that of CMV should be completed in the near future. In this review, the major emphasis will be on the correlation of the published sequence data of representatives of the four virus groups with their known biological properties.

II. GENERAL STRUCTURE OF TRIPARTITE GENOMES

Each of the four encapsidated RNAs of the four virus groups falls into four size classes with RNA 1 varying in length between 3200 and

ROBERT H. SYMONS   •   Department of Biochemistry, The University of Adelaide, Adelaide, South Australia 5000, Australia.

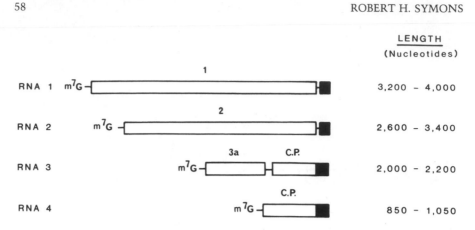

FIGURE 1. General structure of the four viral RNAs of Bromoviruses, Cucumoviruses, Ilarviruses, and alfalfa mosaic virus. The open boxes represent the cistrons on each RNA for
the translation products; protein 1 from RNA 1, protein 2 from RNA 2, protein 3a from
RNA 3, and coat protein (C.P.) from RNA 4. The coat protein cistron in RNA 3 is not
translated *in vitro*. The 3'-terminal closed boxes represent the 140- to 320-residue region
of sequence homology between the four RNAs of each virus.

4000 residues, RNA 2 from 2600 to 3400 residues, RNA 3 from 2000 to
2200 residues, and RNA 4 from 850 to 1050 residues (Fig. 1, Table I). The
exact size of an RNA can only be determined by sequencing of the complete molecule and this has been done for the four RNAs of one strain
of AMV and one of BMV, RNAs 2, 3, and 4 of CMV, RNA 3 of TSV, and
RNA 4 of CCMV (Table I). The variation in molecular weight estimates
from different laboratories for the same RNA determined by gel electrophoresis using RNA molecular weight standards indicates the difficulty
of obtaining more than an approximate molecular weight by this method.
In spite of this problem, there appears to be only limited variation in size
for each of the four RNAs (Fig. 1, Table I).

An m^7G cap structure is present at the 5'-end of each of the four
RNAs of AMV (Pinck, 1975), BMV (Dasgupta *et al.*, 1976), CMV (Symons,
1975), and RNA 4 of CCMV (Dasgupta and Kaesberg, 1982). None of these
RNAs contain a 3'-terminal poly(A) tail. An important feature is the presence of a long 3'-terminal substantially homologous sequence at the end
of each of the four RNAs of each virus (Fig. 1). This varies from about
140 residues for the AMV RNAs (Gunn and Symons, 1980a; Cornelissen
et al., 1983b; Barker *et al.*, 1983) to 250 to 300 residues for the RNAs of
CMV (Symons, 1979), BMV (Gunn and Symons, 1980b; Ahlquist *et al.*,
1981a,b, 1984), CCMV, and BBMV (Ahlquist *et al.*, 1981a). It seems reasonable to assume that the RNAs of all members of the four virus groups
will eventually be shown to contain m^7G caps, long 3'-terminal homologous sequences and to be lacking poly(A) tails.

In vitro translation studies have been carried out with purified RNAs
of AMV, BMV, CMV, and TAV (for review, see Atabekov and Morozov,

TABLE I. Molecular Weight of Viral RNAs

Virus	M_r (× 10^{-6})				References
	RNA 1	RNA 2	RNA 3	RNA 4	
Alfalfa mosaic virus (AMV)	1.24*	0.88*	0.69*	0.30*	Cornelissen et al. (1981a, b), Barker et al. (1983)
Bromoviruses					
Brome mosaic virus (BMV)	1.10*	0.97*	0.72*	0.30*	Ahlquist et al. (1981b, 1984), Dasgupta and Kaesberg (1982)
Cowpea chlorotic mottle virus (CCMV)	1.20	1.07	0.81	0.28*	Lane (1981), Dasgupta and Kaesberg (1982)
Broad bean mottle virus (BBMV)	1.10	1.03	0.90	0.36	Lane (1981)
Cucumoviruses					
Cucumber mosaic virus (CMV)	1.35	1.03*	0.75*	0.35*	Peden and Symons (1973), Gould and Symons (1982), Rezaian et al. (1984)
Tomato aspermy virus (TAV)	1.26	1.10	0.90	0.43	Habili and Francki (1974)
Peanut stunt virus (PSV)	1.3	1.0	0.75	0.35	Lot and Kaper (1976)
Ilarviruses					
Tobacco streak virus (TSV)	1.04	0.98	0.75*	0.3	Lister and Saksena (1976), Cornelissen et al. (1984)
Tulare apple mosaic virus (TAMV)	1.01	0.92	0.74	0.3	Lister and Saksena (1976)
Spinach latent virus (SLV)	1.30	1.18	0.91	0.35	Bos et al. (1980)
Citrus leaf rugose virus (CLRV)	1.1	1.0	0.7	0.3	Gonsalves and Garnsey (1975a)
Citrus variegation virus (CVV)	1.1	1.0	0.7	0.3	Gonsalves and Garnsey (1975b)
Lilac ring mottle virus (LRMV)	1.18	1.13	0.90	0.37	van der Meer and Huttinga (1979)

* Complete sequence determined for RNA from one strain of virus.

1979; van Vloten-Doting and Neeleman, 1982). Each RNA behaves as a monocistronic mRNA with the translation products of RNAs 1, 2, and 4 accounting for most of the coding potential of each RNA. However, RNA 3 is known to be dicistronic (Fig. 1) with the 5'-cistron coding for protein 3a and the silent 3'-cistron coding for the coat protein. RNA 4, a subgenomic RNA, is a very efficient mRNA for the synthesis of coat protein.

III. RNA SEQUENCE AND GENE CONTENT

A. Information Derived from RNAs that Have Been Completely Sequenced

Table II summarizes data obtained from the complete sequence of 13 RNAs. The lengths of the 5'-untranslated regions not counting the m^7G cap vary widely, from 9 residues for BMV RNA 4 and 10 residues for CCMV RNA 4 to 240 residues for AMV RNA 3. The 3'-untranslated regions are, except in the case of AMV RNA 3 and TSV RNA 3, appreciably longer and less variable in length than the 5'-untranslated regions. The 3'-terminal 145-residue homologous sequence for the four RNAs of strain 425 of AMV does not extend into the coding region for any of the RNAs (Table II). However, it appears that the longer regions of 250 to 300 residues showing extensive sequence homology in the BMV, CMV, and CCMV RNAs may extend either up to or just into the coding regions of the coat protein cistron of the RNAs 4 (Table II). The possible biological significance of this 3'-terminal sequence homology is considered in Section IV.C.

There is only one long open reading frame in AMV and BMV RNAs 1 and 2, CMV RNA 2, and in RNA 4 of AMV, TSV, BMV, CCMV, and CMV. The predicted sizes of these gene products correspond to those found on *in vitro* translation of these RNAs (Atabekov and Morozov, 1979; van Vloten-Doting and Neeleman, 1982). In the case of RNA 4 of AMV (Brederode *et al.*, 1980), BMV (Dasgupta and Kaesberg, 1982; Moosic *et al.*, 1983), and CCMV (Rees and Short, 1982), the amino acid sequences of the coat proteins predicted from the nucleotide sequence are in good agreement with those determined by direct sequencing of the coat protein. For CMV RNA 4 (Gould and Symons, 1982), the predicted amino acid sequence of the coat protein is in good agreement with its amino acid composition.

The sequences of RNA 3 of AMV, TSV, BMV, and CMV contain two nonoverlapping reading frames, the one at the 5'-end corresponding in size to the protein 3a found on *in vitro* translation, while the 3'-cistron contains the coat protein gene (Table II). A most unusual feature of BMV RNA 3 is the presence of a poly(A) tract of variable length (16 to 22 residues) in the intercistronic region (Ahlquist *et al.*, 1981b) while poly(A)

TABLE II. Lengths of Various Regions in Viral RNAs

RNA		Total	5'-untranslated	Coding	Intercistronic	Coding	3'-untranslated	References
				Number of nucleotides[a]				
AMV	RNA 1	3644	100	3381	—	—	163	Cornelissen et al. (1983b)
(425)	RNA 2	2593	54	2373	—	—	166	Cornelissen et al. (1983a)
	RNA 3	2037	240	903	49	666	179	Barker et al. (1983)
	RNA 4	881	36	—	—	666	179	Brederode et al. (1980)
TSV	RNA 3	2205	210	870	123	714	288	Cornelissen et al. (1984)
BMV	RNA 1	3234	74	2886	—	—	274	Ahlquist et al. (1984)
	RNA 2	2865	103	2469	—	—	293	Ahlquist et al. (1984)
	RNA 3	2111–2117	91	912	241–247	570	297	Ahlquist et al. (1981b)
	RNA 4	876	9	—	—	570	297	Dasgupta and Kaesberg (1982)
CCMV	RNA 4	824	10	—	—	573	241	Dasgupta and Kaesberg (1982)
CMV	RNA 2	3035	92	2520	—	—	423	Rezaian et al. (1984)
(Q)	RNA 3	2193	94	1002	123	711	263	Gould and Symons (1982)
	RNA 4	1027	53	—	—	711	263	Gould and Symons (1982)

[a] 5'-untranslated region does not include m^{7}G cap. Coding region includes termination codon.

tracts of unknown length and location are also present in CCMV and BBMV RNA 3 (Ahlquist *et al.*, 1981a). These poly(A) tracts in the Bromovirus RNAs may have some unique function since none are present in RNA 3 of CMV, AMV, and TSV.

B. Proportion of Each RNA Containing Protein-Coding Regions

The model in Fig. 1 predicts that the RNAs of the tripartite viruses considered here contain only four cistrons. Since this could be an oversimplification of what happens *in vivo*, it is important to consider the possible existence of other cistrons in these RNAs as well as the formation, *in vivo*, of subgenomic mRNAs other than RNA 4 which would give other polypeptide products.

A first approach is to determine the proportion of each RNA which codes for known polypeptide products. This can be done accurately for those 13 RNAs which have been sequenced and much less accurately for those which have not. In the latter case, the difficulty of determining the molecular weights of both the RNA and its *in vitro* translation product by standard gel electrophoresis techniques means that only an approximate calculation can be made. Data for the RNAs of AMV, TSV, BMV, CCMV, and CMV are given in Table III; there appear to be no data available for BBMV and the Ilarviruses. For RNAs 1 and 2 of most viruses, about 90% is required to code for the *in vitro* translation products and this leaves essentially no RNA to code for an extra polypeptide. The exception at present appears to be RNA 1 of CMV where only about 65% of the RNA is required to code for the synthesis of the M_r-95,000 protein. Even allowing for a long 3'-untranslated region of about 260 residues, a putative intercistronic region, and a 5'-untranslated region, there still remains enough RNA to code for a protein of about M_r 30,000. Complete sequence determination of CMV RNA 1 will eventually show if such calculations are justified.

C. Are There Overlapping Reading Frames on Viral RNAs? Are There Subgenomic RNAs Other Than RNA 4?

Analysis of the sequence data of the 13 RNAs in Table II indicates the potential of these RNAs to code for protein in the two reading frames other than the one used for the known full-length *in vitro* translation products. BMV RNA 3 contains a number of potential AUG initiation codons which could code for short polypeptides (Ahlquist *et al.*, 1981b). With CMV RNA 3, a similar situation exists but the three largest putative AUG-initiated polypeptides are 95, 42, and 33 amino acids long (Gould and Symons, 1982). In AMV RNA 1, the longest AUG-initiated open reading frame can potentially code for 35 amino acids (Cornelissen *et al.*,

TABLE III. Sizes of Translation Products of Viral RNAs

RNA		Length (nucleotides)	Translation product		% residues coding	References
			M_r	No. of amino acids		
AMV (425)	RNA 1*	3644	125,700	1126	92.8	Cornelissen et al. (1983b)
	RNA 2*	2593	89,750	790	91.5	Cornelissen et al. (1983a)
	RNA 3*	2037	32,400	301	44.5 (77.3)[a]	Barker et al. (1983)
	RNA 4*	884	24,380	222	75.7	Brederode et al. (1980)
TSV	RNA 3*	2205	31,740	289	39.5 (71.8)[a]	Cornelissen et al. (1984)
BMV	RNA 1*	3234	109,000	961	89.2	Shih and Kaesberg (1976), Ahlquist et al. (1984)
	RNA 2*	2865	94,200	822	86.2	Shih and Kaesberg (1976), Ahlquist et al. (1984)
	RNA 3*	2111–2117	32,480	303	43.0 (70.0)[a]	Ahlquist et al. (1981b)
	RNA 4*	876	20,900	189	65.1	Dasgupta and Kaesberg (1982)
CCMV	RNA 1	3500	105,000	955	81.0	Davies and Verduin (1979)
	RNA 2	3150	105,000	955	91.0	Verduin (1978)
	RNA 3	2380	32,000	290	37.0	Davies and Verduin (1979)
	RNA 4*	824	21,000	175	70.0	Dasgupta and Kaesberg (1982)
CMV (Q)	RNA 1	4000	95,000	860	65.0	Schwinghamer and Symons (1977)
	RNA 2*	3035	94,300	839	83.0	Rezaian et al. (1984)
	RNA 3*	2193	36,700	333	45.6 (78.1)	Gould and Symons (1982)
	RNA 4*	1027	26,200	236	69.2	Gould and Symons (1982)

* RNAS that have been sequenced.
[a] % for both protein 3a and coat protein.

1983a) while in AMV RNA 2 it is 42 amino acids (Cornelissen *et al.*, 1983b). A similar analysis has not been done for AMV RNA 3 (Barker *et al.*, 1983).

An important unresolved problem is whether or not any potential AUG-initiated polypeptides of any length and derived from any of the three reading frames are synthesized *in vivo* and, if they are, whether this occurs by selective translation of the viral RNAs or after the formation of subgenomic mRNAs. Since the full-length *in vitro* translation products of RNAs 1, 2, and 3 of both CMV and AMV have yet to be detected *in vivo* or have been detected only with difficulty (Nassuth *et al.*, 1981; Samac *et al.*, 1983; A. R. Gould, J. Haseloff, and R. H. Symons, unpublished data), sensitive techniques must be used to search for any of these putative polypeptides *in vivo*. Fortunately, techniques are now becoming available which should allow definitive answers to be obtained. The availability of recombinant DNA clones to specific parts of each viral RNA will allow the selection of potential subgenomic RNAs from plant extracts by hybridization and their translation in cell-free systems. Preliminary evidence with CMV indicates that translatable subgenomic mRNAs exist for CMV RNAs 1 and 2 (Gordon and Symons, unpublished data). Other methods need to be developed for the sensitive and specific detection of low concentrations of the putative polypeptides. Given the nucleotide sequence of the coding region of any subgenomic RNA, and hence the amino acid sequence of the polypeptide, synthetic peptides 10–20 residues long corresponding to a part of the polypeptide can be made and used as antigens (Walter and Doolittle, 1983). Antibodies to such peptides should hopefully provide the means of showing whether these putative polypeptides are present or absent *in vivo*.

Another aspect which should be considered is whether the complementary minus strands of the viral RNAs contain AUG-initiated open reading frames which could code for polypeptides. The complementary strands of RNA 4 of AMV and BMV do have open reading frames which could code for 138 and 118 amino acids, respectively, whereas the longest potential product of CCMV RNA 4 is 39 amino acids (van Vloten-Doting *et al.*, 1982). For the complementary strands of BMV RNA 3, the two longest potential products are 117 and 98 amino acids (Ahlquist *et al.*, 1981b), while the longest product for RNA 1 is 160 amino acids, and for RNA 2, 159 amino acids (Ahlquist *et al.*, 1984). The minus strand of AMV RNA 2 could code for polypeptides of 94, 91, and 60 amino acids (Cornelissen *et al.*, 1983b), and that of AMV RNA 1 for polypeptides of 85 and 78 amino acids (Cornelissen *et al.*, 1983a). The significance of such open reading frames, which have also been noticed in the complementary sequence of TMV RNA (Goelet *et al.*, 1982) and in the noncoding DNA strand of many structural genes (Casino *et al.*, 1981), is unknown. However, as considered above, techniques are now becoming available which allow the direct testing for the presence or absence of the putative polypeptides *in vivo*.

IV. SEQUENCE AND STRUCTURE OF VIRAL RNAs

The complete as well as partial sequence data for a number of viral RNAs allow the comparative analysis of specific parts of an RNA molecule for both sequence and proposed secondary structures. Such comparisons, both between the RNAs of different viruses as well as between the RNAs of different strains of the same virus, should provide us with information on the relation between viral RNA structure and function.

A. 5'-Terminal Sequences

An m^7G cap structure has been shown to be present at the 5'-ends of all RNAs of AMV, BMV, CMV, and RNA 4 of CCMV (see Section II). Unlike many mRNAs of eukaryotes, the two residues adjacent to the m^7G cap of the RNAs that have been sequenced (Table II) are not methylated.

The first 30 residues of the 13 sequenced RNAs are given in Fig. 2. The only common features appear to be the $m^7GpppGU$ sequence at the 5'-end of each RNA with either A or U as the third residue, and a preponderance of A and U in the leader sequences to the initiating AUG (see also Section IV.B). The possible role of such features in the *in vivo* translation of these RNAs is unknown (van Vloten-Doting and Neeleman, 1982). An interesting feature of BMV RNAs 1 and 2 is the substantial sequence homology in the first 42 residues (Fig. 2) with only two mismatches (Ahlquist *et al.*, 1984). The 5'-terminal leader sequences of the RNAs from different strains of the same virus should provide information on conserved and variable residues. Thus, the sequences of the leader sequence are highly conserved in RNAs 1 and 2 of strains S, B, and 425 of AMV with lengths of 101, 99, and 100 residues, respectively, for RNA 1, and 54, 55, and 53 residues for RNA 2 (Cornelissen *et al.*, 1983a,b; Ravelonandro *et al.*, 1983). On the other hand, there are significant differences at the 5'-terminus of AMV RNA 3; the leader sequence is 257 residues long for strain S but only 241 residues for strain B and 240 residues for strain 425. The sequences from residue 39 of strain S and residue 22 of strains B and 425 through to the initiating AUG are highly conserved, whereas the homologous 22 5'-terminal residues of strains B and 425 differ considerably from the 5'-terminal 39 residues of strain S (Barker *et al.*, 1983; Ravelonandro *et al.*, 1983). There are no similar comparative data for the other viral RNAs.

B. Sequences Upstream from Initiating AUG

It appears that initiation of translation of mRNA or eukaryotic ribosomes begins at the AUG nearest the 5'-end of the RNA and it has been

```
                        1              10             20            30
                        |              |              |             |
AMV    RNA 1   m⁷GppppGUUUUUAUCUUACACACGCUUGUGUAAGAU
(425)
       RNA 2   m⁷GppppGUUUUUAUCUUUUCGCGAUUGAAAAGAUAA

       RNA 3   m⁷GppppGUUUUAAAACCAUUUUCAAAAUAUUCCAAU

       RNA 4   m⁷GppppGUUUUUAUUUUUAAUUUUCUUUCAAUUACU


                        |              |              |             |
TSV    RNA 3   m⁷GppppGUAUUCUCCGAGCUUAAGAUACCACUUGCA


                        |              |              |             |
BMV    RNA 1   m⁷GppppGUAGACCACGGAACGAGGUUCAAUCCCUUG

       RNA 2   m⁷GppppGUAAACCACGGAACGAGGUUCAAUCCCUUG

       RNA 3   m⁷GppppGUAAAAUACCAACUAAUUCUCGUUCGAUUC

       RNA 4   m⁷GppppGUAUUAAUAAUGUCGACUUCAGGAACUGGU
                                    ‾‾‾

                        |              |              |             |
CCMV   RNA 4   m⁷GppppGUAAUUUAUCAUGUCUACAGUCGGAACAGG
                                    ‾‾
```

FIGURE 2. 5′-Terminal sequences of the 13 viral RNAs that have been sequenced completely. The four AMV RNAs are from strain 425 and the three CMV RNAs from strain Q. References are given in Table II. The initiation codons for the coat proteins of BMV and CCMV RNA 4 are underlined.

```
                        |              |              |             |
CMV    RNA 2   m⁷GppppGUUUAUUCUCAAGAGCGUAUGGUUCAACCC
(Q)
       RNA 3   m⁷GppppGUAAUCUUACCACUUUCUUUCACGUCGUGU

       RNA 4   m⁷GppppGUUUAGUUGUUCACCUGAGUCGUGUUUUCU
```

suggested that this may be the only signal required (Baralle and Brownlee, 1978; Kozak, 1981). However, attempts have been made to find other recognition signals 5′ to the AUG which may provide recognition sites for ribosome binding. For example, Kozak (1981) has suggested that $^A_GXXAUGG$ is a favored sequence for eukaryotic initiation with either a purine in position −3, or a G in position +4, or both. Sargan *et al.* (1982) have proposed a consensus sequence of AUCACC in the 5′-leader sequence which may bind to the 3′-terminus of 18 S rRNA in the ribosome. Sequences of the 13 viral RNAs up to 20 residues upstream from the initiating AUG are presented in Fig. 3. The sequences do not fit the latter proposal but most of the RNAs conform to the model of Kozak (1981). Since ribosome recognition and binding of an mRNA most likely involves both the sequence and the three-dimensional structure of the RNA, it is perhaps not surprising that a simple, uniform pattern has not emerged.

In the case of AMV RNA 1 of strain 425, the initiating codon is the second AUG from the 5′-terminus. The first AUG at residue 52 is fol-

```
                       -20              -10             -1
                        |                |               |
 AMV     RNA 1   A A U A C U G U G A A G A U U U C A C U A U G A A U G
(425)
         RNA 2   U U C A G U U U A A U C U U U U C A A U A U G U U C A

         RNA 3   U C U U C G U G A G U A A G U U G U A A A U G G A G A

         RNA 4   U U C U U U C A A U U A C U U C C A U C A U G A G U U


 TSV     RNA 3   G A C C C G C C A C U G A A A G G A A G A U G G C G U

         RNA 4   C U C G G A C U U A C C U G A G A U G U A U G A A U A


 BMV     RNA 1   C U U U G U U U U U C A C C A A C A A A A U G U C A A

         RNA 2   U U C U U U C U A C U A U C A C C A A G A U G U C U U

         RNA 3   U G A U A C U G U U U U U G U U C C C G A U G U C U A

         RNA 4           Cap — G U A U U A A U A A U G U C G A


 CCMV    RNA 4           Cap — G U A A U U U A U C A U G U C U A


 CMV     RNA 2   C U A G U C U C U C U U C U G U U A C U A U G A U A A
 (Q)
         RNA 3   G U G U U U A G A U U A C G A A G G U U A U G G C U U

         RNA 4   U U U U G C G U C U C A G U G U G C C U A U G G A C A
```

FIGURE 3. Sequences around the initiation codon (underlined) for proteins in the 13 viral RNAs that have been sequenced completely. References are given in Table II.

lowed by an in-phase termination codon at residue 88 (Cornelissen *et al.*, 1983a). Whether or not this 12-amino-acid polypeptide is produced *in vivo* is unknown. Of interest, however, is an A to U mutation at residue 52 which eliminates this AUG in RNA 1 of strains S and B (Ravelonandro *et al.*, 1983). A similar situation exists in RNA 1 of strain S, but not strains B and 425, of AMV in which an AUG at residue 91 is followed by a UGA termination codon at residue 103, whereas correct initiation is assumed to occur at the AUG at residue 102 (Ravelonandro *et al.*, 1983). For CMV RNA 2, the initiating codon is also the second AUG from the 5'-terminus (Rezaian *et al.*, 1984). The first AUG occurs at residues 19–21 and is followed by an in-phase stop codon at residues 43–45.

C. 3'-Terminal Sequences and Structures

The 3'-termini of viral RNAs are of great interest because they must be the sites of initiation of viral RNA synthesis; also their sequences and tertiary structures will be important in allowing recognition by RNA replicase. A peculiar feature of the Bromoviruses (BMV, CCMV, and BBMV) and the Cucumoviruses is the ability of their RNAs to be aminoacylated on the 3'-terminal hydroxyl by tyrosine in the presence of plant aminoacyl tRNA synthetases (see Hall, 1979). Such a property implies a tRNA-like terminal structure. AMV RNAs cannot be aminoacylated; the RNAs of Ilarviruses have not been tested but it is presumed that they will behave like the AMV RNAs since both groups of viruses require the presence of coat protein for the initiation of viral replication, in contrast to the Bromoviruses and the Cucumoviruses (van Vloten-Doting and Neeleman, 1982).

1. Bromoviruses and Cucumoviruses

A common feature of the four RNAs of BMV, CCMV, BBMV, CMV, and TAV is the long 3'-terminal homologous sequence for each virus (Symons, 1979; Gunn and Symons, 1980b; Wilson and Symons, 1981; Ahlquist et al., 1981a,b 1984; Dasgupta and Kaesberg, 1982). In BMV RNAs, this homology extends for at least 230 residues from the 3'-termini, at least 200 residues for the CCMV and BBMV RNAs, and at least 190 residues for TAV RNAs 1, 3, and 4 (RNA 2 sequence has not been determined; Wilson and Symons, 1981). In the case of the Q strain of CMV, there is an unusual distribution of homologous and nonhomologous sequences (Fig. 4). The first 138 residues of RNAs 1 and 2 are identical to those of RNAs 3 and 4 except for one residue in RNA 1 and three residues in RNA 2. From residue 139 to 270 from the 3'-terminus, RNAs 1 and 2 showed, relative to RNAs 3 and 4, a nonhomologous region of 40 residues, a partially homologous region of 46 residues, and a homologous region of 14 residues which probably extends further. There are 11 residues different between RNAs 1 and 2 (Symons, 1979). This unusual distribution of homologous and nonhomologous sequences is probably not a general feature of CMV RNA since, although the RNAs of P-CMV show a very similar distribution of the 210 3'-terminal residues so far sequenced, the 165 residues sequenced of the four T-CMV RNAs do not. In addition, there is extensive sequence homology between the 3'-terminal regions of Q- and P-CMV RNAs, but much less homology with those of the T-CMV and M-CMV RNAs which are closely similar (R. F. Barker, P. A. Wilson, and R. H. Symons, unpublished data). Hence, the four strains of CMV investigated so far seem to fall into two groups as determined by 3'-terminal sequence homology and by the distribution of homologous and nonhomologous regions between the genomic RNAs of each strain. It is feasible that the Q and P strains may have arisen by reassortment of RNAs following mixed infection by two different parental strains, one of which provided RNAs 1 and 2 and the other RNA 3, and hence RNA 4. Such reassortment of RNAs may occur in vivo in view of the demonstrated formation of pseudorecombinant virus strains by mixing of different CMV RNAs in vitro (Rao and Francki, 1982).

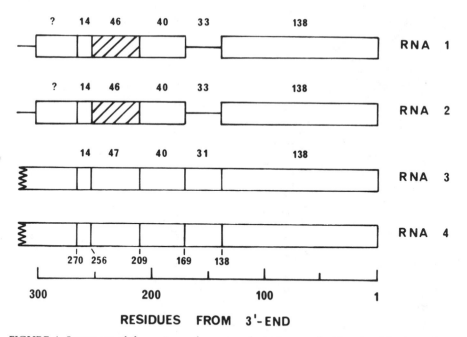

FIGURE 4. Summary of the regions of sequence homology at the 3'-ends of the four RNAs of Q-CMV. The open boxes correspond to regions of complete or almost complete homology while the cross-hatched areas represent partial homology of RNAs 1 and 2 with RNAs 3 and 4. The single horizontal lines represent no homology of RNAs 1 and 2 with RNAs 3 and 4. The numbers above each region indicate the number of residues in that region. The question mark represents a region of homology of about 30 to 40 residues. Data from Symons (1979).

The 3'-terminal 45 residues of RNA 4 of BMV, CCMV, BBMV, Q-CMV, and V-TAV are given in Fig. 5. All these RNAs as well as the others sequenced contain the terminal CCA expected for an RNA that can be aminoacylated. The terminal sequence GAGACCA is conserved in all RNAs except those of TAV. The sequences in Fig. 5 have been arranged to maximize the limited sequence homology outside this terminal region.

The 3'-terminal 170 or so residues of all the Bromovirus and Cucumovirus RNAs sequenced can be arranged in a two-dimensional hydrogen-bonded structure which mimics the familiar cloverleaf pattern of tRNA (Symons, 1979; Gunn and Symons, 1980b; Wilson and Symons, 1981; Ahlquist et al., 1981a). Two such predicted structures, those of RNAs 4 of Q-CMV and V-TAV, are given in Fig. 6. In spite of significant differences in sequence between these RNAs, the two-dimensional models are very similar and include the possible hydrogen bonding of the hairpin of loop E to the 3'-stem (marked with a solid line in Fig. 6) and the sequence in loop F with the region underlined in the variable loop. Similar structures can be drawn for all the sequenced Bromovirus and Cucumovirus RNAs; the ones given by Ahlquist et al. (1981a) are vari-

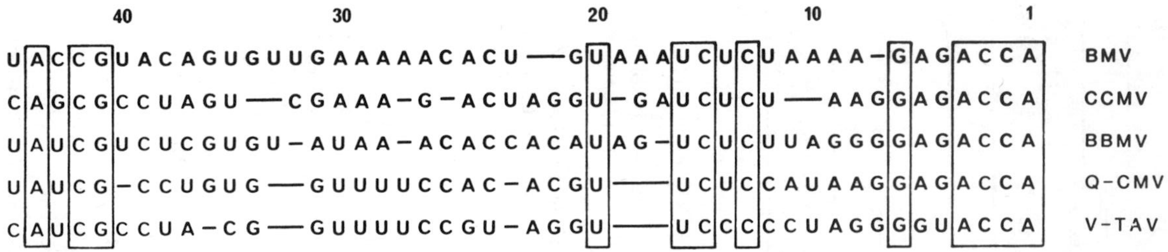

FIGURE 5. Comparison of the 3'-terminal sequences of RNA 4 of five viruses. Sequences have been arranged to show maximum sequence homology (boxed for all five RNAs). References are given in Table II.

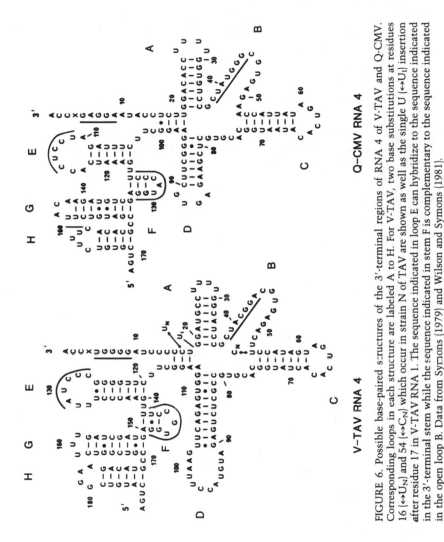

V-TAV RNA 4

Q-CMV RNA 4

FIGURE 6. Possible base-paired structures of the 3'-terminal regions of RNA 4 of V-TAV and Q-CMV. Corresponding loops in each structure are labeled A to H. For V-TAV, two base substitutions at residues 16 (↔U$_N$) and 54 (↔C$_N$) which occur in strain N of TAV are shown as well as the single U (↔U$_I$) insertion after residue 17 in V-TAV RNA 1. The sequence indicated in loop E can hybridize to the sequence indicated in the 3'-terminal stem while the sequence indicated in stem F is complementary to the sequence indicated in the open loop B. Data from Symons (1979) and Wilson and Symons (1981).

ations on the one in Fig. 6 and include the double-stranded region between loop F and the single-stranded stretch in the variable loop. However, the structures proposed by Ahlquist et al. (1981a) do not contain the variable loop B and the 3'-terminal -ACCA is not completely single-stranded, both features of which are characteristic of tRNAs.

An intriguing feature of the proposed structure of the BBMV RNAs is the absence of loop D (Ahlquist et al., 1981a). This difference corresponds exactly to a 20-residue gap when the sequences of the BBMV RNAs are arranged to give maximum homology with the other Bromovirus and Cucumovirus RNAs (Ahlquist et al., 1981a).

The conservation of a very similar base-pairing pattern at the 3'-ends of all the Bromovirus and Cucumovirus RNAs is a strong argument for its physical and biological significance. Whether or not aminoacylation is an essential part of RNA replication remains to be determined. However, the three-dimensional structure of the 3'-ends of the RNAs must play some essential role in the initiation of RNA replication.

2. Alfalfa Mosaic Virus and Ilarviruses

In contrast to the Bromovirus and Cucumovirus RNAs, the 3'-terminal sequences of AMV and TSV, an Ilarvirus, cannot be folded into a tRNA-like structure and they cannot be aminoacylated (Koper-Zwarthoff and Bol, 1980). In addition, they do not contain a 3'-terminal -CCA characteristic of tRNAs (Koper-Zwarthoff and Bol, 1980; Gunn and Symons, 1980a). It is interesting to note that the coat proteins of AMV, TSV, CLRV, and CVV are equally capable of activating their own as well as each other's genomes (van Vloten-Doting, 1975; Gonsalves and Garnsey, 1975c). In addition, the RNAs of AMV and TSV are both able to withdraw protein subunits from intact AMV particles, indicating that the RNAs of the two viruses contain specific sites with high affinity for AMV coat protein (van Vloten-Doting and Jaspars, 1972; van Vloten-Doting, 1975; Houwing and Jaspars, 1978). It has been proposed that the binding of the coat protein to the 3'-terminus of these RNAs is required for proper recognition of the genome fragments by the viral RNA replicase (Houwing and Jaspars, 1978). Hence, there appears to be a fundamental difference in the initiation of replication between the coat protein-dependent and the coat protein-independent viruses.

In spite of the similarities considered so far between the AMV and TSV RNAs, no sequence homology between them could be detected by competition hybridization experiments (Bol et al., 1975) but this technique is not as sensitive as a direct hybridization approach (Gould and Symons, 1983). The 3'-terminal 142 residues of all four AMV RNAs are substantially conserved after which there is not significant homology, except for RNA 3 which contains the complete sequence of RNA 4 at its 3'-end (Cornelissen et al., 1983a,b; Barker et al., 1983). In contrast, there is only 35% sequence homology in the 3'-terminal 140 residues of TSV

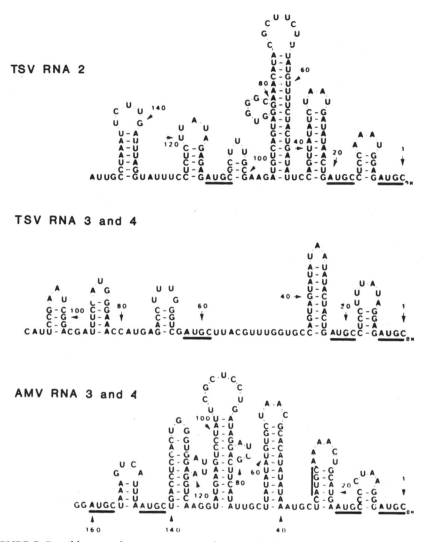

FIGURE 7. Possible secondary structures at the 3'-ends of TSV and AMV RNAs. The conserved AUGC sequences are underlined. Data from Koper-Zwarthoff and Bol (1980).

RNAs 2 and 3, most of which is near the 3'-end (Koper-Zwarthoff and Bol, 1980). There is about 24% sequence homology between the 3'-terminal 140 residues of the AMV RNAs and TSV RNA 3 and even less with TSV RNA 2 (Koper-Zwarthoff and Bol, 1980).

Possible secondary structures at the 3'-termini of TSV RNA 2, TSV RNA 3, and AMV RNA 3 are given in Fig. 7 (Koper-Zwarthoff and Bol, 1980). In spite of the differences in sequence between the RNAs, similar hairpin structures can be formed. In addition to the hairpin structures, the sequence AUGC (underlined) appears conserved, but its significance is unknown.

D. Intercistronic Region of RNA 3

The intercistronic region between the genes coding for protein 3a and the coat protein on RNA 3 varies from 49 residues for AMV to 247 residues for BMV (Table II). The corresponding region has only been sequenced for one of the Ilarviruses, TSV. A unique feature of BMV RNA 3 is the presence of a poly(A) tract of variable length (16 to 22 residues) in this intercistronic region which is absent from RNA 3 of CMV, AMV, and TSV; this aspect has been considered in Section III.B.

The intercistronic regions of BMV, CMV, and AMV, and TSV RNA 3 can be folded into long hairloop structures (Koper-Zwarthoff et al., 1980; Ahlquist et al., 1981b; Gould and Symons, 1982; Cornelissen et al., 1984) which may be important for site-specific recognition if an RNA processing enzyme is involved in the production of RNA 4 by cleavage of RNA 3. Alternatively, it may prevent ribosome binding and hence initiation of protein synthesis at the coat protein cistron. Other secondary structures may be feasible; e.g., the intercistronic region of CMV RNA 3 could form a stable hydrogen-bonded structure with the 5'-untranslated region (Gould and Symons, 1983).

E. Characterization of Viral RNAs by Hybridization Analysis

In spite of the increasing rapidity with which nucleic acid molecules can now be sequenced, the complete sequence determination of a viral genome containing some thousands of residues is still a major undertaking. In the case of tripartite plant viruses, the three genomic RNAs total 8000 to 10,000 residues. Hence, in order to compare the genomes of several strains of the same virus or of related viruses in the same virus group, approaches other than direct sequencing must be used, at least for the present. The technique of hybridization analysis using cDNA is a powerful method for estimating the extent of sequence homology between viral nucleic acids, for probing the finer aspects of the organization of the viral genome, and for detecting and quantifying the presence of minor RNA species. The methodology, scope, and limitations of this technique have been considered in detail by Gould and Symons (1983).

One of the first applications of this technique was the estimation of the sequence homology between the four RNAs of one isolate of CMV (Q-CMV) (Gould and Symons, 1977). It was shown that there was very little homology between RNAs 1, 2, and 3, but that the complete sequence of RNA 4 was contained in RNA 3 and not present in the other two RNAs. This work with CMV was extended to the four RNAs of AMV and similar results were obtained (Gould and Symons, 1978). In addition, it was possible to show by the use of cDNA enriched for the 3'-terminal sequences of RNA 4 of both CMV and AMV that the sequence of RNA 4 was present at the 3'-end of RNA 3. These results were subsequently confirmed by sequencing data (see Section IV.C).

Estimating the sequence homology between the genomes of different members of a group of viruses can provide important information for the classification of these viruses. Thus, cDNA hybridization analysis was applied to the RNAs of four Cucumoviruses, three strains of CMV and one strain of TAV (Gonda and Symons, 1978). Although the corresponding RNAs of the Q and P strains of CMV were indistinguishable, those of the M strain showed only 15 to 30% sequence homology to the Q strain RNAs when the hybridization assay was carried out at high stringency (low salt concentration) and from 26 to 56% sequence homology in a lower stringency assay (high salt concentration). It is rather surprising that two viruses that are both classified as CMV show only about 30% sequence homology between their RNAs. The TAV RNAs showed sequence homology with the corresponding Q-CMV RNAs of 2 to 22% depending on the stringency of the hybridization assay.

In spite of hybridization analysis with labeled cDNA being a useful technique for a comparison of nucleic acids, it has been little used for plant viral RNAs. It requires a minimum of technical apparatus, although the purification of viral RNAs and the actual hybridization analysis are rather laborious. The technique would seem particularly appropriate for an initial characterization of the Ilarviruses and for further characterization of the Cucumoviruses in view of the large number of isolates worldwide.

V. SATELLITE RNAs OF THE CUCUMOVIRUSES

Of all the tripartite plant viruses, only the Cucumoviruses have been found to contain satellite RNAs. This term is used to refer to an RNA that is unable to multiply in cells without the assistance of a specific helper virus, is not necessary for the multiplication of the helper virus, but is encapsidated in helper virus coat protein, and has no appreciable sequence homology with the helper virus genome (Murant and Mayo, 1982). Most field isolates of CMV and PSV do not contain a satellite RNA while there are no reports of a satellite RNA in natural isolates of TAV. In view of the very high infectivity and stability of satellite RNA of CMV, glasshouse stocks of CMV can inadvertently become infected with satellite RNA unless extreme precautions are taken (Mossop and Francki, 1978, 1979).

When RNA isolated from purified CMV is fractionated by electrophoresis on polyacrylamide gels, the four major RNAs are readily detected by staining or by scanning the gel in ultraviolet light (Kaper and West,

1972; Peden and Symons, 1973). In addition, three smaller, minor RNA components are usually found: RNA 4a (M_r approx. 0.25×10^6), RNA 5 (M_r approx. 0.12×10^6), and RNA 6 (M_r approx. 0.05×10^6) (Peden and Symons, 1973; Symons, 1978). Although RNA 4a has not been characterized, RNA 5 usually makes up less than 5% by weight of total viral RNA and consists of specific cleavage products of CMV RNAs 1–4, while RNA 6 contains fragments of viral RNAs as well as host RNA, possibly tRNA (Symons, 1978; Gould et al., 1978). Bands corresponding to RNAs 4a, 5, and 6 can often be seen in published gel patterns of CMV RNAs (e.g., Kaper and West, 1972).

When CMV satellite RNA is also present, it usually migrates slightly slower than RNA 5 on gel electrophoresis (Gould et al., 1978). Since the original RNA 5 is not a satellite RNA, but derived from the viral genomic RNAs, it is ambiguous to refer to CMV satellite RNAs as RNA 5 (Murant and Mayo, 1982) as has been done for CARNA 5, or cucumber mosaic virus-associated RNA 5 which is a satellite RNA (Kaper and Tousignant, 1977).

The sequences of three satellite RNAs of CMV are compared in Fig. 8. The necrosis-inducing (n)CARNA 5 contains 335 residues (Richards et al., 1978) and is one residue larger than the nonnecrotic (1)CARNA 5 and is 93% homologous (Collmer et al., 1983). Sat-RNA contains 336 residues (Gordon and Symons, 1983) and is over 90% homologous with the other two satellite RNAs. Most of the differences between the three molecules occur in the half adjoining the 3'-end and involve mainly base substitutions with some deletions and insertions. All RNAs contain an m^7G cap at the 5'-end. Unlike the CMV RNAs, the satellite RNAs cannot be aminoacylated which is not surprising since they lack the characteristic 3'-terminal CCA of tRNA and their sequence cannot be folded into the cloverleaf tRNA-like structure (Gordon and Symons, 1983). The sequence of a satellite RNA found associated with PSV has not been reported; however, CMV could not support the replication of the PSV satellite RNA and vice versa (Kaper et al., 1978).

The amount of satellite RNA in preparations of CMV is highly variable and depends on the strain of helper CMV, the host plant, and the growth conditions. Different satellite RNA–CMV combinations can lead to either enhancement or attenuation of disease symptoms relative to the satellite-free CMV (see Murant and Mayo, 1982). How the satellite RNAs mediate these effects is not known although attenuation may be due to a decreased synthesis of helper virus. On the basis of their sequences (Fig. 8), the putative polypeptide products vary considerably. Sat-RNA can code for only one polypeptide of 16 amino acids, (1)CARNA 5 for two of 27 and 53 amino acids, and (n)CARNA 5 for three of 27, 41, and 3 amino acids. Although in vitro translation products of CMV satellite RNAs have been reported (Owens and Kaper, 1977; Yamaguchi et al., 1982; Gordon and Symons, 1983), no satellite RNA-specific proteins are known to be produced in vivo.

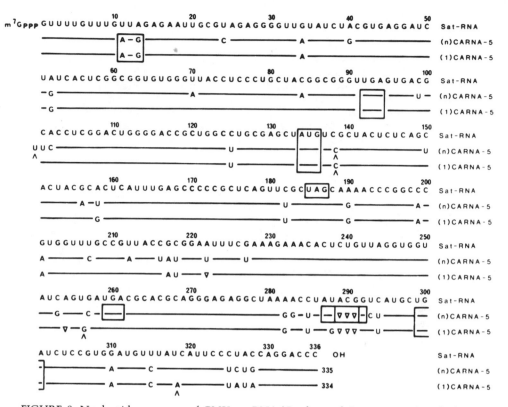

FIGURE 8. Nucleotide sequence of CMV sat-RNA (Gordon and Symons, 1983) with that of (n)CARNA 5 (Richards *et al.*, 1978) and (1)CARNA 5 (Collmer *et al.*, 1983). Only residues differing from sat-RNA are shown while insertions and deletions in (n)CARNA 5 and (1)CARNA 5 are represented by ∧ and ∇, respectively. Possible initiation codons and their corresponding in-phase termination codons are boxed.

Since CMV satellite RNA shows no sequence homology with the genomic RNAs of CMV as determined by hybridization analysis (Gould *et al.*, 1978), its mechanism of replication is of considerable interest. CMV satellite RNA presumably makes use of the machinery required for the replication of the viral RNAs so that some common features would be expected to be present at the 3'-ends of both the viral and satellite RNAs where replication is initiated. It is significant that the proposed 3'-terminal structure of sat-RNA shows some sequence and structural homology with the viral RNAs in that sat-RNA possesses a truncated tRNA-like structure (Gordon and Symons, 1983).

ACKNOWLEDGMENTS.   The author thanks his colleagues in the laboratory for many helpful discussions, Jenny Rosey and Sharon Freund for assistance, and the Australian Research Grants Scheme for financial support.

REFERENCES

Ahlquist, P., Dasgupta, R., and Kaesbert, P., 1981a, Near identity of 3' RNA secondary structure in Bromoviruses and cucumber mosaic virus, *Cell* **23:**183–189.

Ahlquist, P., Lucknow, V., and Kaesberg, P., 1981b, Complete nucleotide sequence of brome mosaic virus RNA 3, *J. Mol. Biol.* **153:**23–38.

Ahlquist, P., Dasgupta, R., and Kaesberg, P., 1984, Nucleotide sequence of the brome mosaic virus genome and its implications for viral replication, *J. Mol. Biol.* **172:**369–383.

Atabekov, J. G., and Morozov, S. Y., 1979, Translation of plant virus messenger RNAs, *Adv. Virus Res.* **25:**1–91.

Baralle, F. E., and Brownlee, G. G., 1978, AUG is the only recognizable signal sequence in the 5' non-coding regions of eucaryotic mRNA, *Nature (London)* **274:**84–87.

Barker, R. F., Jarvis, N. P., Thompson, D. V., Loesch-Fries, L. S., and Hall, T. C., 1983, Complete nucleotide sequence of alfalfa mosaic virus RNA 3, *Nucleic Acids Res.* **11:**2881–2891.

Bol, J. F., Brederode, F. T., Janze, G. C., and Rauh, D. C., 1975, Studies on sequence homology between the RNAs of alfalfa mosaic virus, *Virology* **65:**1–15.

Bos, L., Huttinga, H., and Maat, D. Z., 1980, Spinach latent virus, a new Ilarvirus seed-borne in *Spinacia oleracea*, *Neth. J. Plant Pathol.* **86:**79–98.

Brederode, F. T., Koper-Zwarthoff, E. C., and Bol, J. F., 1980, Complete nucleotide sequence of alfalfa mosaic virus RNA 4, *Nucleic Acids Res,* **8:**2213–2223.

Casino, A., Cipollaro, M., Guerrini, A. M., Mastrocinque, G., Spena, A., and Scarlato, V., 1981, Coding capacity of complementary DNA strands, *Nucleic Acids Res.* **9:**1499–1518.

Collmer, C. W., Tousignant, M. E., and Kaper, J. M., 1983, Cucumber mosaic virus-assiciated RNA 5. X. The complete nucleotide sequence of a CARNA 5 incapable of inducing tomato necrosis, *Virology* **127:**230–234.

Cornelissen, B. J. C., Brederode, F. T., Moormann, R. J. M., and Bol, J. F., 1983a, Complete nucleotide sequence of alfalfa mosaic virus RNA 1, *Nucleic Acids Res.* **11:**1253–1265.

Cornelissen, B. J. C., Brederode, F. T., Veeneman, G. H., van Boom, J. H., and Bol, J. F., 1983b, Complete nucleotide sequence of alfalfa mosaic virus RNA 2, *Nucleic Acids Res.* **11:**3019–3025.

Cornelissen, B. J. C., Janssen, H., Zuidema, D., and Bol, J. F., 1984, Complete nucleotide sequence of tobacco streak virus RNA 3, *Nucleic Acids Res.* **12:**2427–2437.

Dasgupta, R., and Kaesberg, P., 1982, Complete nucleotide sequence of the coat protein messenger RNAs of brome mosaic virus and cowpea chlorotic mottle virus, *Nucleic Acids Res.* **10:**703–713.

Dasgupta, R., Harada, F., and Kaesberg, P., 1976, Blocked 5' termini in brome mosaic virus RNA, *J. Virol.* **18:**260–267.

Davies, J. W., and Verduin, B. J. M., 1979, *In vitro* synthesis of cowpea chlorotic mottle virus polypeptides, *J. Gen. Virol.* **44:**545–549.

Goelet, P., Lomonossoff, G. P., Butler, P. J. G., Akam, M. E., Gait, K. J., and Karn, J., 1982, Nucleotide sequence of tobacco mosaic virus RNA, *Proc. Natl. Acad. Sci. USA* **79:**5818–5822.

Gonda, T. J., and Symons, R. H., 1978, The use of hybridization analysis with complementary DNA to determine the RNA sequence homology between strains of plant viruses: Its application to several strains of Cucumoviruses, *Virology* **88:**361–370.

Gonsalves, D., and Garnsey, S. M., 1975a, Functional equivalence of an RNA component and coat protein for infectivity of citrus leaf rugose virus, *Virology* **64:**23–31.

Gonsalves, D., and Garnsey, S. M., 1975b, Nucleic acid components of citrus variegation virus and their activation by coat protein, *Virology* **67:**311–318.

Gonsalves, D., and Garnsey, S. M,. 1975c, Infectivity of heterologous RNA–protein mixtures from alfalfa mosaic, citrus leaf rugose, citrus variegation, and tobacco steak viruses, *Virology* **67:**319–326.

Gordon, K. H. J., and Symons, R. H., 1983, Satellite RNA of cucumber mosaic virus forms a secondary structure with partial 3'-terminal homology to genomal RNAs. *Nucleic Acids Res.* **11**:947–960.

Gould, A. R., and Symons, R. H., 1977, Determination of the sequence homology between the four RNA species of cucumber mosaic virus by hybridization with complementary DNA, *Nucleic Acids Res.* **4**:3787–3802.

Gould, A. R., and Symons, R. H., 1978, Alfalfa mosaic virus RNA: Determination of the sequence homology between the four RNA species and a comparison with the four RNA species of cucumber mosaic virus, *Eur. J. Biochem.* **91**:269–278.

Gould, A. R., and Symons, R. H., 1982, Cucumber mosaic virus RNA 3: Determination of the nucleotide sequence provides the amino acid sequences of protein 3A and viral coat protein, *Eur. J. Biochem.* **126**:217–226.

Gould, A. R., and Symons, R. H., 1983, A molecular biological approach to relationships among viruses, *Annu. Rev. Phytopathol.* **21**:179–199.

Gould, A. R., Palukaitis, P., Symons, R. H., and Mossop, D. W,. 1978, Characterization of a satellite RNA associated with cucumber mosaic virus, *Virology* **84**:443–455.

Gunn, M. R., and Symons, R. H., 1980a, Sequence homology at the 3'-termini of the four RNAs of alfalfa mosaic virus, *FEBS Lett.* **109**:145–150.

Gunn, M. R., and Symons, R. H., 1980b, The RNAs of Bromoviruses: 3'-terminal sequences of the four brome mosaic virus RNAs and comparison with cowpea chlorotic mottle virus RNA 4, *FEBS Lett.* **115**:77–82.

Habili, N., and Francki, R. I. B., 1974, Comparative studies on tomato aspermy and cucumber mosaic viruses. I. Physical and chemical studies, *Virology* **57**:392–401.

Hall, T. C., 1979, Transfer RNA-like structures in viral genomes, *Int. Rev. Cytol.* **60**:1–26.

Houwing, C. J., and Jaspars, E. M. J., 1978, Coat protein binds to 3' terminal part of RNA 4 alfalfa mosaic virus, *Biochemistry* **17**:2927–2933.

Kaper, J. M., and Tousignant, M. E., 1977, Cucumber mosaic virus-associated RNA 5. I. Role of host plant and helper strain in determining amount of associated RNA 5 with virions, *Virology* **80**:186–195.

Kaper, J. M., and West, C. K., 1972, Polyacrylamide gel separation and molecular weight determination of the components of cucumber mosaic virus RNA, *Prep. Biochem.* **2**:251–263.

Kaper, J. M,. Tousignant, M. E., Diaz-Ruiz, J. R., and Tolin, S. A., 1978, Peanut stunt virus-associated RNA 5: Second tripartite genome virus with an associated satellite-like replicating RNA, *Virology* **88**:166–170.

Koper-Zwarthoff, E. C., and Bol, J. F., 1980, Nucleotide sequence of the putative recognition site for coat protein in the RNAs of alfalfa mosaic virus and tobacco streak virus, *Nucleic Acids Res.* **8**:3307–3318.

Koper-Zwarthoff, E. C., Brederode, F. T., Veeneman, G., van Boom, J. H., and Bol, J. F., 1980, Nucleotide sequences at the 5'-termini of the alfalfa mosaic virus RNAs and the intercistronic junction in RNA 3, *Nucleic Acids Res.* **8**:5635–5647.

Kozak, M., 1981, Possible role of flanking nucleotides in recognition of the AUG initiator codon by eukaryotic ribosomes, *Nucleic Acids Res.* **9**:5233–5252.

Lane, L. C., 1981, Bromoviruses, in: *Handbook of Plant Virus Infections and Comparative Diagnosis* (E. Kurstak, ed.), pp. 333–376, Elsevier/North-Holland, Amsterdam.

Lister, R. M., and Saksena, K. N., 1976, Some properties of tulare apple mosaic and ILAR viruses suggesting grouping with tobacco streak virus, *Virology* **70**:440–450.

Lot, H., and Kaper, J. M., 1976, Physical and chemical differentiation of three strains of cucumber mosaic virus and peanut stunt virus, *Virology* **74**:209–222.

Moosic, J. P., McKean, D. J., Shih, D. S., and Kaesberg, P., 1983, Primary structure of brome mosaic virus coat protein, *Virology* **129**:517–520.

Mossop, D. W., and Francki, R. I. B., 1978, Survival of a satellite RNA *in vivo* and its dependence on cucumber mosaic virus for replication, *Virology* **86**:562–566.

Mossop, D. W., and Francki, R. I. B., 1979, The stability of satellite RNAs *in vivo* and *in vitro*, *Virology* **94**:243–253.

Murant, A. F., and Mayo, M. A., 1982, Satellites of plant viruses, *Annu. Rev. Phytopathol.* **20**:49–70.

Nassuth, A., Alblas, F., and Bol, J. F., 1981, Localization of genetic information involved in the replication of alfalfa mosaic virus, *J. Gen. Virol.* **53**:207–214.

Owens, R. A., and Kaper, J. M., 1977, Cucumber mosaic virus-associated RNA 5. II. *In vitro* translation in a wheat germ protein-synthesis system, *Virology* **80**:196–203.

Peden, K. W. C., and Symons, R. H., 1973, Cucumber mosaic virus contains a functionally divided genome, *Virology* **53**:487–492.

Pinck, L., 1975, The 5′-end groups of alfalfa mosaic virus RNAs are $m^7G^{5'}ppp^{5'}Gp$, *FEBS Lett.* **59**:24–28.

Rao, A. L. N., and Francki, R. I. B., 1982, Distribution of determinants for symptom production and host range on the three RNA components of cucumber mosaic virus, *J. Gen. Virol.* **61**:197–205.

Ravelonandro, M., Godefroy-Colburn, T., and Pinck, L., 1983, Structure of the 5′-terminal untranslated region of the genomic RNAs from two strains of alfalfa mosaic virus, *Nucleic Acids Res.* **11**:2815–2826.

Rees, N. W., and Short, M. N., 1982, The primary structure of cowpea chlorotic mottle virus coat protein, *Virology* **119**:500–503.

Rezaian, M. A., Williams, R. H. V., Gordon, K. H. J., Gould, A. R., and Symons, R. H., 1984, Nucleotide sequence of cucumber mosaic virus RNA 2 reveals a translation product significantly homologous to corresponding proteins of other viruses, *Eur. J. Biochem.* **143**:277–284.

Richards, K. E., Jonard, G., Jacquemond, M., and Lot, H., 1978, Nucleotide sequence of cucumber mosaic virus associated RNA 5, *Virology* **89**:395–408.

Samac, D. A., Nelson, S. E., and Loesch-Fries, L. S., 1983, Virus protein synthesis in alfalfa mosaic virus infected alfalfa protoplasts, *Virology* **131**:455–462.

Sargan, D. R., Gregory, S. P., and Butterworth, P. H. W., 1982, A possible novel interaction between the 3′-end of 18S ribosomal RNA and the 5′-leader sequence of many eukaryotic messenger RNAs, *FEBS Lett.* **147**:133–136.

Schwinghamer, M. W., and Symons, R. H., 1977, Translation of the four major RNA species of cucumber mosaic virus in plant and animal cell-free systems and in toad oocytes, *Virology* **79**:88–108.

Shih, D. S., and Kaesberg, P., 1976, Translation of the RNAs of brome mosaic virus: The monocistronic nature of RNA 1 and RNA 2, *J. Mol. Biol.* **103**:77–88.

Symons, R. H., 1975, Cucumber mosaic virus RNA contains 7-methylguanosine at the 5′-terminus of all four RNA species, *Mol. Biol. Rep.* **2**:277–285.

Symons, R. H., 1978, The two-step purification of ribosomal RNA and plant viral RNA by polyacrylamide gel electrophoresis, *Aust. J. Biol. Sci.* **31**:25–37.

Symons, R. H., 1979, Extensive sequence homology at the 3′-terminus of the four RNAs of cucumber mosaic virus, *Nucleic Acids Res.* **7**:825–837.

van der Meer, F. A., and Huttinga, H., 1979, Lilac ring mottle virus, *CMI/AAB Descriptions of Plant Viruses* No. 201.

van Vloten-Doting, L., 1975, Coat protein is required for infectivity of tobacco streak virus: Biological equivalence of the coat proteins of tobacco streak virus and alfalfa mosaic virus, *Virology* **65**:215–225.

van Vloten-Doting, L., and Jaspars, E. M. J., 1972, The uncoating of alfalfa mosaic virus by its own RNA, *Virology* **48**:699–708.

van Vloten-Doting, L., and Jaspars, E. M. J., 1977, Plant covirus systems: Three components systems, in: *Comprehensive Virology* (H. Fraenkel-Conrat and R. Wagner, eds.), Volume 11, pp. 1–53, Plenum Press, New York.

van Vloten-Doting, L., and Neeleman, L., 1982, Translation of plant virus RNA's, in: *Nucleic Acids and Proteins in Plants* (B. Parthier and D. Boulter, eds.), pp. 337–367, Springer-Verlag, Berlin.

van Vloten-Doting, L., Dubelaar, M., and Bol, J. F., 1982, Open reading frame in the minus strand of two plus type RNA viruses, *Plant Mol. Biol.* **1:**155–158.

Verduin, B. J. M., 1978, Degradation of cowpea chlorotic mottle virus ribonucleic acid *in situ*, *J. Gen. Virol.* **39:**131–147.

Walter, G., and Doolittle, R. F., 1983, Antibodies against synthetic peptides, in: *Genetic Engineering: Principles and Methods* (J. K. Setlow and A. Hollaender, eds.), Volume 5, pp. 61–69, Plenum Press, New York.

Wilson, P. A., and Symons, R. H., 1981, The RNAs of Cucumoviruses: 3'-terminal sequence analysis of two strains of tomato aspermy virus, *Virology* **112:**342–345.

Yamaguchi, K., Hidaka, S., and Miura, K., 1982, Relationship between structure of the 5' non-coding region of viral mRNA and efficiency in the initiation step of protein synthesis in a eukaryotic system, *Proc. Natl. Acad. Sci. USA* **79:**1012–1016.

CHAPTER 4

Virus Multiplication

R. HULL AND A. J. MAULE

I. INTRODUCTION

The multiplication of a plant virus covers all the events from the entry of the virus particles into the initially infected cells, via the spread of the virus through the plant, to the formation of complete virus particles ready for transmission to other plants. Because the tripartite viruses occur in relatively high concentrations and because their divided genomes offer experimental advantages, more is known about their multiplication than that of many other viruses.

The study of plant virus multiplication necessitates having *in vivo* and *in vitro* systems in which the various stages can be dissected. It is only with advances in techniques that a deeper understanding of the mechanisms involved can be obtained. Therefore, we consider the various systems which are available and which have been used.

There are several stages in virus multiplication which are logically, but not necessarily, sequential. The virus particles have first to enter the cell and to be uncoated. There is some initial translation of the input RNA (plus-strand RNA) to give gene products essential for RNA replication. This is followed by RNA replication, first by the formation of RNA complementary to input RNA (minus-strand RNA) and then the formation of new plus-strand RNA. It is likely that RNA replication amplifies the mRNAs of gene products (e.g., that of coat protein) needed in the later stages leading finally to encapsidation of the newly synthesized plus-strand RNA.

The aim of this chapter is to review what is known about the multiplication of viruses with tripartite genomes and to highlight areas where knowledge is lacking but where progress is likely to be made in the near future. There have been previous reviews on the subject (see van Vloten-

R. HULL AND A. J. MAULE • John Innes Institute, Norwich NR4 7UH, England.

Doting and Jaspars, 1977; Lane, 1979): this chapter is intended to mainly cover the findings which have been reported since these reviews.

II. EXPERIMENTAL SYSTEMS

A. Whole Plants

Before the advent of protoplast techniques in the late 1960s and eukaryotic *in vitro* translation systems in the early 1970s, studies of the multiplication of plant viruses relied upon the inoculation of whole plants or detached leaves and monitoring the increase in virus-specific products or infectious virus as the infection progressed. This approach was, on the whole, reliable and predictable, and much basic information relating to virus accumulation and spread was obtained. Moreover, although the newer techniques have provided the opportunity to break down the complex interaction between host and virus into more manageable and defined areas, in the 1980s the intact plant still provides the mainstay for experimentation.

The extent to which a virus multiplies in the tissues of a plant does not depend solely upon the competence of individual cells, but also involves factors such as tissue specificity, antagonism from host defenses, efficiency of virus spread, host physiology, and environmental conditions. Consequently, studies of virus multiplication in whole plants have been mainly restricted to the analysis of virus- or infection-specific macromolecules formed in susceptible host tissues, and normally those in which systemic spread and extensive virus accumulation has occurred. The progressive nature of infections in intact tissues will mean that, at best, these analyses will indicate the presence of all the components associated with virus replication and spread at any one time, and probably in quatitites greater than those obtained using single-cell systems. However, a fundamental disadvantage to this approach is that, owing to the asynchrony characteristic of progressive infection, the correlation of particular molecular species with the temporal stages of replication in individual cells is not possible.

Attempts to introduce a "degree of synchrony" by differential temperature treatment of infected plants (Dawson and Schlegel, 1973; W. O. Dawson et al., 1975) have been moderately successful but there have been few examples of its application to the study of replication of the tripartite virus (Dawson and Schlegel, 1973; Roberts and Wood, 1981a). The treatment depends upon the systemic spread of the virus from parts of the host plant held at a permissive temperature to other parts held at a nonpermissive low temperature and the synchronous multiplication of virus upon transfer back to the permissive. Unfortunately, in practice this method is only applicable to plants with a large internodal length (e.g., *Nicotiana tabacum* L. cv. Xanthi). Similar transfers of whole systemically-infected plants or leaves, or even smaller tissue pieces from a low-holding to a high-permissive temperature (Dawson and Lung, 1976) also give improved synchrony. However, in the majority of work involving the kinetics of formation of virus-specific products, synchronous infections have been achieved using protoplasts.

Tissue specificity, seen for instance with the phloem-limited Luteoviruses, could severely limit the use of whole plants for studying virus multiplication. However, for the tripartite viruses considered here, this does not apply since none of the viruses appear to have tissue restrictions in their susceptible or tolerant hosts.

The situation with regard to resistant hosts is less straightforward. For those reacting to hypersensitivity [e.g., cucumber mosaic virus (CMV) in cowpea], virus multiplication in a few cells is necessary for the host to recognize the infection and to react by stimulating necrosis in affected cells. Clearly the whole plant would not be the favored experimental system here. Resistance may also be associated with some inherent property of the plant limiting virus replication in individual cells (Motoyoshi and Oshima, 1977; Beier et al., 1979; Maule et al., 1980c). The problem here is the amount of virus-specific material available rather than the number of cells infected and such plants are probably best avoided when examining replication per se.

Despite the reservations mentioned above, it should be remembered that there is no alternative to the whole plant when trying to correlate virus multiplication with the physiology of disease and its relationship to the situation in the field, especially since none of the tripartite viruses examined cause symptoms to be expressed in protoplasts.

Finally, for the scientist wishing to use whole plants to study multiplication of the tripartite viruses, it is encouraging to note that he is assisted not only by their mechanical transmissibility but also by the infectious nature of their naked RNA. Thus, it is possible to attempt pseudorecombination experiments between different strains or mutants of the same virus (Fulton, 1970; Roosien and van Vloten-Doting, 1983) and between different viruses (Bancroft, 1972; Habili and Francki, 1974; Mossop and Francki, 1977, 1979) allowing biochemical and physiological aspects of virus multiplication to be mapped at least to the genomic segments.

The fine mapping of viral gene functions using mutants is only possible with whole plants when the mutants have a permissive temperature (ts mutants) (Roosien and van Vloten-Doting, 1982) or are nonlethal. The observations that plants can be infected with naked DNA from Geminiviruses (ssDNA) (Goodman, 1977a,b; Bock et al., 1978) and Caulimoviruses (dsDNA) (Shepherd et al., 1968) and that cDNA copies of poliovirus RNA are infectious for animal cell lines (Racaniello and Baltimore, 1981) perhaps indicate that in the future, site-directed mutagenesis of cDNA copies of RNA viruses may be possible for the analysis of gene function. Very recently, Alquist et al. (1984) have shown that RNA transcripts of cDNAs to BMV RNAs are infectious. Thus, even if cDNAs themselves are not infectious, transcription from them should overcome this problem. (Pseudorecombinants and virus mutants are discussed in detail in Chapter 5.)

B. Protoplasts

The infection of isolated protoplasts can be achieved with both virus particles and isolated virus RNA (Table I). With the exception of BMV (Okuno *et al.*, 1977; Furusawa and Okuno, 1978), all the tripartite viruses require some chemical mediator [e.g., poly-L-ornithine (PLO), polyethylene glycol (PEG)] to stimulate infection of protoplasts (Table I). These chemicals operate at least partly by overcoming the charge constraints to close interaction between the negatively'charged plasmalemma and similarly charged virus or viral RNA. For BMV, the surface charge is less negative to the extent that it will complex with cowpea chlorotic mottle virus (CCMV) to give enhanced infections of tobacco protoplasts (Watts *et al.*, 1981, 1984). They probably also operate by causing significant perturbations of the protoplast plasmalemma so stimulating uptake and infection. From this it is important to note that the mechanism of protoplast infection probably bears little relation to natural infection of cells in the plant, and extrapolation from protoplast experiments with regard to the earliest stages in the infection process should be treated with caution (see Section III).

In the absence of symptoms, virus infection of protoplasts is usually monitored as the proportion of protoplasts infected (immunofluorescent staining of viral coat protein, usually as virus particles) and/or, in a population of protoplasts, the accumulation of some virus-specific product (RNA, protein, or infectious virus). These two approaches provide different information, although they have been used interchangeably by some workers. Together they can give an estimate of the average extent of virus multiplication in individual cells. Although the comparison between tissue and isolated protoplasts has only rarely been made, it is likely that, with regard to virus accumulation, protoplasts behave in quantitative terms much as the cells in the tissue from which they were derived (Motoyoshi *et al.*, 1973).

The principal advantage of protoplasts over intact tissue for studying virus replication is that they are infected in a synchronous fashion and, in theory at least, show virus multiplication synchronously. In a situation where cell-to-cell contact is denied, the former must be true, but the latter precept is based upon the assumption that all the cells in the parent tissue were equally competent and that they were equally affected by the trauma of protoplast isolation, neither of which is likely to be correct. Despite this, protoplast populations do show a "cycle" of virus replication representing the average (see Section IV.F). This information is

TABLE I. Infection of Protoplasts by Tripartite Genome Viruses

| Inoculum | | | | |
Virus	Viral RNA	Protoplast host	Uptake inducer	References[a]
AMV (strain 425)	ND[b]	Cowpea	PLO[c]	1–7
AMV-derived ts mutants	ND	Cowpea	PLO	7
	+	Cowpea	None	3
		Cowpea	PLO	4
AMV (strain 425)	+	Alfalfa	PEG	8
TSV	ND	Cowpea	PLO	4
BMV (ATCC 66)	ND	Barley	None	9–12
"	ND	Barley	PLO	9, 11
"	ND	Barley	PLO + osmotic shock	13–15
"	+	Barley	PLO	13, 15
BMV (ATCC 66)	ND	Chenopodium hybridum	PLO + osmotic shock	13
"	ND	Tobacco	PLO + osmotic shock	13
"	ND	Radish	None	10, 12
"	ND	Wheat	None	10
"	ND	Oats	None	10
"	ND	Maize	None	10
"	ND	Turnip	None	12
BMV (Russian)	+	Barley	Protamine sulfate	16, 17
BMV (V5)	ND	Tobacco	None	18
"	ND	Tobacco	PLO	18–21
"	ND	Tobacco	PEG	21
"	+	Tobacco	PLO	18
CCMV (type)	ND	Tobacco	PLO	22–25
"	ND	Tobacco	PEG	21, 26
CMV (Y)	ND	Tobacco	PLO	27–29
CMV (O)	ND	Tobacco	PLO	29
CMV (Q)	ND	Cowpea	PLO	30
CMV ?	ND	Cowpea	PLO	31
"	ND	Cowpea	None	31
CMV (W)	ND	Cucumber	PLO	32, 33
	+	Cucumber	PEG	34

[a] References: 1, Alblas and Bol (1977); 2, Alblas and Bol (1978); 3, Nassuth et al. (1981); 4, Nassuth and Bol (1983); 5, Nassuth et al. (1983a); 6, Nassuth et al. (1983b); 7, Sarachu et al. (1983); 8, Samac et al. (1983); 9, Okuno et al. (1977); 10, Furusawa and Okuno (1978); 11, Okuno and Furusawa (1978a); 12, Maekawa et al. (1981); 13, Okuno and Furusawa (1979); 14, Okuno and Furusawa (1978b); 15, Okuno and Furusawa (1978c); 16, Loesch-Fries and Hall (1980); 17, Kiberstis et al. (1981); 18, Motoyoshi et al. (1974a); 19, Sakai et al. (1979); 20, Watts and Dawson (1980); 21, Sakai et al. (1983); 22, Motoyoshi et al. (1973a); 23, Motoyoshi et al. (1973b); 24, Motoyoshi et al. (1974b); 25, Sakai et al. (1977); 26, Dawson et al. (1978); 27, Otsuki and Takebe (1973); 28, Honda et al. (1974); 29, Takanami et al. (1977); 30, Gonda and Symons (1979); 31, Koike et al. (1977); 32, Coutts and Wood (1976); 33, Maule et al. (1980a); 34, Maule et al. (1980b).
[b] ND, not determined; +, positive infection.
[c] PLO, poly-L-ornithine; PEG, polyethylene glycol.

essential when trying to characterize virus mutants at the physiological level or to interpret data following the use of inhibitors.

Experiments with inhibitors of protein synthesis (cycloheximide, chloramphenicol) and RNA synthesis (actinomycin D, cordycepin), which initially used tissue pieces *in vitro*, are now most frequently performed with protoplasts. This ensures that all the infected cells are exposed to the inhibitor equally, both in time and in concentration (see Section IV.E).

A further advantage conferred by the absence of cell-to-cell contact in protoplast systems is the ability to overcome some host resistance barriers. Hence, the replication of CMV (Gonda and Symons, 1979) and AMV (Alblas and Bol, 1977) have been examined in cowpea protoplasts, a plant demonstrating hypersensitive resistance in inoculated leaves with these viruses. In some cases, it is also possible to extend the host range of the virus; thus, BMV (strain ATCC 66) will infect protoplasts of *Raphanus sativus*, a plant considered to be a nonhost for this virus (Furusawa and Okuno, 1978). In addition to providing a means for studying mechanisms of resistance, this approach allows the worker to compare mechanisms of replication operating in a range of host/protoplast systems.

C. *In Vitro* Systems

1. Translation

The elucidation of mechanisms of viral replication depends upon being able to identify the origins of particular viral gene products and to correlate these with *in vivo* functions. Using *in vitro* translation systems, viral RNA-coded protein products have been characterized for many of the tripartite viruses (Table II) and in many cases it has been possible to assign these to particular RNA species. In the majority of these studies, cell-free translations have been performed using the rabbit reticulocyte lysate, wheat embryo or wheat germ extract systems, with less frequent use being made of ascites cell extracts and *Xenopus* oocyte systems. We could find no evidence for more faithful translation of plant viral RNAs in plant-derived systems compared with eukaryotic systems generally although it is clear that different systems will yield different spectra of protein products. The development of a pea embryo extract active in translation of added mRNA (Penmans *et al.*, 1980) provides the opportunity to use a homologous system for those plant viruses having peas as a host, although this has not yet been exploited.

The relative advantages and disadvantages of the reticulocyte lysate, wheat germ extract, and oocyte systems with regard to plant viral RNAs have been reviewed (Atabekov and Morozov, 1979; Davies, 1979) and will not be covered in detail here, but it might be of value to consider some general points. First, for many of the truly *in vitro* systems listed (not

TABLE II. Translation Products of Tripartite Genome Viruses

Virus	RNA	Cell-free extract	Major in vitro translation products (M_r)	Ref.[a]	Possible in vivo equivalent (M_r)	Ref.[a]	Known function in vivo	Host[b]	Ref[a]
AMV	1	RR, WG[c]	58K, 62K, 115–120K	1, 2	25K		—	Alfalfa	6
	2	RR, WG	95–100K	1–3	90K		—	Alfalfa	
	3	RR, WG	35K	1–3	32K		—	Alfalfa	
	4	RR, WG, KA	25K	1–4	25K		Coat protein	Cowpea, alfalfa	5
TSV	1 + 2	WG	120K, 100K, 50–84K	3	?		—	—	7
	3	WG, RR	34K	3	?		—	—	
	4	WG	28K	3	28K		Coat protein	—	
BMV	1	WG, WE	110–120K	8, 9	110–120K		—	Barley	9, 12, 13
	2	WG, WE	100–110K	8, 9	100–110K		—	Tobacco	
	3	WG, WE	34–35K	9, 10	35–36K		—	C. hybridum	
	4	WG, WE	19.5–20K	9–11	19.5–20K		Coat protein	Tobacco	
CCMV	1	WG, RR	105K	3, 14	100K		—	Tobacco	15
	2	WG, RR	105K	3, 14	100K		—	Tobacco	
	3	WG, RR	32–35K		34–36K		—	Tobacco	
	4	WG, RR	19K		19K		Coat protein	Tobacco	
BBMV	1	—	—		—		—	—	
	2	—	—		—		—	—	
	3	WG	30–35K	11	—		—	—	11
	4	WG	21K	11	—		Coat protein	—	
CMV	1	RR, WG, WE	105K?	16	21K		—	—	11
	2	RR, WG, WE	120K	16	—		—	—	
	3	RR, WG	34–39K	16, 17	—		—	—	
	4	RR, WG, WE, O	24–25K	16, 17	25K		Coat protein	Tobacco (Pl), cowpea, cucumber	18–21
	5	WG	5.2K, 3.8K	22	—		—	—	

[a] References: 1, van Tol and van Vloten-Doting (1979); 2, Gerlinger et al. (1977); 3, Davies (1979); 4, Mohier et al. (1975); 5, Nassuth et al. (1983a); 6, Samac et al. (1983); 7, Ghabrial and Lister (1974); 8, Shih and Kaesberg (1976); 9, Kiberstis et al. (1981); 10, Shih and Kaesberg (1973); 11, Davies and Kaesberg (1974); 12, Sakai et al. (1979); 13, Okuno and Furusawa (1979); 14, Davies and Verduin (1979); 15, Sakai et al. (1977); 16, Schwinghamer and Symons (1977); 17, Schwinghamer and Symons (1975); 18, Ziemiecki and Wood (1975); 19, Gonda and Symons (1979); 20, Roberts and Wood (1981b); 21, M. Boulton (personal communication); 22, Owens and Kaper (1977).

[b] All host protoplasts except Pl (= plant).

[c] RR, rabbit reticulocyte lysates; WG, wheat germ extract; WE, wheat embryo extract; O Xenopus oocytes; KA, Krebs ascites cell extract.

oocytes) the fidelity of translation of individual mRNAs and the nature of the products formed will depend upon the ion concentration or, more exactly, upon the balance between K^+ and Mg^{2+}, being particularly important for large polycistronic RNAs (Davies, 1979). Second, and of particular importance when translating total viral RNAs rather than separated components and when trying to relate *in vitro* products with those formed *in vivo*, is the ability of individual messages to perform competitively.

The use of single RNA species is a convenient way of assigning *in vitro* protein products to particular viral genomic messages. However, for the larger RNAs 1 and 2 of the tripartite viruses and particularly where they are not encapsidated in structurally different particles (e.g., Bromoviruses and Cucumoviruses), this can best be achieved in combination with the elegant techniques of hybrid-arrested translation (Paterson *et al.*, 1977) or hybrid-selected translation (Ricciardi *et al.*, 1979). The *in vitro* protein products of BMV RNAs have thus been assigned to their genomic segments (P. A. Kiberstis, personal communication). An extension of these techniques using cDNA clones specific to particular regions of the genomic RNA in the hybrid-arrest system could give the exact coding position of the protein product, but with the reservations that the cDNA does not block ribosome binding and that its presence may possibly prevent translation of any downstream coding regions.

For those viruses where extensive sequence data have been obtained and potential protein-coding regions identified (see Chapter 3) the origin of a particular *in vitro* translation product can be found by reference to its molecular weight on denaturing gels; the N-terminal protein sequence can then be aligned by the genetic code to the 5'-end of the appropriate gene.

The successful correlation between a potential viral gene, gene product, and functional protein *in vivo* is difficult to achieve and for the tripartite viruses so far has only really been achieved with viral coat protein for AMV, BMV, and CMV (Table II) and in part by extrapolation to other members of the same virus groups. Ideally, after assigning an *in vitro* product to a potential gene, the relationship between the *in vitro* product and its proposed *in vivo* counterpart should be ascertained. This is possible by correlating charge and M_r on two-dimensional gels or M_r alone on SDS denaturing gels, demonstrating antigenic similarity by immunoprecipitation, and finally, by establishing protein sequence homology by tryptic peptide mapping. The worker in this area has also to remember that in the absence of some endogenous protein-processing mechanisms in *in vitro* translation systems, he may still have difficulty in correlating his *in vivo* and *in vitro* protein products.

2. Transcription

Transcription of plant viral RNAs *in vitro* has only recently been attempted, apparently awaiting clarification as to the nature of RNA-dependent RNA polymerase activities responsible for *in vivo* replication of viral RNA. For cowpea mosaic virus (CPMV) and BMV, that confusion has largely been resolved by the isolation of membrane-bound enzymes that are virus-coded and show template specificity (Bujarski *et al.*, 1982; Dorssers *et al.*, 1983) and for which *in vitro* transcription systems are in current use (Miller and Hall, 1983). The technique has been further refined for the BMV replicase by nuclease digestion of endogenous template in the enzyme preparations such that they will initiate and copy added template (Miller and Hall, 1983). This modification makes the system equivalent to *in vitro* translation, and with it provides the opportunity to apply hybrid-arrest transcription for analysis of structural features on BMV RNAs (Ahlquist *et al.*, 1984). In theory, with precisely defined cDNA clones, it should be possible to examine the significance of both primary and theoretical secondary RNA structures on RNA replication.

III. VIRUS ENTRY AND UNCOATING

Although virus entry and uncoating are of paramount importance in the establishment of a successful infection, there is virtually nothing known about the subject; much of what is known has recently been reviewed by de Zoeten (1981) and Shaw (1984). The nucleic acids of all the viruses discussed in this volume are infectious on their own and so, at least under experimental mechanical inoculation conditions, coat protein is not a necessity for cell entry. Because of the lack of knowledge of these early events, it is not known whether infection is initiated by particles which enter the cell and then uncoat or by particles uncoating at the plasmalemma and passing the RNA into the cell. Inoculation of leaves or cells is usually with such large numbers of particles that it is not possible to determine if it is the few particles which behave differently which really initiate infection. However, assuming that infection results from at least a proportion of the particles which enter the cell, there is some information on entry of particles into protoplasts.

The work by Burgess *et al.* (1973) and Bancroft *et al.* (1975) with CCMV indicates that plant protoplasts do not take up virus particles by active pinocytotic mechanisms. For successful virus entry there has to be damage to the plasmalemma and, provided the virus particles have the correct charge, they pass into the cell by passive means. The polycations listed as uptake inducers in Table I are known to damage plasma membranes and also have the correct charge. Virus uptake by pinocytosis and stimulated by polycations has been suggested for CMV (Honda *et al.*, 1974) and is discussed further in relation to Bromoviruses by Watts *et al.* (1981) who showed to the contrary that an inhibitor of energy-dependent reactions, sodium azide, did not inhibit uptake of virus. However, it should be borne in mind that the mechanism by which PEG mediates virus uptake is not fully understood and is likely to be different from that of polycations. If the observations with protoplasts extend to cells within the leaf, virus particles will only enter those cells partially damaged by the vector or during mechanical inoculation.

There is a considerable amount of knowledge concerning the conditions needed *in vitro* for the disassembly of the viruses covered in this review (see Chapter 2). Whether this has any relevance to uncoating *in vivo* is open to question. Durham *et al.* (1977) have suggested that calcium binding sites in the particles may control disassembly. The *in vitro* experiments take no account of the possible involvement of membranes or, say, of ribosomes, interacting with the RNAs of relaxed particles (Wilson, 1984). The only *in vivo* study of the uncoating of one of the viruses under review, BMV (Kurtz-Fritsch and Hirth, 1972), showed that in barley leaves there was a rapid phase of uncoating (17% of the particles in the first 20 min) followed by a much slower phase (40% of the particles in the first 8 hr). Interestingly, BMV particles also uncoated in the leaves of a non-host, cabbage, which indicates that host specificity operates at a stage later than uncoating.

One further feature that needs to be borne in mind when discussing virus particle uncoating is that cells in the later stages of infection contain large numbers of coated (progeny) particles. That these remain coated must indicate that there is a spatial separation between the region where uncoating takes place and that where virus assembly and storage occur.

IV. VIRUS REPLICATION

A. Initial Translation

That the input RNA is translated to give products needed for RNA replication can be inferred from two observations. The genomic RNAs of AMV and Ilarviruses need coat protein or the monocistronic gene for coat protein (RNA 4) for infection to be successfully initiated (Bol *et al.*, 1971; Mohier *et al.*, 1974; Gonsalves and Garnsey, 1974, 1975a,b,c; van Vloten-Doting, 1975; Huttinga and Mosch, 1976; Bos *et al.*, 1980; Thomas *et al.*, 1983). It is thought, at least in the case of AMV, that the binding of a few coat protein molecules to the 3'-terminal sequence, which is homologous in RNAs 1, 2, and 3, is required for the proper recognition of these RNAs by the viral replicase (Houwing and Jaspars, 1978; Koper-Zwarthoff and Bol, 1980; Smit *et al.*, 1981). This phenomenon is discussed more fully in Section IV.C.2. Second, when RNAs 1 + 2 of BMV or B and M components of AMV are inoculated into protoplasts, the RNAs replicate on their own (but without the production of virus particles) (Kiberstis *et al.*, 1981; Nassuth *et al.*, 1981); yet RNA 3 of either virus, which has a similar 3' sequence to RNAs 1 and 2 from that virus, cannot replicate independently. This indicates that products from RNAs 1 and 2 are involved in replication.

B. Proteins Involved in Replication

There is considerable discussion about the enzyme involved in the replication of plant viral RNAs which can be resolved into two opposing viewpoints. One is that only host enzymes mediate the replication and the other is that the virus contributes at least some subunits to the RNA replicase. The situation is complicated by the fact that the normal low level of host polymerase activity is enhanced considerably on virus infection. The whole subject is critically discussed by Hall *et al.* (1982). Based mainly on their observations on the BMV polymerase system, they proposed a model for the RNA-dependent RNA-polymerase replication complex in infected plants. This complex comprises a membrane-bound virus-coded polypeptide(s) which confers specificity on the viral RNA and non-membrane-bound, host-coded polypeptides which have nonspecific polymerase activity. They suggest that many of the extraction procedures used for the isolation of RNA polymerases release the host-coded moiety and that this could give rise to incorrect observations. It is only with the use of selective detergents that the full replication complex is released from membranes. In the BMV replicase system, dodecyl-β-ᴅ-maltoside is used in the purification and stabilization of the enzymatic activity (Bujarski *et al.*, 1982).

As far as tripartite genome viruses are concerned, there are several lines of evidence which suggest that virus-coded functions are involved in the RNA replicase. First, as noted earlier (Section IV.A) the fact that RNAs 1 and 2 of AMV (+ coat protein) and of BMV can replicate on their own (Nassuth *et al.*, 1981; Kiberstis *et al.*, 1981) suggests that some of their gene products are involved directly in replication. This has led Nassuth and Bol (1983) to propose the involvement of the P1/P2 complex (gene products of RNAs 1 and 2) in the replication of AMV. Second, in relatively pure preparations of RNA polymerase from BMV-infected plants, there is a polypeptide of the same electrophoretic mobility in gels and with the same tryptic peptide map as the *in vitro* translation product of BMV RNA 1; no such polypeptide species is found in polymerase preparations from mock-inoculated leaves (Hall *et al.*, 1982). In preparations of RNA polymerase from CMV-infected plants, there is also a polypeptide species of similar size to one of the translation products of CMV RNAs 1 and 2 and which is not found in preparations from healthy tissue (Kumarasamy and Symons, 1979). However, Gordon *et al.* (1982) reported that, based on evidence from differences in translation products of various CMV strains and from tryptic peptide mapping, this polypeptide is not a virus-coded protein. In discussion they comment that the polymerase from CMV-infected plants has a lack of template specificity and that this might be conferred by a virus-coded product(s) which they did not extract. In view of the structural features in common between BMV and CMV RNAs (see Section IV.C.2) it would be surprising if these viruses differed markedly in the basic composition of their replicase complexes. Very recently, amino acid homologues have been shown between the putative RNA polymerases of two of the tripartate genome viruses (AMV and BMV) and also between these viruses and two unrelated viruses (Kamer and Argos, 1984; Haseloff *et al.*, 1984).

As noted in Section IV.A, the genomic RNAs of AMV and at least some of the Ilarviruses need coat protein or the monocistronic gene for coat protein (RNA 4) for successful infection. The coat proteins of AMV and of tobacco streak (TSV), citrus leaf rugose, and citrus variegation viruses are equally capable of activating each other's genomes as well as their own (van Vloten-Doting, 1975; Gonsalves and Garnsey, 1975c). The interaction of coat protein with RNA has been most extensively studied in AMV (for review see Jaspars, 1984).

In AMV RNAs 1 and 4 (and presumably RNAs 2 and 3 which have similar 3' noncoding sequences) there is a high-affinity binding site for coat protein covering the 3'-terminal 68 bases (Houwing and Jaspars, 1980, 1982; Stoker *et al.*, 1980; Zuidema *et al.*, 1983a). There are also some internal sites for coat protein binding which could possibly be involved in virus assembly (Zuidema *et al.*, 1983a).

Zuidema *et al.* (1983b) show that the high-affinity binding of coat protein to RNA is abolished by mild trypsin treatment which just removes 25 amino acids at the N-terminus. Fukuyama *et al.* (1983) point out that the sequence of the N-terminal 36 amino acid residues obeys the requirements of a polypeptide chain structure which associates with RNA. It has been suggested (Houwing and Jaspars, 1978; Nassuth and Bol, 1983) that the coat protein bound at the 3'-end of the plus strand is the replicase recognition site. However, since this sequence is not found at the 3'-end of the minus strand, it would imply that plus-strand formation might utilize a different mechanism.

C. RNA Replication

1. Mechanisms

Most, if not all, of the production of RNA in healthy plant cells is from DNA templates. However, there is no evidence that RNA viruses (at least those being considered here) involve DNA directly in the replication of their nucleic acid. All available evidence suggests that RNA replication is via minus-strand RNA. Although the actual molecular biology of this process is poorly understood, some clues have appeared. Virus-specific dsRNA is found in cells infected with AMV (Pinck and Hirth, 1972; Mohier *et al.*, 1974), BMV (e.g., Philips *et al.*, 1974; Bastin and Kaesberg, 1976), CMV (e.g., Kaper and Diaz-Ruiz, 1977; Takanami *et al.*, 1977), and CCMV (Bancroft *et al.*, 1975). Whatever the significance of dsRNA is (see below), its presence indicates that minus-strand RNA is formed.

dsRNAs are found in two different forms, replicative form (RFs) in which the molecule is completely base-paired, and replicative intermediates (RIs) in which part of the molecule is single-stranded. RIs are rarely reported as the identification of double-stranded molecules usually depends upon RNase treatment; however, RIs have been reported for CMV (Takanami et al., 1977). The rest of the discussion below relates to RFs.

For all the viruses listed above, there are dsRNA forms of the three genomic RNAs. There is some dispute as to whether there are dsRNA forms of RNA 4 of some of the viruses. Philipps et al. (1974) and Bancroft et al. (1975) did not find any dsRNA 4 in BMV-infected tissue whereas Bastin and Kaesberg (1976) and Loesch-Fries and Hall (1980) did. Similarly, Bancroft et al. (1975) reported no dsRNA 4 to be present in cells infected with CCMV whereas Dawson and Dodds (1982) found a molecular species of the expected size. However, with other viruses there is not such a conflict of evidence. Pinck and Hirth (1972) reported that AMV did not have a minus strand for RNA 4, an observation subsequently confirmed by blotting techniques (see Nassuth and Bol, 1983). CMV, on the other hand, appears to consistently have dsRNA 4 (Kaper and Diaz-Ruiz, 1977; Takanami et al., 1977; Bar-Joseph et al., 1983; Rosner et al., 1983).

The above reports assume that molecules of the expected size actually are dsRNA 4. Whether they are or not, there is evidence that has been used to suggest that for at least some viruses, RNA 4 is derived from RNA 3 by nucleolytic cleavage. However, the possibility of internal initiation for the transcription of RNA 4 should not be ignored. The absence of minus strand of the size of AMV RNA 4 has been noted above and the possible site for nucleolytic cleavage in RNA 3 has been suggested by Koper-Zwarthoff et al. (1980). For BMV and CMV, the genomic RNA has a sequence at the 5'-end complementary to the 3' region proposed as a recognition site for polymerase binding (see Section IV.C.2). The 5' regions of RNA 4 of these viruses do not have similar complementary sequences. A site for nucleolytic cleavage has been proposed from the sequence data on CMV RNA 3 (Gould and Symons, 1982). BMV, BBMV, and CCMV RNAs 3 contain internal oligo(A) sequences (Ahlquist et al., 1981a,b). In BMV RNA 3 the oligo(A) sequence is in the intercistronic region about 20 nucleotides upstream of the position of the 5'-terminus of RNA 4 (Ahlquist et al., 1981b). From secondary structure considerations the oligo(A) sequence was considered as being involved in the production of RNA 4 from RNA 3 (Ahlquist et al., 1981b). If RNA 4 is cleaved from RNA 3, the 5' portion of RNA 3 (RNA 3a) would be left. Gonda and Symons (1979) suggested that for CMV, RNA 3a is degraded into fragments which are large enough to hybridize to probes. Virion RNAs which migrate in gels between RNAs 3 and 4 have been reported for BMV (Lane and Kaesberg, 1971), CMV (Kaper and West, 1972; Marchoux et al., 1973; Lot et al., 1974), AMV (Bol and Lak-Kaashoek, 1974), and TSV (Clark and

Lister, 1971). With the recent development of probing techniques the nature of these RNAs could be examined to determine if they have sequences in common with RNA 3a.

Thus, there is considerable doubt about the functional significance of dsRNA 4. There should also be a word of caution about using the sizes of other dsRNA species in the analysis of the replication mechanisms of a virus. Dawson and Dodds (1982) reported seven discrete dsRNA bands from plants infected with CCMV instead of the three or possibly four expected. Similarly, Bar-Joseph *et al.* (1983) and Rosner *et al.* (1983) found bands of dsRNA in CMV-infected plants which did not correspond to the genomic RNAs, RNA 4, or the satellite CARNA-5. For neither of these viruses was it shown by hybridization that the extra bands contained viral sequences. It should be noted that some of the dsRNAs found in cells infected with Sindbis virus are considered to be derived from the replicative intermediates of larger RNAs by the ribonuclease treatment used in extraction (Simmons and Strauss, 1972). Thus, the significance of these extra dsRNA species is not clear.

There has been some debate as to whether viral dsRNAs actually exist in the cell or whether they are artifacts of extraction (see Bishop and Levintow, 1971). However, since they are extracted by a range of techniques it would seem likely that dsRNA molecules exist in the cell. They appear to be rather stable molecules with a low rate of turnover (Dawson and Dodds, 1982) and accumulate during infection.

This then raises the question as to whether the wholly double-stranded molecules (RFs) are actively involved in replication or whether they are molecules which have finished replication. Although we can be certain that the inoculum plus-strand is the template for the formation of, at least some of, the minus-strand RNA, there is less certainty about the template of the new plus-strand RNA. There are two mechanisms commonly proposed for RNA-dependent RNA replication (Fig. 1): (1) semiconservative in which the minus strand is the template and plus strands are displaced by the later-initiated plus strands; in this case dsRNA would be molecules which had finished replication; (2) conservative in which the dsRNA is the template with asymmetric transcription of new plus-strand RNA and the original RNA still remaining base-paired. These two mechanisms and others have been discussed extensively with regard to RNA phage and animal viruses (see Weissmann and Ochoa, 1967; Stavis and August, 1970; Bishop and Levintow, 1971; Agol, 1980). The consensus of opinion was originally that the majority of replication is semiconservative but more recent observations (Meyer *et al.*, 1978) have provided evidence suggesting conservative replication of poliovirus RNA. Thus, we feel that this question should be reexamined especially as the conservative mechanism resembles the normal transcription of RNA from dsRNA.

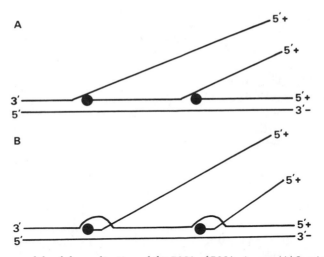

FIGURE 1. Two models of the replication of the RNA of RNA viruses. (A) Semiconservative method copying the minus strand with plus strands being displaced by later-initiated plus strands. (B) Conservative method involving asymmetric transcription of dsRNA. ●, RNA polymerase.

2. Nucleic Acid Structure and Replication

The mechanisms of RNA-dependent RNA replication suggest that the sequences at the 3'-ends of the plus and minus strands of RNA should be important in the binding of the polymerase molecule and the initiation of synthesis of the complementary strand. Since it is likely that the same polymerase is involved for each of the genomic RNAs of any given virus, one might expect sequence or structural similarities at the ends of these RNAs. When the sequences of the plus-strand RNAs for any one virus are examined (see Chapter 3), similarities are found at the 3'-ends of all four RNA species. For RNA 3 of both BMV and CMV the 5'− and 3'− terminal sequences contain extensive regions of complementary sequence (Dasgupta *et al.*, 1980; Gould and Symons, 1982). As noted by Dasgupta *et al.* (1980), this complementarity would be reflected as homology in the 3'-ends of plus and minus strands and could be evidence for common recognition sites for the same polymerase in both strands. No such complementarity is found between the 3'- and 5'-ends of RNA 4 of these viruses; it will be interesting when sequence data become available to see if the 5'-termini of RNAs 1 and 2 also contain sequences complementary to their 3'-ends. However, liquid hybridization shows that although there is 3'-terminal homology between CMV RNAs 1, 2, and 3, there is no other detectable homology between these RNAs (Gould and Symons, 1977). In AMV there is no such complementarity between the 3'- and 5'-terminal sequences of any of the RNAs (Koper-Zwarthoff *et al.*, 1980).

There are some interesting features in the 3'-terminal regions of the

RNAs of tripartite viruses. The RNAs of the Bromoviruses and of CMV and tomato aspermy virus have an amino acid-accepting region (for tyrosine) (for reviews see Hall, 1979; Haenni et al., 1982). In discussing the possible biological significance of the tRNA-like structures found in the RNAs of certain plant viruses, Hall (1979) comments on their possible involvement in RNA replication. The infectivity of BMV RNA was greatly reduced when its tyrosylation ability had been destroyed by chemical modification (Kohl and Hall, 1977). However, in a reappraisal of this experiment, Kiberstis and Hall (1983) suggested that the loss of infectivity probably resulted from nonspecific acylation of the RNA. Some recent elegant work by Ahlquist et al. (1984) suggests that the 3' region of BMV RNA is involved in polymerase binding. Using hybrid-arrested RNA synthesis (blocking of in vitro RNA synthesis by hybridization of specific cDNA fragments) they showed that polymerase binding requires at least some of the 3'-terminal 39 bases but not more that 153 bases. Using two different approaches, Rietveld et al. (1983) and Joshi et al. (1983) have estimated the size of the RNA involved in the tRNA structure to be 133 nucleotides for BMV and 114–118 nucleotides for BBMV. Although there are differences in the primary sequences of the 3'-ends of the various viruses, their secondary structures are very similar (Ahlquist et al., 1981a; Rietveld et al., 1983), raising the possibility that there might be similarities in the mechanism of polymerase binding throughout the Bromo- and Cucumoviruses. As noted above, there is homology in the 3'-ends of the plus and minus strands of RNA 3 of both BMV and CMV (Dasgupta et al., 1980; Gould and Symons, 1982); it would be of interest to know if the 3'-termini of the minus strand also had tRNA-like structures.

The satellite RNAs of CMV can form secondary structures, the 3'-terminal region of which has some features resembling those of the genomic CMV RNAs (Gordon and Symons, 1983). Since the satellite RNAs are completely dependent upon the genomic RNAs for replication, this can be taken as further evidence that the secondary structure of the 3'-end of the RNA is important for polymerase recognition.

A model for the role of aminoacylation of the viral RNA in transcription has been proposed by Hall and Wepprich (1976). This model is supported by the finding that BMV RNAs are aminoacylated in vivo (Loesch-Fries and Hall, 1982). The fact that BMV RNA 4 is aminoacylated has been used to suggest its autonomous replication (Loesch-Fries and Hall, 1982).

D. Further Translation Products

Apart from the proteins involved in replication which are discussed in Section IV.B, various other virus-coded proteins have been identified. The major translation products of RNAs of these viruses are summarized in Table II. In most cases where there is information, the same (or similar) products are found in vivo as those produced in vitro translation. However, AMV RNA 1 gives several products *in vitro* which led van Tol *et al.* (1980) to suggest that there might be internal weak termination codons. But the sequence of RNA 1 (Cornelissen *et al.*, 1983) does not show these termination codons, only a long open reading frame for a protein of M_r 125,685. Furthermore, only the full-size protein has been detected in alfalfa protoplasts (Samac *et al.*, 1983). Thus, in this case, it seems likely that the small *in vitro* translation products from AMV RNA 1 are artifactual.

E. Inhibitors

The use of metabolic inhibitors in virus multiplication studies has been directed toward two goals; first, the suppression of synthesis of host components which interfere with the detection of virus-specific products formed *in vivo*, and second, as a means of identifying biochemical mechanisms important in virus replication.

For the suppression of host RNA synthesis, actinomycin D is most commonly used, either alone or in conjunction with UV light treatment. Thus, host RNA synthesis has been reduced by 90% in tobacco (Takanami *et al.*, 1977) and cucumber (Boulton *et al.*, 1985) protoplasts. UV light treatment is effective in reducing host protein synthesis, allowing the detection of virus-specific proteins in BMV-infected barley (Kiberstis *et al.*, 1981) and CMV-infected cucumber protoplasts (M. Boulton, personal communication). Chloramphenicol is effective in reducing host chloroplast protein synthesis while leaving virus-specific synthesis in the cytoplasm unaffected (Alblas and Bol, 1977; Barnett and Wood, 1978).

It is important to note that while these broad-range inhibitors are useful, if not essential, in characterizing virus-specific products, their action may not always be predictable and the spectrum of products observed may not be representative of the "natural" infection. For example, using UV light to depress host protein synthesis in AMV-infected alfalfa protoplasts, Samac *et al.* (1983) noted a coincidental depression of synthesis of AMV RNAs 1 and 2 relative to RNAs 3 and 4.

As with many other RNA viruses, the sensitivity of the replication of tripartite viruses to inhibitors of DNA transcription has been extensively studied. Unfortunately, with the exception of AMV replication in cowpea protoplasts (Nassuth *et al.*, 1983b), few attempts have been made to characterize the phenomenon beyond gross measurement of synthesized virus (e.g., infectivity) or host susceptibility (e.g., percentage of protoplasts infected). For AMV, treatment of cowpea protoplast with actinomycin D immediately after inoculation reduced the production of infectious virus by 90%, although addition of the inhibitor at later times [i.e., 6 hr postinoculation (p.i.)] did not affect virus multiplication (Alblas and Bol, 1977). This, and similar observations made with some other plant RNA viruses led to the proposal that a host-coded factor was important for the earliest stages of virus replication. However, a comprehensive study of variation in the sensitivity of protoplasts to actinomycin D with time in culture, has not been made.

In what way this putative host-coded factor could assist in AMV replication has not been determined, although its importance in the production of both (+)- and (−)-strand RNA has been suggested (Nassuth *et al.*, 1983). The possibility that the factor could be a subunit of an RNA-dependent polymerase complex was discussed. For (+) RNA, it was interesting that RNA 3 was inhibited significantly less than other genomic RNAs, and that the production of coat protein was comparatively insensitive to actinomycin D. Similar observations have been made for the Comovirus CPMV multiplying in cowpea protoplasts (Rottier *et al.*, 1979). With BMV-infected barley protoplasts (Maekawa *et al.*, 1981), it was observed that actinomycin D was effective in reducing infectious virus before it reduced viral coat protein, detected by immunochemical staining. Furthermore, treatment of alfalfa protoplasts with UV irradiation, also a transcriptional inhibitor, prior to inoculation with AMV, resulted in depressed synthesis of RNAs 1 and 2 relative to RNAs 3 and 4 (Samac *et al.*, 1983). These results are in accord with host involvement in the viral RNA transcription process (Nassuth *et al.*, 1983b) and a spatial or functional separation from a translationally active pool of viral RNA, particularly that coding for viral coat protein.

Recently, it has been suggested that some RNA viruses which are inhibited by actinomycin D or cordycepin require a host-coded factor for the uncoating of the virion (Mayo and Barker, 1983). The authors extrapolated from this to suggest that it would be true for all RNA viruses.

Partial inhibition of BMV replication in barley protoplasts by actinomycin D has been reported (Maekawa *et al.*, 1981), a phenomenon which was again restricted to the earliest stages of infection. Other work showed that treatment applied only 1 hr p.i. gave rise to apparently normal infections (Loesch-Fries and Hall, 1980). Actinomycin D failed to inhibit the replication of CCMV in tobacco protoplasts (Bancroft *et al.*, 1975) although the time of administration was not reported. UV irradiation of barley protoplasts prior to inoculation with BMV resulted in levels of replication which permitted the detection of proteins specific for all the genomic viral RNAs (Kiberstis *et al.*, 1981). Clearly, virus dependence upon host-coded factors has yet to be proven for the Bromoviruses. However, the similarity between Bromoviruses and Cucumoviruses in a 3′-RNA structure potentially important for replication (Section IV.C.2), and the insensitivity of CMV replication to actinomycin D in cucumber plants (Barnett and Wood, 1978; Coutts *et al.*, 1978) and in tobacco (Takanami *et al.*, 1977) and cucumber (Boulton *et al.*, 1985) protoplasts, makes a host-coded function important in RNA transcription unlikely. This is particularly supported by the work with cucumber protoplasts (Boulton *et al.*, 1985) where treatment with actinomycin D failed to inhibit CMV replication when administered either 1 hr before or immediately after inoculation with virus particles or naked viral RNA. If CMV is dependent upon a host factor(s), then it must be a constitutive element or a long-lived mRNA.

Inhibitors of other areas of host metabolism have not yet provided convenient systems for studying the details of virus replication. Cyclo-heximide, an inhibitor of translation of 80 S ribosomes, has been iden-tified as a general inhibitor of virus growth (Barnett and Woods, 1978; Nassuth et al., 1983b). Similarly, in a survey of antibiotics appropriate for preventing microbial contamination of cultures of CCMV-infected tobacco protoplasts (Bancroft et al., 1975), tetracycline and amphotericin B were also found to be potent inhibitors of virus growth. Tetracycline is an inhibitor of translation whereas amphotericin B is a polyene anti-biotic which binds to membrane cholesterol.

F. Time Courses of Replication

Since information concerning time courses requires synchronous replication, much of it is derived from studies using protoplasts. In one case the differential-temperature system described in Section II.A for syn-chronizing replication in plants has been used (Dawson and Schlegel, 1973). Table III is a summary of much of the information available about time courses of replication of tripartite viruses. The overall picture is that detectable synthesis of viral RNA starts about 6–10 hr p.i. with dsRNA sometimes appearing before ssRNA. Proteins tend to be detected at about the same time or slightly later than the mRNA. One anomaly has been reported in AMV-infected protoplasts where the coat protein was detected several hours before its mRNA 4 (Nassuth et al., 1983a). However, this may only reflect differences in the sensitivities of the detection tech-niques.

The relative rate of accumulation of dsRNA is much lower than that of ssRNA. For instance, Loesch-Fries and Hall (1980) reported that at 8 hr p.i., 38% of the BMV specific label was in double-stranded mol-ecules whereas at 22 hr p.i. it was only 2%. In general, RNAs 3 and 4 accumulate at a greater rate than RNAs 1 and 2. However, some care must be taken in comparing rates of accumulation of different RNA spe-cies as it has to be assumed that they have the same specific radioactivity. Some evidence that this might not be so was reported by Bancroft et al. (1975). The relationship between the accumulation of different RNA spe-cies and their encapsidation is discussed in Section V.

The most detailed information on time courses is that on AMV in cowpea protoplasts (Nassuth et al., 1983a; Nassuth and Bol, 1983). Nas-suth and Bol (1983) show that the balance of synthesis of (+) and (−) strands is altered in infection initiated with RNAs 1 + 2 when compared with those with RNAs 1 + 2 + 3. In the RNA 1 + 2 infections, there is a reduced amount of (+) strand made and a greatly increased amount of (−) strand.

TABLE III. Time Courses of Replication

Virus	Host	dsRNA Species	dsRNA First detected	dsRNA Maximum[a]	ssRNA Species	ssRNA First detected	ssRNA Maximum[a]	Protein Species	Protein First detected	Protein Maximum[b]	Ref.[c]
BMV	Barley PP[d]	1	6 hr[b]	—	1	6 hr	—	—[e]	—	—	1
		2	6 hr	36 hr	2	6 hr	36 hr	—	—	—	
		3	8 hr		3	10 hr		—	—	—	
		4	8 hr		4	10 hr		—	—	—	
CCMV	Tobacco PP	—	6 hr	—	—	6 hr	—	CP[f]	10 hr	—	2
	Cowpea Pl	1–3	8 hr	20 hr	1–4	8 hr	~40 hr	—	—	—	3
	Tobacco PP	—	6 hr	—	—	6 hr	—	100K	27 hr	44 hr	2, 4
								34K	12 hr	20 hr	
								CP	16 hr	72 hr	
CMV	Tobacco PP	1–4	—	—	—	6 hr	~30 hr	CP	6 hr	20 hr	5
	Cowpea PP	—	—	—	—	8–10 hr	40 hr	CP	8–10 hr	44 hr	6
	Cucumber PP	1–3	12 hr	—	1–4 + 5[g]	6–7 hr	~40 hr	CP	9 hr	33 hr	7
AMV	Cowpea PP	1–3[h]	3–6 hr	42 hr	1	9 hr	42 hr	—	—	—	8
					2	15 hr	42 hr	—	—	—	
					3	6 hr	42 hr	—	—	—	
					4	15 hr	42 hr	—	—	—	
	Alfalfa PP	—	—	—	1	—	25 hr	CP	6–9 hr	42 hr	9
		—	—	—	2	—	25 hr	126K	5 hr	20 hr	
		—	—	—	3	5 hr	25 hr	90K	9 hr	20 hr	
		—	—	—	4	—	25 hr	32K	5 hr	9 hr	
								CP	5 hr	<28 hr	

[a] Maximum amount synthesized (not max rate).
[b] Hours postinoculation.
[c] References: 1, Loesch-Fries and Hall (1980); 2, Bancroft et al. (1975); 3, Dawson (1978); 4, Sakai et al. (1977); 5, Takanami et al. (1977); 6, Gonda and Symons (1979); 7, M. Boulton (personal communication); 8, Nassuth et al. (1983a); 9, Samac et al. (1983).
[d] PP, protoplasts; Pl, plant.
[e] —, information not given.
[f] CP, coat protein.
[g] CMV RNA 5 = CARNA 5.
[h] Detected as minus strand.

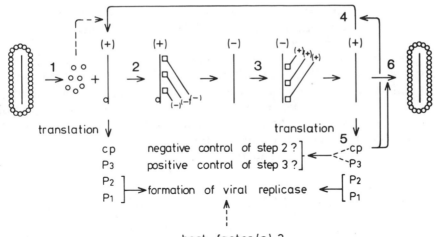

FIGURE 2. Scheme proposed by Nassuth and Bol (1983) for the replication cycle of AMV. The tripartite genome is represented by a single line. ○, virus coat protein; □, RNA polymerase; P_1, P_2, and P_3 are the translation products of viral RNAs 1, 2, and 3, respectively. Numbers 1–6 refer to steps in the replication cycle mentioned in the text and discussed more fully in Nassuth and Bol (1983). Reproduced by kind permission of Dr. J. F. Bol and Academic Press.

From these and other observations, Nassuth and Bol (1983) have proposed a model (Fig. 2) suggesting controls at various steps of virus replication. The first control (Fig. 2, step 1) is the need for virus coat protein subunits for the proper recognition of input (+) RNA by the replicase (Fig. 2, step 2) (see Section IV.B). Some of the newly synthesized (+) RNA is translated and it is suggested that the newly synthesized coat protein associates with the replication complex, thus turning off (−) RNA production. A third control that Nassuth and Bol (1983) suggest is that the gene product from RNA 3, P_3, is involved in switching to the asymmetric production of (+) RNA (Fig. 2, step 3).

G. Joint Infections of Viruses and Virus Strains

The interactions between strains of the same virus and between different viruses have been relatively little studied. Hull and Plaskitt (1970) reported on the interaction of two strains of AMV as revealed by electron microscopy of thin sections. The two strains were differentiated by the structure of the particle aggregation bodies. In joint infections or in sequential infections separated by short time intervals, the aggregation bodies merged. With longer intervals between sequential infections, the strains either occupied different parts of the cell, or different cells, or the second strain was excluded altogether (cross-protection). Thus, it appears that once a strain has become established in even a small part of a cell, another strain cannot multiply there. What the controlling process is has not yet been established.

Watts and Dawson (1980) and Sakai et al. (1983) studied the interactions between BMV and CCMV in tobacco protoplasts. BMV dominated the infection and prevented synthesis of infectious CCMV, though CCMV protein was produced. Cross-protection between the two viruses occurred if more than 8 hr elapsed between inoculations. There was no evidence for pseudorecombinants being formed between the two viruses, nor was there any transcapsidation.

V. ASSEMBLY IN RELATION TO VIRUS REPLICATION

For the tripartite viruses, the encapsidation of viral nucleic acid effectively removes (+) RNA from the biochemically active pool. In the correct ionic conditions, in vitro polymerization of coat protein subunits is a spontaneous process, and there is no evidence for or against the suggestion that a different process operates in vivo. It seems that the efficiency and rate at which encapsidation occurs will depend upon the relative and absolute concentrations of both RNA and coat protein at the correct intracellular site; for maximum efficiency, the relative rates of synthesis of viral RNA and viral coat protein should be finely tuned. Clearly this does not always happen, since the occurrence of empty capsids for many viruses and a surfeit of naked viral RNA (Gonda and Symons, 1979) for CMV, indicates imbalance in the two processes. In vitro assembly studies show that, at least for some of the viruses, viruslike particles can be assembled around polyanions other than viral nucleic acid (see Lebeurier et al., 1969; Hull, 1970; Hull and Read, 1970); there is no evidence of any polyanions in purified virus preparations. This suggests that the replication of the RNA, the translation of coat protein, and the assembly of the virus particles are closely integrated both temporally and spatially.

In a variety of in vitro translation systems, viral coat protein is produced from subgenomic RNA 4 (see Section IV). In vivo, the accumulation of RNA 4 in AMV-infected cowpea protoplasts (Nassuth et al., 1983a) and BMV-infected barley protoplasts (Loesch-Fries and Hall, 1980), to become the most abundant RNA species, is commensurate with its role as messenger for the most abundant viral protein. However, in the former case, a detailed comparison of the time courses of synthesis of RNAs 3 and 4 with coat protein and infectivity shows RNA 3 to have a closer correlation with coat protein synthesis (Nassuth et al., 1983a). With barley protoplasts, inoculation with large amounts of BMV RNA 4 failed to show any coat protein messenger activity in vivo (Kiberstis et al., 1981). For CMV (Q strain)-infected cowpea protoplasts, the results were equivocal, the proportion of RNA 3 to RNA 4 remaining as 1:1 over the period with maximum rates of coat protein synthesis (Gonda and Symons, 1979). However, for CMV infections of tobacco (Takanami *et al.*, 1977) and cucumber (M. Boulton *et al.*, 1985) protoplasts, RNA 3 predominates throughout the infection.

For both (+) RNAs 3 and 4, and other genomic RNAs, their concentration at the site of assembly or translation must be primarily determined at the stage of transcription from the (−) RNA. An intriguing control mechanism depending upon negative feedback on (−) RNA synthesis by RNA 3 or one of its products, has been proposed for AMV in cowpea protoplasts (Nassuth and Bol, 1983) and was discussed in detail in Section IV.B.

If for the tripartite viruses the efficiency of encapsidation is the same for different RNA species, then one would expect the relative proportions of the RNAs present in virions to reflect the respective proportions at the site of assembly. However, for the *in vitro* encapsidation of BMV RNA, the rate of accumulation in capsids is greater for RNAs 3 and 4 than for 1 and 2 (Herzog and Hirth, 1978). Furthermore, in BMV-infected barley protoplasts, the proportions of encapsidated RNAs did not reflect the proportions of total viral RNAs *in vivo* (Loesch-Fries and Hall, 1980). It is perhaps significant that the proportions of encapsidated RNA did vary over the period of infection and did at no time correlate with the proportions present in the virus inoculum purified from plants. This may represent a fundamental difference in the processes in plants and protoplasts. In AMV-infected cowpea protoplasts, there is also a disparity between total viral RNAs and encapsidated RNAs. Late in infection approximately 100% RNA 1, 80% RNAs 2 and 3, and 35% RNA 4 of total RNA is present in capsids (Nassuth *et al.*, 1983a).

To date, studies on viral RNA replication have taken little account of the possible existence of different intracellular pools of RNA, perhaps separately committed to translation or destined for encapsidation. If this were so, proportional differences between total viral RNA and encapsidated RNA would not be surprising; indeed, for AMV-infected cowpea protoplasts, the low percentage of RNA 4 in capsids might be expected if it is accepted that RNA 4 is functioning as the coat protein message *in vivo*.

VI. CELLULAR LOCATION OF RNA REPLICATION

Identification of intracellular sites active in either RNA replication or the virus encapsidation process has not been achieved for any of the tripartite viruses. In fact, with the exception of TYMV and its close association with chloroplasts (Matthews, 1973), this is true for most plant viruses. The difficulty seems to be in the demonstration of a correlation between ultrastructural observations and biochemical functions. For the tripartite viruses the presence of virus particles has been noted for many intracellular sites but only rarely (de Zoeten and Schlegel, 1967; Lastra and Schlegel, 1975) has this been linked with autoradiography in an attempt to characterize the biochemical components involved. Perhaps, with the advent of *in situ* hybridization for specific viral RNA sequences, it may be possible to add more biochemical evidence to the extensive ultrastructural information.

It is likely that the detailed interaction between host and virus will be specific and thus prevent extrapolation of information between hosts and even between viruses in the same group. Hence, in some recent work, Sakai *et al.* (1983) provided indirect evidence from mixed infections of tobacco protoplasts that two related viruses (CCMV and BMV) replicated at different intracellular sites.

The effects of infection by tripartite genome viruses on the ultrastructure of the cell, and the distribution of these viruses within the cells are discussed in detail in Chapter 6. Three general points are pertinent to this chapter: (1) Particles of the viruses under discussion are found evenly distributed through the cytoplasm, and occasionally in various organelles. (2) None of the viruses has a definite association with any specific organelle. (3) There are membranous structures within infected cells which have led to the suggestion that they are sites of virus replication.

At the biochemical level, translation of both AMV (Alblas and Bol, 1977) and CMV coat protein (Barnett and Wood, 1978; M. Boulton, personal communication) has been shown to be resistant to chloramphenicol and susceptible to cycloheximide, arguing against chloroplasts and mitochondria having a major role in the replication of these viruses.

For RNA replication specifically, there may be a correlation between the occurrence of a BMV-coded RNA-dependent RNA polymerase associated with membranes (Bujarski *et al.*, 1982), and the proliferation of endoplasmic reticulum (Motoyoshi *et al.*, 1973a; Burgess *et al.*, 1974). This has yet to be confirmed.

VII. CELL-TO-CELL SPREAD

Identification of factors important in the cell-to-cell spread of plant viruses has depended upon the physical mapping of virus strains or mutants which produce smaller than wild-type lesions or limited systemic spread, and on their examination in conjunction with virus multiplication studies in intact tissue and protoplasts. The best characterized systems so far are for the Ls_1 (Nishiguchi *et al.*, 1978, 1980) and Ni2519 (Zimmern and Hunter, 1983) mutants of TMV which are temperature-sensitive in cell-to-cell movement. Analysis of virus-specific proteins formed *in vivo* and *in vitro* showed both mutations to reside in the virus-coded 30K polypeptide (Leonard and Zaitlin, 1982; Zimmern and Hunter, 1983). Of greater significance is the observation that with Ls_1 the restriction of virus spread at the nonpermissive temperature can be overcome by complementation either with the temperature-resistant strain TMV-*vulgare* or with PVX (Taliansky *et al.*, 1982a). Furthermore, in a wider study, it was possible to complement the spread of avirulent viruses (including BMV) in resistant hosts by preinfection with a virulent unrelated virus. From this, the authors postulate that mechanisms controlling cell-to-cell spread may be common within large groups of viruses (Taliansky *et al.*, 1982b); it would be interesting to know whether there was any antigenic and functional similarity between the TMV 30K protein and nonstructural proteins from other unrelated viruses.

The determinant(s) for systemic spread of CCMV has been mapped to genomic RNA 3. The ts mutant used showed specific changes in the amino acid composition of viral coat protein resulting in increased thermolability (Bancroft *et al.*, 1972; Bancroft and Lane, 1973). The instability of the virus particle at high nonpermissive temperatures and low efficiency with which naked RNA moves from cell to cell probably account for the phenotype of this mutant. However, it was not excluded (Bancroft and Lane, 1973) that ts RNA 3 contained multiple mutations covering other genes.

One could speculate that the 35K protein coded for at the 5'-end of RNA 3 could function in a manner similar to the 30K protein from TMV. In some similar work, complementation studies between wild-type CCMV (strain T) and a systemic symptom-deficient variant (strain M) isolated following passage through beans, mapped the phenotype to RNA 3 (Kuhn and Wyatt, 1979). The authors suggest that the factor responsible resides in the 35K gene of RNA 3, a theory supported by the finding that strain M contains only 10% of RNA 3 present in wild-type virus (Wyatt and Kuhn, 1979).

In these discussions it has been assumed that the viral agent active in cell-to-cell spread is the mature virus particle. Unlike tobacco rattle virus RNA 1 which will replicate, spread, and cause symptoms in the absence of coat protein (Sänger, 1968), the tripartite viruses all appear to require coat protein or the coat protein gene. For AMV and the Ilarviruses this is also a prerequisite for RNA replication (see Section IV.B). BMV RNAs 1 + 2 appear to replicate alone in barley protoplasts (Kiberstis *et al.*, 1981) whereas RNAs 1 + 2 + 3 are required for a competent infection in whole plants. Similarly, CMV RNAs 1 + 2 + 3 are required for infection of plants (Peden and Symons, 1973).

As with viruses from other groups, the Cucumovirus, chrysanthemum aspermy virus has been observed in the plasmodesmata of infected tissues (Lawson and Hearon, 1970).

VIII. CONCLUSIONS

Although much has been discovered about some stages of the multiplication of tripartite genome viruses, there are still stages about which little is known. Some of these have been noted in this chapter and it is these areas which will probably attract more attention in the future. Over the last few years there has been a great expansion of the technical "tool bag" which can be used to tackle these problems. We can expect that techniques which involve specific probes to individual RNA species or even certain regions of RNA and probes to specific proteins (synthetic oligopeptides) will widen the understanding of gene products for which no function is at present known. However, there is a word of caution especially about the noncritical application of the present *in vitro* techniques. A plant cell is a very complex structure and it is likely that features not generally considered are involved in virus multiplication. Thus, it would seem likely that membranes are involved at various stages, e.g., uncoating, cell-to-cell transport, virus assembly. We feel that future developments of *in vitro* techniques for the investigation of cellular events should take into account at least some aspects of the cellular complexity.

ACKNOWLEDGMENTS.   We thank Drs. L. S. Loesch-Fries, E. M. J. Jaspars, P. Ahlquist, and J. F. Bol for preprints of manuscripts, and Dr. P. Kiberstis and Miss M. Boulton for permitting us to quote unpublished information. We are indebted to Mrs. M. Hobbs for preparation of the manuscript and to Drs. J. W. Davies and T. M. A. Wilson for critical reading of the manuscript.

REFERENCES

Agol, V. I., 1980, Structure, translation and replication of picornaviral genomes, *Prog. Med. Virol.* **26:**119.

Ahlquist, P., Dasgupta, R., and Kaesberg, P., 1981a, Near identity of 3' RNA secondary structure in Bromoviruses and cucumber mosaic virus, *Cell* **23:**183.

Ahlquist, P., Luckow, V., and Kaesberg, P., 1981b, Complete nucleotide sequence of brome mosaic virus RNA 3, *J. Mol. Biol.* **153:**23.

Ahlquist, P., Bujarski, J. L., Kaesberg, P., and Hall, T. C., 1984, Localisation of the replicase recognition site within brome mosaic virus RNA by hybrid-arrested RNA synthesis, *Plant Mol. Biol.* **3:**37.

Ahlquist, P., French, R., Janda, M., and Loesch-Fries, L. S., 1984, Multicomponent RNA plant virus infection derived from cloned viral cDNA, *Proc. Natl. Acad. Sci. USA* (in press).

Alblas, F., and Bol, J. F., 1977, Factors influencing the infection of cowpea mesophyll protoplasts by alfalfa mosaic virus, *J. Gen. Virol.* **36:**175.

Alblas, F., and Bol, J. F., 1978, Coat protein is required for infection of cowpea protoplasts with alfalfa mosaic virus, *J. Gen. Virol.* **41:**653.

Atabekov, J. G., and Morozov, S. Y., 1979, Translation of plant virus messenger RNAs, *Adv. Virus Res.* **25:**1.

Bancroft, J. B., 1972, A virus made from parts of the genome of brome mosaic and cowpea chlorotic mottle viruses, *J. Gen. Virol.* **14:**223.

Bancroft, J. B., and Lane, L. C., 1973, Genetic analysis of cowpea chlorotic mottle and brome mosaic viruses, *J. Gen. Virol.* **19:**381.

Bancroft, J. B., Rees, M. W., Dawson, J. R. O., McClean, G. D., and Short, M. N., 1972, Some properties of a temperature-sensitive mutant of cowpea chlorotic mottle virus, *J. Gen. Virol.* **16**:69.

Bancroft, J. B., Motoyoshi, F., Watts, J. W., and Dawson, J. R. O., 1975, Cowpea chlorotic mottle and brome mosaic viruses in tobacco protoplasts, in: 2nd John Innes Symposium: "Modification of the Information Content of Plant Cells", pp. 133–160, North-Holland, Amsterdam.

Bar-Joseph, M., Rosner, A., Moscovitz, M., and Hull, R., 1983, A simple procedure for the extraction of double-stranded RNA from infected plants, *J. Virol. Methods* **6**:1.

Barnett, A., and Wood, K. R., 1978, Influence of actinomycin D, ethephon, cycloheximide and chloramphenicol on the infection of a resistant and a susceptible cucumber cultivar with cucumber mosaic virus (Price's No. 6 strain), *Physiol. Plant Pathol.* **12**:257.

Bastin, M., and Kaesberg, P., 1976, A possible replicative form of brome mosaic virus RNA 4, *Virology* **72**:536.

Beier, H., Bruening, G., Russell, M. L., and Tucker, C. L., 1979, Replication of cowpea mosaic virus in protoplasts isolated from immune lines of cowpeas, *Virology* **95**:165.

Bishop, J. M., and Levintow, L., 1971, Replicative forms of viral RNA, structure and function, *Prog. Med. Virol.* **13**:1.

Bock, K. R., Guthrie, E. J., and Meredith, G., 1978, Distribution, host range, properties and purification of cassava latent virus, a Geminivirus, *Ann. Appl. Biol.* **90**:361.

Bol, J. F., and Lak-Kaashoek, M., 1974, Composition of alfalfa mosaic virus nucleoproteins, *Virology* **60**:476.

Bol, J. F., van Vloten-Doting, L., and Jaspars, E. M. J., 1971, A functional equivalence of top component *a* RNA and coat protein in the initiation of infection by alfalfa mosaic virus, *Virology* **46**:73.

Bos, L., Huttinga, H., and Maat, D. Z., 1980, Spinach latent virus, a new Ilarvirus seed-borne in *Sinacea oleracea, Neth. J. Plant Pathol.* **86**:79.

Boulton, M. I., Maule, A. J., and Wood, K. R., 1985, Effect of actinomycin D and u.v. irradiation on the replication of cucumber mosaic virus in protoplasts isolated from resistant and susceptible cucumber cultivars, *Phys. Plant. Path.* (in press).

Bujarski, J. J., Hardy, S. F., Miller, W. A., and Hall, T. C., 1982, Use of dodecyl-β-D-maltoside in the purification and stabilization of RNA polymerase from brome mosaic virus-infected barley, *Virology* **119**:465.

Burgess, J., Motoyoshi, F., and Fleming, E. N., 1974, Structural changes accompanying infection of tobacco protoplasts with two spherical viruses, *Planta* **117**:133.

Clark, M. F., and Lister, R. M., 1971, Preparations and some properties of the nucleic acid of tobacco streak virus, *Virology* **45**:61.

Cornelissen, B. J. C., Brederode, F. T., Moormann, R. J. M., and Bol, J. F., 1983, Complete nucleotide sequence of alfalfa mosaic virus RNA 1, *Nucleic Acids Res.* **11**:1253.

Coutts, R. H. A., and Wood, K. R., 1976, Investigations on the infection of cucumber mesophyll protoplasts with cucumber mosaic virus, *Arch. Virol.* **52**:307.

Coutts, R. H. A., Barnett, A., and Wood, K. R., 1978, Aspects of the resistance of cucumber plants and protoplasts to cucumber mosaic virus, *Ann. Appl. Biol.* **89**:336.

Dasgupta, R., Ahlquist, P., and Kaesberg, P., 1980, Sequence of the 3' untranslated region of brome mosaic virus coat protein messenger RNA, *Virology* **104**:339.

Davies, J. W., 1979, Translation of plant virus ribonucleic acids in extracts from eukaryotic cells, in: *Nucleic Acids in Plants* (T. C. Hall and J. W. Davies, eds.), Volume 2, CRC Press, Boca Raton, Fl.

Davies, J. W., and Kaesberg, P., 1974, Translation of virus mRNA: Protein synthesis directed by several virus RNAs in a cell-free extract from wheat germ, *J. Gen. Virol.* **25**:11.

Davies, J. W., and Verduin, B. J. M., 1979, *In vitro* synthesis of cowpea chlorotic mottle virus polypeptides, *J. Gen. Virol.* **44**:545.

Dawson, J. R. O., Dickerson, P. E., King, J. M., Sakai, F., Trim, A. R. H., and Watts, J. W., 1978, Improved methods for infection of plant protoplasts with viral ribonucleic acid, *Z. Naturforsch.* **33c**:548.

Dawson, W. O., 1978, Time-course of cowpea chlorotic mottle virus RNA replication, *Intervirology* **9**:119.

Dawson, W. O., and Dodds, J. A., 1982, Characterization of sub-genomic dsRNAs from virus infected plants, *Biochem. Biophys. Res. Commun.* **107**:1230.

Dawson, W. O., and Lung, M. C. Y., 1976, Synchronisation of cowpea chlorotic mottle virus replication in cowpea leaves, *Intervirology* **7**:284.

Dawson, W. O., and Schlegel, D. E., 1973, Differential temperature treatment of plants greatly enhances multiplication rates, *Virology* **53**:476.

Dawson, W. O., Schlegel, D. E., and Lung, M. C. Y., 1975, Synthesis of tobacco mosaic virus in intact tobacco leaves systemically inoculated by differential temperature treatment, *Virology* **65**:565.

de Zoeten, G. A., 1981, Early events in plant virus infection, in: *Plant Diseases and Vectors* (K. F. Harris and K. Maramorosch, eds.), pp. 221–239, Academic Press, New York.

de Zoeten, G. A., and Schlegel, D. E., 1967, Nuclear and cytoplasmic uridine-^{3}H incorporation into virus-infected plants, *Virology* **32**:416.

Dorssers, L., van der Meer, J., van Kammen, A., and Zabel, P., 1983, The cowpea mosaic virus RNA replication complex and the host encoded RNA-dependent RNA polymerase–template complex are functionally different, *Virology* **125**:155.

Durham, A. C. H., Hendry, D. A., and von Wechmar, M. B., 1977, Does calcium ion binding control plant virus disassembly?, *Virology* **77**:524.

Fukuyama, K., Abdel-Mequid, S. S., Johnson, J. E., and Rossmann, M. G., 1983, The structure of a $T = 1$ aggregate of alfalfa mosaic virus coat protein seen at 4.5Å resolution, *J. Mol. Biol.* **167**:873.

Fulton, R. W., 1970, The role of particle heterogeneity in infection by tobacco streak virus, *Virology* **41**:288.

Furusawa, I., and Okuno, T., 1978, Infection with BMV of mesophyll protoplasts isolated from five plant species, *J. Gen. Virol.* **40**:489.

Gerlinger, P., Mohier, E., Le Meur, M. A., and Hirth, L., 1977, Monocistronic translation of alfalfa mosaic virus RNAs, *Nucleic Acids Res.* **4**:813.

Ghabrial, S. A., and Lister, R. M., 1974, Chemical and physiochemical properties of two strains of tobacco streak virus, *Virology* **57**:1.

Gonda, T. J., and Symons, R. H., 1979, Cucumber mosaic virus replication in cowpea protoplasts: Time course of virus, coat protein and RNA synthesis, *J. Gen. Virol.* **45**:723.

Gonsalves, D., and Garnsey, S. M., 1974, Infectivity of the multiple nucleoprotein and RNA components of citrus leaf rugose virus, *Virology* **61**:343.

Gonsalves, D., and Garnsey, S. M., 1975a, Functional equivalence of an RNA component and coat protein for infectivity of citrus leaf rugose virus, *Virology* **64**:23.

Gonsalves, D., and Garnsey, S. M., 1975b, Nucleic acid components of citrus variegation virus and their activation by coat protein, *Virology* **67**:311.

Gonsalves, D., and Garnsey, S. M., 1975c, Infectivity of heterologous RNA–protein mixtures from alfalfa mosaic, citrus leaf rugose, citrus variegation and tobacco streak viruses, *Virology* **67**:319.

Goodman, R. M., 1977a, Infectious DNA from a whitefly transmitted virus of *Phaseolus vulgaris*, *Nature (London)* **266**:54.

Goodman, R. M., 1977b, Single-stranded DNA genome in a whitefly-transmitted plant virus, *Virology* **83**:171.

Gordon, K. H. J., and Symons, R. H., 1983, Satellite RNA of cucumber mosaic virus forms a secondary structure with partial 3′-terminal homology to genomal RNAs, *Nucleic Acids Res.* **11**:947.

Gordon, K. H. J., Gill, D. S., and Symons, R. H., 1982, Highly purified cucumber mosaic virus-induced RNA-dependent RNA polymerase does not contain any full length translation products of the genomic RNAs, *Virology* **123**:284.

Gould, A. R., and Symons, R. H., 1977, Determination of the sequence homology between the four RNA species of cucumber mosaic virus by hybridization analysis with complementary DNA, *Nucleic Acids Res.* **4**:3787.

Gould, A. R., and Symons, R. H., 1982, Cucumber mosaic virus RNA 3: Determination of the nucleotide sequence provides the amino acid sequences of protein 3A and viral coat protein, *Eur. J. Biochem.* **126:**217.

Habili, N., and Francki, R. I. B., 1974, Comparative studies on tomato aspermy and cucumber mosaic viruses. III. Further studies on relationship and construction of a virus from parts of two viral genomes, *Virology* **61:**443.

Haenni, A.-L., Joshi, S., and Chapeville, F., 1982, tRNA-like structures of RNA viruses, *Prog. Nucleic Acid Res. Mol. Biol.* **27:**85.

Hall, T. C., 1979, Transfer RNA-like structures in viral genomes, *Int. Rev. Cytol.* **60:**1.

Hall, T. C., and Wepprich, R. K., 1976, Functional possibilities for aminoacylation of viral RNA in transcription and translation, *Ann. Microbiol. (Inst. Pasteur).* **127A:**143.

Hall, T. C., Miller, W. A., and Bujarski, J. J., 1982, Enzymes involved in the replication of plant viral RNAs, *Adv. Plant Pathol.* **1:**179.

Haseloff, J., Goelet, P., Zimmern, D., Ahlquist, P., Dasqusta, R., and Kaesberg, P., 1984, Striking similarities in amino acid sequence among nonstructural proteins encoded by RNA viruses that have dissimilar genomic organization, *Proc. Natl. Acad. Sci. USA* **81:**4358.

Herzog, M., and Hirth, L., 1978, *In vitro* encapsidation of the four RNA species of brome mosaic virus, *Virology* **86:**48.

Honda, Y., Matsui, C., Otsuki, Y., and Takebe, I., 1974, Ultrastructure of tobacco mesophyll protoplasts inoculated with cucumber mosaic virus, *Phytopath.* **64:**30.

Houwing, C. J., and Jaspars, E. M. J., 1978, Coat protein binds to the 3'-terminal part of RNA 4 of alfalfa mosaic virus, *Biochemistry* **17:**2927.

Houwing, C. J., and Jaspars, E. M. J., 1980, Preferential binding of 3'-terminal fragments of alfalfa mosaic virus RNA 4 to virions, *Biochemistry* **19:**5261.

Houwing, C. J., and Jaspars, E. M. J., 1982, Protein binding sites in nucleation complexes of alfalfa mosaic virus RNA 4, *Biochemistry* **21:**3408.

Hull, R., 1970, Studies on alfalfa mosaic virus. III. Reversible dissociation and reconstitution studies, *Virology* **40:**34.

Hull, R., and Plaskitt, A., 1970, Electron microscopy on the behaviour of two strains of alfalfa mosaic virus in mixed infections, *Virology* **42:**773.

Hull, R., and Read, D. B., 1970, Alfalfa mosaic virus: Reversible dissociation and reconstitution, *Annu. Rep. John Innes Inst.* 1969, p. 48.

Huttinga, H., and Mosch, W. H. M., 1976, Lilac ring mottle virus: A coat protein-dependent virus with a tripartite genome, *Acta Hortic.* **59:**113.

Jaspars, E. M. J., 1984, Interaction of alfalfa mosaic virus nucleic acid and protein, in: *Molecular Plant Virology*, Volume I (J. W. Davies, ed.), CRC Press, Boca Raton, Fla.

Joshi, R. L., Joshi, S., Chapeville, F., and Haenni, A.-L., 1983, tRNA-like structures of plant viral RNAs: Conformational requirements for adenylation and aminoacylation, *EMBO J.* **2:**1123.

Kamer, G., and Argos, P., 1984, Primary structural comparison of RNA-dependent polymerases from plant, animal and bacterial viruses, *Nuc. Acids Res.* **12:**7269.

Kaper, J. M., and Diaz-Ruiz, J. R., 1977, Molecular weights of the double-stranded RNAs of cucumber mosaic virus strain S and its associated RNA 5, *Virology* **80:**214.

Kaper, J. M., and West, C. K., 1972, Polyacrylamide gel separation and molecular weight determination of the components of cucumber mosaic virus RNA, *Prep. Biochem.* **2:**251.

Kiberstis, P. A., and Hall, T. C., 1983, N-acetyl-tyrosine at the 3' end of brome mosaic virus RNA has little effect on infectivity, *J. Gen. Virol.* **64:**2073.

Kiberstis, P. A., Loesch-Fries, L. S., and Hall, T. C., 1981, Viral protein synthesis in barley protoplasts inoculated with native and fractionated brome mosaic virus RNA, *Virology* **112:**804.

Kohl, R. J., and Hall, T. C., 1977, Loss of infectivity of brome mosaic virus RNA after chemical modification of the 3'- or 5'-terminus, *Proc. Natl. Acad. Sci. USA* **74:**2682.

Koike, M., Hibi, T., and Yova, K., 1977, Infection of cowpea mesophyll protoplasts with cucumber mosaic virus, *Virology* **83**:413.

Koper-Zwarthoff, E. C., and Bol, J. F., 1980, Nucleotide sequence of the putative recognition site for coat protein in the RNAs of alfalfa mosaic virus and tobacco streak virus, *Nucleic Acids Res.* **8**:3307.

Koper-Zwarthoff, E. C., Brederode, F. T., Keeneman, G., van Boom, J. H., and Bol, J. F., 1980, Nucleotide sequence at the 5'-termini of the alfalfa mosaic virus RNAs and the intercistronic junction in RNA 3, *Nucleic Acids Res.* **8**:5635.

Kuhn, C. W., and Wyatt, S. D., 1979, A variant of cowpea chlorotic mottle virus obtained by passage through beans, *Phytopathology* **69**:621.

Kumarasamy, R., and Symons, R. H., 1979, Extensive purification of the cucumber mosaic virus-induced RNA replicase, *Virology* **96**:622.

Kurtz-Fritsch, C., and Hirth, L., 1972, Uncoating of two spherical plant viruses, *Virology* **47**:385.

Lane, L. C., 1979, The nucleic acids of multipartite, defective and satellite plant viruses, in: *Nucleic Acids in Plants*, Volume 2 (T. C. Hall and J. W. Davies, eds.), p. 65, CRC Press, Boca Raton, Fla.

Lane, L. C., and Kaesberg, P., 1971, Multiple genetic components in bromegrass mosaic virus, *Nature New Biol.* **232**:40.

Lastra, J. R., and Schlegel, D. E., 1975, Viral protein synthesis in plants infected with broad bean mottle virus, *Virology* **65**:16.

Lawson, R. H., and Hearon, S., 1970, Subcellular localisation of chrysanthemum aspermy virus in tobacco and chrysanthemum leaf tissue, *Virology* **41**:30.

Lebeurier, G., Wurtz, M., and Hirth, L., 1969, Auto-assemblage de RNA extracts de trois types de virus différents et des sous-unités protéique du Virus de la Mosaïque de la Luzerne, *C. R. Acad. Sci.* **268**:2002.

Leonard, D. A., and Zaitlin, M., 1982, A temperature-sensitive strain of tobacco mosaic virus defective in cell to cell movement generates an altered viral-coded protein, *Virology* **117**:416.

Loesch-Fries, L. S., and Hall, T. C., 1980, Synthesis, accumulation and encapsidation of individual brome mosaic virus RNA components in barley protoplasts, *J. Gen. Virol.* **47**:323.

Loesch-Fries, L. S., and Hall, T. C., 1982, *In vivo* aminoacylation of brome mosaic and barley stripe mosaic virus RNAs, *Nature (London)* **298**:771.

Lot, H., Marchoux, G., Marrou, J., Kaper, J. M., West, C. K., van Vloten-Doting, L., and Hull, R., 1974, Evidence for three functional RNA species in several strains of cucumber mosaic virus, *J. Gen. Virol.* **22**:81.

Maekawa, K., Furusawa, I., and Okuno, T., 1981, Effects of actinomycin D and ultraviolet irradiation on multiplication of brome mosaic virus in host and non-host cells, *J. Gen. Virol.* **53**:353.

Marchoux, G., Douine, L., Lot, H., and Esvan, C., 1973, Identification et estimation du poids moléculaire des acides ribonucléiques du virus de la mosaïque du concombre (VMC souche D), *C.R. Acad. Sci. Ser. D* **277**:1409.

Matthews, R. E. F., 1973, Induction of disease by viruses with special reference to turnip yellow mosaic virus, *Annu. Rev. Phytopathol.* **11**:147.

Maule, A. J., Boulton, M. I., Edmunds, C., and Wood, K. R., 1980a, Polyethylene glycol-mediated infection of cucumber protoplasts by cucumber mosaic virus RNA, *J. Gen. Virol.* **47**:199.

Maule, A. J., Boulton, M. I., and Wood, K. R., 1980b, An improved method for the infection of cucumber leaf mesophyll protoplasts with cucumber mosaic virus, *Phytopathol. Z.* **97**:118.

Maule, A. J., Boulton, M. I., and Wood, K. R., 1980c, Resistance of cucumber protoplasts to cucumber mosaic virus: A comparative study, *J. Gen. Virol.* **51**:271.

Mayo, M. A., and Barker, H., 1983, Effects of actinomycin D on infection of tobacco protoplasts by four viruses, *J. Gen. Virol.* **64**:1775.

Meyer, J., Lundquist, R. E., and Maizel, J. V., 1978, Structural studies on the RNA component of the poliovirus replication complex. II. Characterization by electron microscopy and autoradiography, *Virology* **85**:445.

Miller, W. A., and Hall, T. C., 1983, Use of micrococcal nuclease in the purification of highly template dependent RNA-dependent RNA polymerase from brome mosaic virus-infected barley, *Virology* **125**:236.

Mohier, E., Pinck, L., and Hirth, L., 1974, Replication of alfalfa mosaic virus RNAs, *Virology* **58**:9.

Mohier, E., Hirth, L., Le Meur, M.-A., and Gerlinger, P., 1975, Translation of alfalfa mosaic virus RNAs in mammalian cell-free systems, *Virology* **68**:349.

Mossop, D. W., and Francki, R. I. B., 1977, Association of RNA 3 with aphid transmission of cucumber mosaic virus, *Virology* **81**:177.

Mossop, D. W., and Francki, R. I. B., 1979, The stability of satellite viral RNAs *in vivo* and *in vitro*, *Virology* **94**:243.

Motoyoshi, F., and Oshima, N., 1977, Expression of genetically controlled resistance to tobacco mosaic virus infection in isolated tomato leaf mesophyll protoplasts, *J. Gen. Virol.* **34**:499.

Motoyoshi, F., Bancroft, J. B., Watts, J. W., and Burgess, J., 1973a, The infection of tobacco protoplasts with cowpea chlorotic mottle virus and its RNA, *J. Gen. Virol.* **20**:177.

Motoyoshi, F., Bancroft, J. B., and Watts, J. W., 1973b, A direct estimate of the number of cowpea chlorotic mottle virus particles absorbed by tobacco protoplasts that become infected, *J. Gen. Virol.* **21**:159.

Motoyoshi, F., Bancroft, J. B., and Watts, J. W., 1974a, The infection of tobacco protoplasts with a variant of brome mosaic virus, *J. Gen. Virol.* **25**:31.

Motoyoshi, F., Watts, J. W., and Bancroft, J. B., 1974b, Factors influencing the infection of tobacco protoplasts by cowpea chlorotic mottle virus, *J. Gen. Virol.* **25**:245.

Nassuth, A., and Bol, J. F., 1983, Altered balance of the synthesis of plus- and minus-strand RNAs induced by RNAs 1 and 2 of alfalfa mosaic virus in the absence of RNA 3, *Virology* **124**:75.

Nassuth, A., Alblas, F., and Bol, J. F., 1981, Localization of genetic information involved in the replication of alfalfa mosaic virus, *J. Gen. Virol.* **53**:207.

Nassuth, A., ten Bruggencate, G., and Bol, J. F., 1983a, Time course of alfalfa mosaic virus and coat protein synthesis in cowpea protoplasts, *Virology* **125**:75.

Nassuth, A., Alblas, F., van der Geest, A. J. M., and Bol, J. F., 1983b, Inhibition of alfalfa mosaic virus RNA and protein synthesis by actinomycin D and cycloheximide, *Virology* **126**:517.

Nishiguchi, M., Motoyoshi, F., and Oshima, N., 1978, Behavior of a temperature sensitive strain of tobacco mosaic virus in tomato leaves and protoplasts, *J. Gen. Virol.* **39**:53.

Nishiguchi, M., Motoyoshi, F., and Oshima, N., 1980, Further investigation of a temperature sensitive strain of tobacco mosaic virus; its behaviour in tomato leaf epidermis, *J. Gen. Virol.* **46**:497.

Okuno, T., and Furusawa, I., 1978a, Modes of infection of barley protoplasts with brome mosaic virus, *J. Gen. Virol.* **38**:409.

Okuno, T., and Furusawa, I., 1978b, The use of osmotic shock for the inoculation of barley protoplasts with brome mosaic virus, *J. Gen. Virol.* **39**:187.

Okuno, T., and Furusawa, I., 1978c, Factors influencing the infection of barley mesophyll protoplasts with brome mosaic virus RNA, *J. Gen. Virol.* **41**:63.

Okuno, T., and Furusawa, I., 1979, RNA polymerase activity and protein synthesis in brome mosaic virus-infected protoplasts, *Virology* **99**:218.

Okuno, T., Furusawa, I., and Hiruki, C., 1977, Infection of barley protoplasts with brome mosaic virus, *Phytopathology* **67**:610.

Otsuki, Y., and Takebe, I., 1973, Infection of tobacco mesophyll protoplasts by cucumber mosaic virus, *Virology* **52**:433.

Owens, R. A., Kaper, J. M., 1977, Cucumber mosaic virus-associated RNA 5. II. *In vitro* translation in a wheat germ protein-synthesis system, *Virology* **80**:196.

Paterson, B. M., Roberts, B. E., and Kuff, E. L., 1977, Structural gene identification and mapping by DNA: mRNA hybrid-arrested cell-free translation, *Proc. Natl. Acad. Sci. USA* **74:**4370.

Peden, K. W., C., and Symons, R. H., 1973, Cucumber mosaic virus contains a functionally divided genome, *Virology* **53:**487.

Penmans, W. J., Carlier, A. R., and Schreurs, J., 1980, Cell-free translation of exogenous mRNA in extracts from dry pea primary axes, *Planta* **147:**302.

Philipps, G., Gigot, C., and Hirth, L., 1974, Replicative forms and viral RNA synthesis in leaves infected with alfalfa mosaic virus, *Virology* **60:**370.

Pinck, L., and Hirth, L., 1972, The replicative RNA and the viral RNA synthesis rate in tobacco infected with alfalfa mosaic virus, *Virology* **19:**413.

Racaniello, V. R., and Baltimore, D., 1981, Cloned poliovirus complementary DNA is infectious in mammalian cells, *Science* **214:**916.

Ricciardi, R. P., Miller, J. S., and Roberts, B. E., 1979, Purification and mapping of specific mRNAs by hybridization-selection and cell-free translation, *Proc. Natl. Acad. Sci. USA* **76:**4927.

Rietveld, K., Pleij, C. W. A., and Bosch, L., 1983, Three-dimensional models of the tRNA-like 3′ termini of some plant viral RNAs, *EMBO J.* **2:**1079.

Roberts, P. L., Wood, K. R., 1981a, Methods for enhancing the synchrony of cucumber mosaic virus replication in tobacco plants, *Phytopathol. Z.* **102:**114.

Roberts, P. L., and Wood, K. R., 1981b, Protein synthesis in cucumber mosaic virus-infected leaves, pre-infected cells and protoplasts, *Arch. Virol.* **70:**115.

Roosien, J., and van Vloten-Doting, L., 1982, Complementation and interference of U.V. induced Mts mutants of alfalfa mosaic virus, *J. Gen. Virol.* **63:**189.

Roosien, J., and van Vloten-Doting, L., 1983, A mutant of alfalfa mosaic virus with an unusual structure, *Virology* **126:**155.

Rosner, A., Bar-Joseph, M., Moscovitz, M., and Mevarech, M., 1983, Diagnosis of specific viral RNA sequences in plant extracts by hybridization with a polynucleotide kinase-mediated, ^{32}P-labelled, double-stranded RNA probe, *Phytopathology* **73:**699.

Rottier, P. J. M., Rezelman, G., and van Kammen, A., 1979, The inhibition of cowpea mosaic virus replication by actinomycin D. *Virology* **92:**299.

Sakai, F., Watts, J. W., Dawson, J. R. O., and Bancroft, J. B., 1977, Synthesis of proteins in tobacco protoplasts infected with cowpea chlorotic mottle virus, *J. Gen. Virol.* **34:**285.

Sakai, F., Dawson, J. R. O., and Watts, J. W., 1979, Synthesis of proteins in tobacco protoplasts infected with brome mosaic virus, *J. Gen. Virol.* **42:**323.

Sakai, F., Dawson, J. R. O., and Watts, J. W., 1983, Interference in infections of tobacco protoplasts with two Bromoviruses, *J. Gen. Virol.* **64:**1347.

Samac, D. A., Nelson, S. F., and Loesch-Fries, L. S., 1983, Virus protein synthesis in alfalfa mosaic virus infected alfalfa protoplasts, *Virology* **131:**455.

Sänger, H. L., 1968, Characteristics of tobacco rattle virus: Evidence that its two particles are functionally defective and mutually complementing, *Mol. Gen. Genet.* **101:**346.

Sarachu, A. N., Nassuth, A., Roosien, J., van Vloten-Doting, L., and Bol, J. F., 1983, Replication of temperature sensitive mutants of alfalfa mosaic virus in protoplasts, *Virology* **125:**64.

Schwinghamer, M. W., and Symons, R. H., 1975, Fractionation of cucumber mosaic virus RNA and its translation in a wheat embryo cell-free system, *Virology* **63:**252.

Schwinghamer, M. W., and Symons, R. H., 1977, Translation of the four major RNA species of cucumber mosaic virus in plant and animal cell-free systems and in toad oocytes, *Virology* **79:**88.

Shaw, J. G., 1984, Early events in plant virus infection, in: *Molecular Plant Virology*, Volume 2 (J. W. Davies, ed.), CRC Press, Boca Raton, Fla.

Shepherd, R. J., Wakeman, R. J., and Romanko, R. R., 1968, DNA in cauliflower mosaic virus, *Virology* **36:**150.

Shih, D. S., and Kaesberg, P., 1973, Translation of brome mosaic viral ribonucleic acid in a cell free system derived from wheat embryo, *Proc. Natl. Acad. Sci. USA* **70:**1799.

Shih, D. S., and Kaesberg, P., 1976, Translation of the RNAs of brome mosaic virus: The monocistronic nature of RNA 1 and RNA 2, *J. Mol. Biol.* **103**:77.

Simmons, D. T., and Strauss, J. H., 1972, Replication of Sindbis virus. II. Multiple forms of double-stranded RNA isolated from infected cells, *J. Mol. Biol.* **71**:615.

Smit, C. H., Roosien, J., van Vloten-Doting, L., and Jaspars, E. M. J., 1981, Evidence that alfalfa mosaic virus infection starts with three RNA–protein complexes, *Virology* **112**:169.

Stavis, R. L., and August, J. T., 1970, The biochemistry of RNA bacteriophage replication, *Annu. Rev. Biochem.* **39**:527.

Stoker, K., Koper-Zwarthoff, E. C., Bol, J. F., and Jaspars, E. M. J., 1980, Localization of a high affinity binding site for coat protein on the 3'-terminal part of RNA 4 of alfalfa mosaic virus, *FEBS Lett.* **121**:123.

Takanami, Y., Kubo, S., and Imaizumi, S., 1977, Synthesis of single- and double-stranded cucumber mosaic virus RNAs in tobacco mesophyll protoplasts, *Virology* **80**:376.

Taliansky, M. E., Malyshenko, S. I., Pshennikova, E. S., Kaplan, I. B., Ulanova, E. F., and Atabekov, J. G., 1982a, Plant virus-specific transport function. I. Virus genetic control required for systemic spread, *Virology* **122**:318.

Taliansky, M. E., Malyshenko, S. I., Pshennikova, E. S., and Atabekov, J. G., 1982b, Plant virus-specific transport function. II. A factor controlling virus host range, *Virology* **122**:327.

Thomas, B. J., Barton, R. J., and Tuszynski, A., 1983, Hydrangea mosaic virus, a new Ilarvirus from *Hydrangea macrophylla* (Saxifragaceae), *Ann. Appl. Biol.* **103**:261.

van Tol, R. G. L., and van Vloten-Doting, L., 1979, Translation of alfalfa mosaic virus RNA 1 in the mRNA-dependent translation system from rabbit reticulocyte lysates, *Eur. J. Biochem.* **93**:461.

van Tol, R. G. L., van Gemeren, R., and van Vloten-Doting, L., 1980, Two leaky termination codons on AMV RNA 1, *FEBS Lett.* **118**:67.

van Vloten-Doting, L., 1975, Coat protein is required for infectivity of tobacco streak virus: Biological equivalence of the coat proteins of tobacco streak virus and alfalfa mosaic virus, *Virology* **65**:215.

van Vloten-Doting, L., and Jaspars, E. M. J., 1977, Plant covirus systems: Three component systems, in: *Comprehensive Virology*, Volume 11 (H. Fraenkel-Conrat and R. R. Wagner, eds.), Plenum Press, New York.

Watts, J. W., and Dawson, J. R. O., 1980, Double infection of tobacco protoplasts with brome mosaic virus and cowpea chlorotic mottle virus, *Virology* **105**:501.

Watts, J. W., King, J. M., 1984, The effect of charge on infection of tobacco protoplasts by Bromoviruses, *J. Gen. Virol.* **65**:1709.

Watts, J. W., Dawson, J. R. O., and King, J. M., 1981, The mechanism of entry of viruses into plant protoplasts, *Ciba Found. Symp.* **80**:56.

Weissmann, C., and Ochoa, S., 1967, Replication of phage RNA, *Prog. Nucleic Acid Res. Mol. Biol.* **6**:353.

Wilson, T. M. A., 1984, Cotranslational disassembly of tobacco mosaic virus *in vitro*, *Virology* **137**:255.

Wyatt, S. D., and Kuhn, C. W., 1979, Replication and properties of cowpea chlorotic mottle virus in resistant cowpeas, *Phytopathology* **69**:125.

Ziemiecki, A., and Wood, K. R., 1975, Changes in the soluble protein constitution of cucumber cotyledons following infection with two strains of cucumber mosaic virus, *Physiol. Plant Pathol.* **7**:79.

Zimmern, D., and Hunter, T., 1983, Point mutation in the 30K open reading frame of TMV implicated in temperature-sensitive assembly and local lesion spreading of mutant Ni 2519, *EMBO J,* **2**:1893.

Zuidema, D., Bierhuizen, M. F. A., Cornelissen, B. J. C., Bol, J. F., and Jaspars, E. M. J., 1983a, Coat protein binding sites on RNA 1 of alfalfa mosaic virus, *Virology* **125**:361.

Zuidema, D., Bierhuizen, M. F. A., and Jaspars, E. M. J., 1983b, Removal of the N-terminal part of alfalfa mosaic virus coat protein interferes with the specific binding to RNA 1 and genome activation, *Virology* **129**:255.

CHAPTER 5

Virus Genetics

L. van Vloten-Doting

1. INTRODUCTION

In the last 10 years our knowledge about the organization and expression of RNA plant viruses has increased considerably (Matthews, 1981; van Vloten-Doting *et al.*, 1983). The complete base sequence of a number of plant viruses is known. These studies combined with *in vitro* translation studies (Atabekov and Morozov, 1979; Davies and Hull, 1982; van Vloten-Doting and Neeleman, 1982) revealed the number and size of the primary translation products of the different virus RNAs. However, our knowledge about the function of virus-coded proteins is lagging behind. One reason for this is that genetic studies with RNA viruses are hampered by the apparent lack of recombination taking place at the RNA level (King *et al.*, 1982). Moreover, compared to RNA bacterial or animal viruses, very little genetic work has been done with RNA plant viruses, mainly because plants are less amenable for genetic studies than bacterial or animal cells (e.g., there are no replica plating methods for the easy selection of conditional lethal mutants, etc.).

The genetic analysis of plant viruses with tripartite genomes started about 15 years ago with the construction of hybrids (later called pseudorecombinants) obtained by test tube exchange of RNAs or nucleoprotein components between different viruses or strains (see Section II). The information which can be gained by this approach is rather limited. In addition, the analysis of plants or protoplasts infected with only one or two genomic RNA segments can reveal the location of information required for RNA and/or nucleoprotein multiplication (see Section III). However, the only way to obtain a detailed understanding of all steps involved in virus multiplication is the study of conditional lethal mu-

L. VAN VLOTEN-DOTING • Department of Biochemistry, University of Leiden, Leiden, The Netherlands

tants. Therefore, the main body of this chapter will be devoted to the description of the production, maintenance, and analysis of this type of mutant.

II. ANALYSIS OF PSEUDORECOMBINANTS CONSTRUCTED FROM GENOME PARTS OF NATURALLY OCCURRING VIRUS STRAINS

In this chapter only the general principles and conclusions drawn from the study of pseudorecombinants will be discussed. The majority of experiments of this type were done in the first half of the 1970s and the results obtaind with tripartite genome viruses have been reviewed (van Vloten-Doting and Jaspars, 1977). For the Bromo- and Cucumoviruses, exchange of genetic material is only possible by manipulating isolated RNAs, but with the Ilarviruses,* exchanges can also be made using separated nucleoprotein particles.

Pseudorecombinants (also called hybrids) consist of a combination of genome parts originating from different strains or viruses. Combinations of genetic material from different strains of the same virus have always been found to be viable.

With the Bromoviruses it was also found that the combination brome mosaic virus (BMV) RNA 1 and RNA 2 plus cowpea chlorotic mottle virus (CCMV) RNA 3 did produce virus while the reverse combination did not (Bancroft, 1972). With the Cucumviruses cucumber mosaic virus (CMV) and tomato aspermy virus (TAV), it was found that most combinations of RNA 1 plus RNA 2 from one virus with RNA 3 from another virus were viable. However, exchange of RNAs 1 and 2 yielded no viable pseudorecombinants (Rao and Francki, 1981, and references therein). With the Ilarviruses, no successful construction of intervirus pseudorecombinants has been reported, although limited attempts involving one strain of tobacco streak virus (TSV) and one of AMV have been made (van Vloten-Doting, 1975).

In all cases where this was attempted, it was found that starting from the pseudorecombinants, the original viruses could be reconstructed when appropriate RNA segments were combined. This indicates that the different virus RNAs replicate true to type.

In all pseudorecombinants the serotype is determined by RNA 3, in accordance with the location of the coat protein cistron on this RNA. Aphid transmissibility of CMV also mapped on RNA 3 (Mossop and Francki, 1977). Most of the other markers used were differences in symptoms on a variety of hosts (see also Matthews, 1981). On certain hosts the symptoms induced by some Bromoviruses, Cucumoviruses, and Il-

---

* Throughout this chapter, alfalfa mosaic virus (AMV) will be considered to be a member of the Ilarvirus group, as proposed by van Vloten-Doting et al. (1981).

arviruses appear to be determined by RNA 3, RNA 2, or occasionally RNA 1, whereas others are induced by a combination of RNA 2 and RNA 3 (Lane, 1979; Rao and Francki, 1982). Since the molecular basis of symptom development is unknown, these data cannot be interpreted any further at present.

III. ANALYSIS OF INCOMPLETE INFECTIONS

Individual genome segments from both AMV and BMV and all possible combinations thereof have been inoculated into protoplasts. When BMV infection was studied in barley protoplasts, a mixture of RNA 1 and RNA 2 was sufficient to induce viral RNA synthesis. Since the viral RNA was detected by "Northern blots" using cloned viral sequences as probe, no distinction could be made between plus- and minus-strand synthesis (Kiberstis *et al.*, 1981). Comparable amounts of P1 and P2 (translation products of RNA 1 and RNA 2, respectively) were produced in the complete and incomplete infections, suggesting that the amount of plus strands synthesized was similar in both cases (Kiberstis, personal communication). If this assumption is correct, asymmetric RNA synthesis (much more plus strands than minus strands) is an intrinsic property of the BMV replicase.

AMV infection has been studied in cowpea protoplasts. Viral RNA synthesis was detected on "Northern blots" using end-labeled virion RNA for the detection of minus-strand RNA and labeled cDNA for the detection of plus-strand RNA. Viral RNA synthesis was only found in protoplasts inoculated with the complete mixture or with mixtures containing B + M (RNA 1, RNA 2, and coat protein). In incomplete infections, more minus strand and less plus strand was synthesized when compared to the RNA synthesized in complete infections (Fig. 1 and Nassuth and Bol, 1983). Incomplete infections, much more plus-strand RNA than minus-strand RNA was synthesized, whereas in incomplete infections both types of RNA were produced in roughly equal amounts. Apparently, an RNA 3-coded product (3a or coat protein) is involved in the regulation of plus- and minus-strand RNA synthesis. Nassuth and Bol (1983) favored the coat protein in this role, since it is already known that the coat protein of AMV plays a role differing from that of BMV, and they assumed that it is less likely that the function of the 3a proteins of these two viruses also differed.*

In an attempt to prove the role of coat protein in the regulation of RNA synthesis, protoplasts were infected with a mixture of TSV and AMV B + M. No pseudorecombinants could be constructed from these

---

* This assumption is less convincing now that it is known (Cornelissen and Bol, 1984) that the amino acid sequence homology between Bromo- and Ilarviruses is restricted to RNAs 1 and 2.

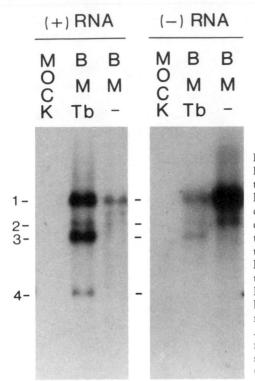

FIGURE 1. Detection of viral plus-strand RNA and minus-strand RNA. RNA extracted from cowpea protoplasts inoculated with different mixtures of AMV nucleoproteins (as indicated in figure) was denatured by glyoxal treatment, electrophoresed on a 1.5% agarose gel, and transferred to DBM paper. The blot was hybridized to [32P]-AMV cDNA to obtain the left panel. After autoradiography, the label was washed off and the blot was hybridized to [32P]-AMV RNA to obtain the right panel. The positions of plus-strand AMV RNAs 1, 2, 3, and 4 are indicated in the margin. Quantititative comparison is only allowed within each panel. Courtesy of Drs. A. Nassuth and J. F. Bol.

viruses but the coat proteins have been shown to be interchangeable in genome activation (van Vloten-Doting, 1975). No effect of the addition of TSV (which was replicated in the protoplasts) was observed on the balance of plus and minus AMV RNAs synthesized. Assuming that the replication of TSV and AMV did take place in the same protoplasts, these results indicate that none of the TSV-coded proteins could substitute for the AMV RNA 3-coded function lacking in B + M-infected protoplasts.

IV. ANALYSIS OF MUTANTS

A. Type of Mutants

1. Mutants Differing in Symptoms, Host Range, Vector Transmission, or Structural Properties

The first mutants of plant viruses to be recognized were symptom mutants (Matthews, 1981). Atypical symptoms are frequently observed in systemically infected plants. A number of single-lesion transfers at high dilution are normally sufficient to obtain a virus isolate which induces symptoms differing from those induced by the parental strain. Less frequently, a systemic infection will be observed in plants which normally react hypersensitively to the virus under investigation (e.g., Yarwood, 1979). From these systemically infected leaves, a new virus mutant can be isolated. A high percentage of AMV symptom mutants thus isolated showed a thermosensitive character (van Vloten-Doting et al., 1980).

Spontaneous or artificially induced mutants may differ in host range and/or vector transmissibility. Virus isolates which are maintained for a long time on one particular host will sometimes lose their capability to infect some other host species. Similarly, prolonged mechanical transmission of a virus may result in the loss of vector transmission. For wound tumor virus (WTV, Reoviridae, genus Phytoreovirus) which has 12 genome segments of dsRNA, it has been shown that loss of the ability to multiply and be transmitted by leafhoppers is coupled with the loss of two of its RNA segments (Reddy and Black, 1977). Several virus isolates of the Cucumo- and Ilarviruses have been kept on host plants of one type for several years by mechanical inoculation but there are no reports about any loss of genome parts, suggesting that none are exclusively devoted to information required for vector transmission. This is in accordance with the observation that aphid transmissibility of CMV is located on RNA 3, which contains the information for the 3a and the coat proteins (Mossop and Francki, 1977).

Upon analysis of virus nucleoprotein on sucrose gradients, polyacrylamide gels, or in the electron microscope, strains or mutants with deviating structural properties are sometimes encountered (Section IV.J.1). All these type of mutants described above are useful in determining the genetic basis of one particular characteristic (e.g., vector transmission or viral architecture) but the usefulness for general genetic analysis is limited.

2. Thermosensitive Mutants

The use of conditional lethal mutants for plant viruses is restricted to thermosensitive or thermoresistant ones. Fortunately, these types of mutants can be found for all essential functions. The molecular basis for the thermosensitive character involves a small change; e.g., variation of just one nucleotide or one amino acid may affect the conformational stability at higher or lower temperature and thus affect the function of the RNA and/or protein. The majority of thermosensitive (ts) mutations are expressed at the protein level, but some are expressed at the RNA level (for plant viruses, e.g., Taliansky et al., 1982a; Roosien et al., 1983b).

Loss of a particular function due to a ts mutation in a virus-coded protein does not necessarily imply that the mutated virus-coded protein itself is responsible for that function. It can also indicate that the mutant protein at the nonpermissive temperature can no longer associate with the appropriate host- or virus-coded macromolecules.

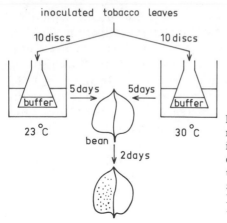

FIGURE 2. Scheme representing the determination of ts character of virus production in tobacco. Discs were punched at random out of five tobacco leaves directly after inoculation, and placed in preincubated buffer. After 5 days, virus production was measured by local lesion assay. From Roosien and van Vloten-Doting (1982).

B. Determination of ts Character

A ts defect may be expressed as a failure to induce virus symptoms or to produce progeny virus particles at the nonpermissive temperature. Since we do not know which steps of the virus multiplication process are required for symptom induction, part of the ts mutations may escape detection when only symptom formation is studied. For instance, incomplete infections, leading to the multiplication and spread of naked viral RNA, may sometimes induce the same symptoms as complete infections. The best studied examples of this phenomenon are the so-called stable and unstable infections by tobacco rattle virus (TRV, a Tobravirus with a bipartite genome; Lister, 1966, 1968).

When whole plants are compared, one should be aware that plants differ substantially in their susceptibility to virus infections and that the temperature within a leaf may be considerably lower than the surrounding temperature (Robinson, 1973). Both problems can be circumvented by comparing the virus production in two sets of leaf discs taken from one set of inoculated leaves, and incubated in solutions kept at different temperatures (Fig. 2). Unfortunately, discs from some plants (e.g., bean and cowpea) will not survive in buffer (Roosien and van Vloten-Doting, unpublished results) but another way to circumvent the problems mentioned above is by the use of protoplasts (Sarachu et al., 1983). A disadvantage of the use of protoplasts is that ts mutations involving virus spread will escape detection (Nishiguchi et al., 1978).

Virus replication is an intimate interplay between host- and virus-coded functions. Virus mutants may show a ts character on one host but a wild-type (wt) phenotype on others (Robinson, 1973; van Vloten-Doting et al., 1980; Roosien and van Vloten-Doting, 1982). Some complex mutants may even express different ts mutations in different hosts (Sarachu

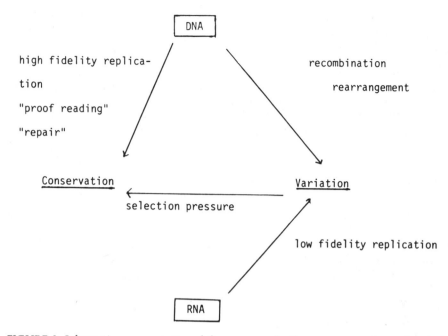

FIGURE 3. Schematic representation of the processes leading to conservation and variation of DNA and RNA genomes.

et al., 1983; Section IV.H.2). These observations show that results obtained with one host may not be extrapolated to others.

C. Spontaneous Mutations

1. Estimation of Spontaneous Mutation Frequency

In the last 5 years it has become apparent that RNA replication takes place with less fidelity than DNA replication (Taniguchi *et al.*, 1978; Holland *et al.*, 1982; Prabhakar *et al.*, 1982, and references therein). This difference may be due to the fact that RNA replicating enzymes can start *de novo* synthesis which means that chain elongation cannot be dependent upon perfect base pairing between template and growing chains. Thus, RNA replication lacks the proofreading mechanism which operates during DNA replication. Conservation of a particular base sequence will depend on selection pressure (Fig. 3). For bacteriophage Qβ (± 4000 bases), it has been estimated that in a multiply passaged population, each viable genome differs in one or two bases from the "average" parental population (Domingo *et al.*, 1978). For the animal virus Coxsackie B, it has been found that the frequency of antigenic mutations (mutations expressed at the protein level) is on the order of 10^{-4} (Prabhakar *et al.*, 1982). Since neutral, silent, and lethal mutations do not contribute to the antigenic variation, the rate of base substitutions must be much higher.

No estimates of mutation rates are available for RNA plant viruses. However, it is reasonable to assume that the rate of base substitution during plant viral RNA replication will be of the same order as that observed for RNA bacterial and animal viruses. Moreover, high variability of RNA plant viruses has been observed. For instance, Kunkel (1940) observed that 1 out of every 200 lesions induced by tobacco mosaic virus (TMV, a Tobamovirus with a monopartite genome with ssRNA of 6395 bases; Goelet et al., 1982) contained a distinguishable mutant, indicating that the rate of base substitution is very high.

Different viruses vary quite widely in the rate at which they give rise to new strains (Matthews, 1981); e.g, among the viruses with a tripartite genome, much less variation is observed with CCMV than with AMV (W. O. Dawson, personal communication). This difference may be attributed to the possibility that some viral replicases are less error-prone than others. However, one should be aware that the spontaneous mutation frequency observed is usually very much lower than the frequency of base substitution during RNA replication. After the random base substitutions which take place during RNA replication, there is a process of selection, eliminating all lethal mutations and, depending on how the experiment is performed, most of the less vigorously replicating ones. Moreover, the number of mutants detected depends on the method of detection, but even when the same method is used, the number of mutants observed cannot be correlated directly with the number of base substitutions.

2. Nonrandom Distribution of Spontaneous Mutations over the Genome?

There is a remarkable difference in reversion rate between different mutants. Nearly all mutants selected by host shift will revert to wt phenotype upon propagation in the original host (Section IV.F.1). Some artificially induced mutants isolated directly after the mutagenic treatment will quickly revert to the wt phenotype (Section IV.G) while others will be quite stable. This difference may suggest that the base substitutions which arise spontaneously during RNA replication are not distributed randomly over the genome. Nonrandom distribution could arise when there is a "context" effect during RNA replication; e.g., the primary (or secondary) structure of the RNA template determines whether transcription will be more or less faithful. In the low-fidelity areas, base substitutions take place frequently and changes in selection pressure (e.g., host shift) allow the reproducible selection (Donis-Keller et al., 1981) of better-adapted virus variants. If the original selection pressure is applied (return to original host), the original sequence is again selected (see Fig. 4). Artificially induced mutations will be present in the low- as well as in

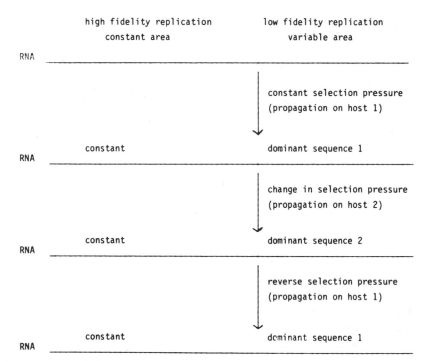

FIGURE 4. Schematic representation of the effect of variation in fidelity of transcription on a virus genome propagated under different selection pressures. The variation in fidelity could be generated by the local sequence and/or structure of the template ("context" effect). The virus genome is thought to consist of a large number of high- and low-fidelity areas.

the high-fidelity regions. However, only the latter class will give rise to stable mutants.

3. Consequences of the High Mutation Frequency

The high mutation rate observed in RNA viruses has a number of consequences for the maintenance and handling of plant viruses. (1) All stock material of wt virus and mutants should be kept without an opportunity to replicate (see Section IV.D). (2) Determination of nucleotide sequence should preferentially be performed on RNA (which gives the average sequence of all RNA molecules present) instead of on cloned cDNA (which gives the sequence of a limited number of amplified RNA molecules). (3) Cloned DNA of RNA viruses might be unsuitable as vectors for gene transfer (van Vloten-Doting, 1983). (4) Cross-protection of crops by mild strains is risky because the mild strain could give rise to virulent strains. The risk is higher for vector-transmissible strains with a broad host range than for nontransmissible viruses with a narrow host range.

D. Stability of Mutants

1. Stability during Virus Propagation

During propagation of viruses, some base substitution will always occur. It is conceivable that especially during the first multiplication cycle after the mutational event (either spontaneous or artificially induced), additional mutations enhancing the viability of the variant may accumulate (see also Section IV.F.1). There will even be some kind of selection for more vigorously replicating mutants because during the sequential single-lesion transfer required to obtain a pure mutant isolate, the lesions appearing first will usually be the ones selected.

Indications that such an accumulation of mutations may indeed take place are as follows: (1) the presence of two amino acid substitutions in the coat protein of a nitrous acid-induced CCMV ts mutant from which only one is consistent with the mutagenic action (Rees and Short, 1982); (2) the presence of five amino acid substitutions in the coat protein of another nitrous acid-induced CCMV structural mutant (Bancroft *et al.*, 1976); again, only part of the changes are consistent with the mutagenic action; (3) the observation of Dawson (1978) that CCMV ts mutants become stable only after numerous sequential single-lesion transfers; (4) the finding that some of the AMV ts mutants carry ts mutations in more than one component (Sarachu *et al.*, 1983; Section IV.H.2).

Theoretically, a mutation in the viral replicase (subunit) leading to a less faithful transcription may increase the "spontaneous" mutation rate. During propagation the characteristics of the mutants will vary continuously. Convincing evidence for the existence of such mutants is difficult to obtain but one AMV mutant [Mts 3(s)t,b,c] is suspected of being such a "mutator mutant" (see Section IV.H.2).

2. Stability during Storage

As argued above (Section IV.C.3), virus isolates should be maintained without an opportunity to replicate if their original properties are to be retained. For viruses which lose their viability when stored as nucleoprotein in buffer, preservation as nucleoprotein in 50% glycerol (Lister, personal communication) or dried leaf material (Bos, 1969) can be recommended. With AMV mutants, we normally infect a number of plants with it and store the virus from these plants. Several well-infected leaves are dried over $CaCl_2$ at 4°C. The dried leaf material is divided into small parcels which are then kept over $CaCl_2$ at -20°C. Analysis of mutant characteristics should preferably be done with virus cultivated from this dried leaf material, avoiding numerous transfers from one plant to another. In this way the genetic variability between different batches of one particular mutant (or of wt virus) is kept to a minimum. Most AMV mutants were found to keep their viability for several years at -20°C, and no changes in properties of the mutant were observed after storage. Only one mutant lost its ts character after storage at $-20°C$ for more than 4 years. In this case we found that material which had been kept in culture was still thermosensitive, suggesting that the original dried leaf material contained besides the mutant some wt or revertant virus, which survived while the ts mutant was inactivated (van Vloten-Doting et al., 1980).

E. Induction of Mutations

Only results obtained with viruses with tripartite genomes are discussed. A more general discussion as well as information about the mechanism of mutagenesis is given by Matthews (1981).

1. Mutagenesis by Chemicals

Bancroft and co-workers (Bancroft et al., 1971, 1972; Bancroft and Lane, 1973) reported the isolation of a number of mutants after treatment of CCMV and BMV RNA with nitrous acid at pH 4.8 to a survival level between 10^{-2} and 10^{-3}. Mutations on all genomic RNAs were observed and from the eight mutants isolated, at least two carried mutations on two genome segments.

Hartmann et al. (1976) reported the isolation of mutants after treatment of AMV nucleoprotein with 0.04 M N-methyl-N'-nitro-N-nitrosoguanidine (NNG) at pH 7.0 to a survival level of 10^{-2} (24 hr at 25°C). Mutants were selected on the basis of aberrant symptoms and were found to be unstable. The three stable isolates carried mutations in RNA 1 and/or RNA 3. van Vloten-Doting et al. (1980) also used NNG to inactivate AMV but apparently without induction of mutations.

Dawson (1978) also used NNG for the induction of mutants of CCMV grown in cowpea. A survival level of 20% was found for CCMV under conditions which resulted in survival of only 1% for AMV. ts mutants were selected on soybean by the method developed by Dawson and Jones (1976) (see Section IV.F.3). During purification of isolates by sequential single-lesion transfer, most of the mutants reverted to the wt phenotype. Only three stable mutants were obtained, two of which mapped on RNA 1 and one on RNA 3. These mutants were derived from the same initial lesion, indicating that either this lesion contained more than one mutant and/or that additional mutations accumulated during propagation of these isolates in soybean (the parent virus was cultivated in cowpea plants). Even in the absence of mutagenic treatment, host shift often results in the appearance of mutants (see Section IV.F.1). In view of the fact that in the two studies reporting positive results with NNG, the number of mutants isolated was very low and hence additional factors may have played a role in selection of mutants, it is doubtful whether NNG should be used as a mutagen for RNA plant viruses.

2. Mutagenesis by UV Light

UV light was found to be mutagenic for AMV nucleoprotein (van Vloten-Doting et al., 1980; Roosien and van Vloten-Doting, 1982). Separated nucleoprotein components were irradiated until survival levels between 10^{-1} and 10^{-4} were reached. The UV inactivation increased with increasing RNA target size (J. Roosien, unpublished results). Mutations could be induced in all the RNA components. Normally, stable isolates were obtained but unstable isolates were also obtained when a host shift was employed during single-lesion transfer. From the symptoms induced by irradiated virus, local lesions were picked at random. Depending on the experiment, between 10 and 60% of the isolates contained ts mutations. The mutations always mapped on the component treated with UV light, confirming the mutagenic action of UV light.

3. Mutagenesis by Manipulation of Viral cDNA

Manipulation of base sequences at the DNA level has the great advantage that all types of mutations (base substitutions, insertions, as well as deletions) can be introduced at will on any predetermined site in the genome. Full-length DNA copies of some viral RNAs are already available. "Polio DNA" was reported to be infectious (Racaniello and Baltimore, 1981) but there are no reports as yet of infectious DNA from any RNA plant virus.

If the DNA itself is not infectious, this does not impair the usefulness of these clones. With present-day techniques it must be possible to obtain precise transcripts from the cloned sequences. The viral cDNA can be inserted after a strong promoter such that the first base transcribed coincides with the first base of the viral sequence. Termination of transcription at the last base of the viral sequence may be more difficult and trimming of the primary transcript may be necessary. Investigations of the biological properties of these transcripts will (hopefully) reveal how the structure of the viral RNA and the encoded proteins are related to their functions. Serious problems with experiments of this type could be caused by the rather high mutation rate during RNA replication. To permit the conclusion that a certain effect is due to a particular mutation introduced, one has to exclude the presence of additional spontaneous mutations. This could mean that the base sequence of the progeny of each mutant should be checked.

F. Selection of Mutants

1. Selection by Host Shift

For a large number of plant viruses including AMV (Oswald *et al.*, 1955; Roosien and van Vloten-Doting, 1983), CMV (Yarwood, 1970, 1979) and CCMV (Kuhn and Wyatt, 1979; Wyatt and Kuhn, 1979), host passage effects (also called host adaptation) have been reported. One of the major questions in the study of host passage effect is whether it is due to host-directed mutagenesis or to random mutation followed by selection by the particular host. Although there is as yet no final proof to distinguish between these possibilities, evidence in favor of the latter is accumulating.

My working hypothesis is that mutations take place at random, but the type and number of mutations that are expressed are selected by the host. There is ample evidence that during RNA replication, a certain amount of base substitution always takes place (see Section IV.C), some of which will inactivate the virus whereas others will change one or more of the virus characters. In the plant there will be competition between these mutants and the parental strain. If the parental strain is well adapted to this host, it can be expected that it will successfully outcompete most mutants. However, upon transmission of the virus from this host to another, some of the mutants may have a better chance of competing successfully with the parental strain. The molecular basis for this selection is unknown but there are some indications that speed of systemic movement might be an important factor. During several multiplication cycles, such positive mutations may accumulate and a gradual change in character of the isolate is observed. (Donis-Keller *et al.*, 1981). Roosien and van Vloten-Doting (1982) found that after host shift, from tobacco to beans for four transfers and back to tobacco, 8 of 50 AMV isolates were thermosensitive. However, 7 of the 8 ts isolates reverted to wt phenotype after three transfers on tobacco.

2. Selection by Gradual Temperature Shift

Franck and Hirth (1976) selected a thermoresistant isolate of AMV by passing the virus a number of times through tobacco discs which were incubated at increasing temperature (see Table I). The first temperature regime was close to the nonpermissive temperature for the wt. At each temperature the time course of accumulation of infectious virus and of viral antigen was determined. Virus was always transferred to the next set of discs when the infectivity had reached its maximum. The isolate obtained in this way could hardly multiply at 22°C (optimal temperature for the parental strain), while it could multiply reasonably well at 34°C (highest temperature employed during selection). As expected, the isolate contained a mixture of mutants because it induced two different types

TABLE I. Scheme for Selection of Temperature-Resistant
Mutants of AMV by Gradual Temperature Shift Employed
by Franck and Hirth (1972)[a]

Number of transfers	Incubation temperature °C of tobacco leaf discs	Days between transfers
6	28	8
10	30	6
5	33	4
7	34	3

[a] Inoculum from the last set of tobacco discs was used to start single-lesion purification of mutant lines.

of local lesions in *Vigna catjang* (Franck, 1978). For further genetic work, serial single-lesion transfers would be required. Mutants obtained in this way may be expected to contain several mutations because at each transfer the best adapted mutant will dominate and an accumulation of advantageous mutations will be selected.

3. Selection by Differential Development of Symptoms

Dawson (1978) used the following method to select ts mutants from a mutagen-treated preparation of CCMV. Soybean leaves inoculated with the treated virus (cultivated in cowpea) were incubated for 2 days at 35°C (permissive temperature for wt and nonpermissive for ts mutants). All lesions developed at 35°C (wt phenotype) were marked. Plants were then transferred to 25°C; all lesions newly developed at this temperature might be expected to contain ts mutants. Potential ts isolates were passaged through several single-lesion transfers on soybean at 25°C, while the ts character was checked on both soybean and cowpea.

This procedure is an adaptation of the method developed by Dawson and Jones (1976) for TMV. With TMV they found that only 25 of 225 potential isolates contained stable ts mutants. According to Dawson (1978) the procedure worked even less satisfactorily with CCMV. Most isolates were unstable (lost their ts character) but some stable isolates were obtained after numerous sequential single-lesion transfers. Finally, only the progeny from one of all the original lesions assayed, yielded ts virus.

This poor result (less than 1% of all potential isolates contained ts virus) makes it doubtful whether it is worthwhile to put so much effort into this type of selection method. Direct screening of all isolates obtained after mutagenic treatment and avoiding host shifts, might be more profitable.

G. Reversion of Mutations

Although there are several reports about unstable virus mutants, there is very little information available about reversion, except for mutants selected by host shift (Section IV.F.1). Determination of reversion frequency of plant viruses is very laborious. To enhance the selection pressure for reversion, Roosien and van Vloten-Doting (1982) cultured mutant virus at, or close to, the nonpermissive temperature.

In the first experiment the ts character of virus present in local lesions developed at the nonpermissive temperature was assayed. We used an AMV RNA 3 ts mutant [Tbts 7(uv)t,b], in bean leaves. However, in plants kept continuously at the nonpermissive temperature the virus production was below the level of detection. The procedure adopted was as follows: plants were inoculated and incubated for 18 hr at the permissive temperature (local lesion development takes about 48 hr) and prior to the shift to the nonpermissive temperature the bean leaves were immersed in water of 50°C for 45 sec to delay the necrotic host reaction (Wu et al., 1969). After the warm water treatment, plants were incubated for an additional 72 hr at the nonpermissive temperature to allow lesion development. Plants infected with the ts mutant developed at 30°C produced less than 10% of the number of lesions observed at 23°C whereas there was no significant difference in the numbers of lesions observed in plants infected with wt. The progeny from 10 of the local lesions induced by the mutant at the nonpermissive temperature was analyzed and found to be as thermosensitive as the parent isolate.

In a second experiment, Roosien and van Vloten-Doting (1982) compared the ts character of mutant nucleoprotein isolated from primarily infected tobacco leaves kept at 23 and 30°C. The higher temperature is nonpermissive for mutant multiplication, but due to the difference in temperature between an intact leaf and its surroundings (Robinson, 1973), some virus multiplication is still possible. If a significant amount of reversion did take place, the virus isolated from plants held at 30°C should be less thermosensitive than that from plants at 23°C. However, no significant differences were found between the virus isolated from plants kept at 23 and 30°C (Table II). It is difficult to make a quantitative estimate of the reversion rate from this type of experiment.

That phenotypic reversion of artificially induced mutations can take place was demonstrated by the third experiment, in which the ts character of mutant virus isolated from systemically infected tobacco plants kept at 23 and 30°C was analyzed. From Table III it can be seen that the ts character of all three preparations of the mutant grown at 23°C was comparable. However, two from the three preparations grown at 30°C showed an intermediate ts character indicating that at least some virus had become thermoresistant.

TABLE II. Comparison of ts Character of Mutant Virus Grown in Tobacco Plants Kept at 23 and 30°C

| Mutant | Inoculum virus grown in plants kept at | Virus production[a] in tobacco leaf discs incubated at | | $\dfrac{\text{Virus production at 30°C}}{\text{virus production at 23°C}} \times 100$ |
		23°C	30°C	
Mts 6(uv)t	23°C	263	6	2
	30°C	255	8	3
Mts 7(uv)t,b	23°C	472	2	0
	30°C	165	0	0
Mts 8(uv)t	23°C	289	5	2
	30°C	157	2	1

[a] Determined at 5 days p.i. by local lesion assay on bean leaves (see Section IV.B and Fig. 2).

Phenotypic reversion of the ts character may be due to real reversion of the first base substitution or to a second mutation suppressing the effect of the first mutation. The fact that mutant preparations I and III grown at 30°C still showed a ts character on bean while their mutant character on tobacco (on which they were selected) was partly lost, suggests that the reversion was due to acquisition of such a suppression mutation.

TABLE III. Selection for Revertants of Tbts 7(uv)t,b[a]

| | | ts character | |
Inoculum	Expt.	On tobacco leaf discs[b] $\dfrac{\text{virus prod. at 30°C}}{\text{virus prod. at 23°C}} \times 100$	On bean[c] $\dfrac{\text{virus prod. at 30°C}}{\text{virus prod. at 23°C}} \times 100$
Mutant preparation isolated from tobacco plants kept at 23°C	I	4	1
	II	4	2
	III	6	0
Mutant preparation isolated from tobacco plants kept at 30°C	I	34	10
	II	3	2
	III	51	5
Wild-type preparation from tobacco plants kept at 23°C	IV	84	38
Wild-type preparation from tobacco plants kept at 30°C	IV	72	24

[a] Roosien and van Vloten-Doting (unpublished results).
[b] Determined at 5 days p.i. by local lesion assay on bean leaves (see Section IV.B and Fig. 2).
[c] Determined by direct local lesion assay on bean plants, preincubated, inoculated, and developed at 30 or 23°C.

H. Assignment of Mutations to RNA Components

1. Assignment by Construction of Pseudorecombinants (See Also Section II)

This method is the most rigorous, but also the most laborious. It requires the purification of pure RNA and/or nucleoprotein components of both wt and all mutants to be analyzed. In principle, six pseudorecombinants have to be constructed from each mutant and wt. However, often when there are already some indication on which RNA the mutation is located, only the combination containing the suspected mutant RNA and the complementing two wt RNAs as well as the reverse combination are analyzed (Bancroft and Lane, 1973; Roosien and van Vloten-Doting, 1983).

2. Assignment by Supplementation Test

In the supplementation test the characteristics of the mutant virus (lesion size or ts character, etc.) are compared to those of the mutant virus supplemented separately with each of the wt RNAs or nucleoproteins. wt phenotype will only be obtained when the wt RNA species added corresponds to the RNA species carrying the mutation. Bancroft and Lane (1973) used mixtures of a phenol extract of mutant-infected leaves with purified wt CCMV RNAs. Dawson (1981) avoided the necessity of extracting RNA from the mutant-infected leaves by a procedure of sequential inoculation. Each half leaf of a local lesion host was first inoculated with one purified wt RNA component followed by a second inoculation of the leaves with sap from a mutant-infected leaf.

Supplementation tests of AMV mutants on intact plants have been performed by inoculating purified mutant nucleoprotein or sap from mutant-infected leaves supplemented with single wt nucleoprotein components (van Vloten-Doting et al., 1980; Roosien and van Vloten-Doting, 1982). Phenol extraction of AMV mutant-infected leaves is unnecessary, since the genome parts of AMV can be separated as nucleoprotein particles. At 0°C these particles will retain their infectivity in leaf sap for at least 1 hr (unpublished results).

The supplementation test is much easier than construction of pseudorecombinants but the data are sometimes less clear-cut. One complicating factor is that the wt RNA species must always be present in excess over the mutant RNA species. Unfortunately, the concentration of the latter is often not known. In Table IV results of supplementation tests with two AMV RNA 2 ts mutants are shown. The actual amount of supplementation (numbers in italic) is extremely variable. This variation may be due to the fact that in experiment 1 the amount of purified wt component added was insufficient and a large part of the infection centers contained only mutant components. Such a situation will increase the number of local lesions at 23°C and decrease the percentage of supple-

TABLE IV. Supplementation of Two AMV RNA 2 TS Mutants on Bean

	Local lesions on bean[a] on:					
	Expt. 1			Expt. 2		
	23°C	30°C	(30/23) × 100	23°C	30°C	(30/23) × 100
Mts 7(uv)t,b[b]	260	8	3	47	1	1
Mts 7(uv)t,b + wt B[c]	193	9	5	81	0	0
Mts 7(uv)t,b + wt M[c]	253	78	31	174	164	94
Mts 7(uv)t,b + wt Tb[c]	601	24	4	410	6	1
Mts 11(uv)b[b]	203	11	5	945	107	11
Mts 11(uv)b + wt B[c]	205	3	2	375	60	16
Mts 11(uv)b + wt M[c]	541	101	19	1075	1160	108
Mts 11(uv)b + wt Tb[c]	691	7	1	738	195	26

[a] Bean plants were preincubated, inoculated, and developed at 23 or 30°C.
[b] Mutant inoculum consisted of homogenate from systemically infected leaf.
[c] Purified wt components were added to a final concentration of 2 μg/ml.

mentation. Both mutants shown in Table IV were induced by UV irradiation of M, and therefore it is unlikely that the low percentage of supplementation found in experiment 1 was due to the presence of additional mutations in other components. Spontaneous mutants may sometimes contain ts mutations on more than one component. When in such a case supplementation tests are only performed with single components, one of the ts mutations may be missed or the ts mutation may be assigned to the wrong component. An example of such a complicated mutant is given in Table V. The ts mutation present in a spontaneous AMV mutant and expressed in tobacco and bean was assigned to RNA 2 (van Vloten-Doting et al., 1980). In later work (Sarachu et al., 1983), it was found that this mutant expressed a ts mutation on RNA 1 and one on RNA 3 in tobacco, while ts mutations on RNA 2 and RNA 3 were expressed on bean. This discrepancy could be explained by assuming that the wt M preparation used in the first study was contaminated with B and Tb. Alternatively, additional mutations may have been acquired during propagation of this mutant. The original mutation in RNA 2 might have had an effect on the fidelity of transcription (Section IV.D.1) since the 1983 preparation carried at least one ts mutation in each component.

3. Assignment by Complementation Test

Besides the construction of pseudorecombinants and the supplementation test, Bancroft and Lane (1973) used a complementation test for the assignment of mutations. In this test the symptoms induced by a mixture of two mutants were compared to the symptoms produced by each of the mutants alone and those of wt. If wt symptoms did appear, it was concluded that the mutations present in these two mutants were

TABLE V. Supplementation of a Spontaneous AMV Mutant [Mts 3(s)t,b,c] on Different Hosts

	Virus production at 30°C (%)				
	Tobacco leaf discs[a]		Bean plants[b]		
Wt component added	van Vloten-Doting et al. (1980)	Sarachu et al. (1983)	van Vloten-Doting et al. (1980)	Sarachu et al. (1983)	Cowpea protoplasts[a]
None	0	0	0	1	0
B	2	0	1	3	7
M	19	0	58	24	4
Tb	2	0	1	19	1
B + M	—	0	—	17	101
B + Tb	—	28	—	7	20
M + Tb	—	1	—	84	1

[a] Virus production (measured by local lesion assay on bean) at 30°C as percentage of the corresponding virus production at 23°C (tobacco) or 25°C (cowpea).
[b] Number of local lesions induced on bean plants preincubated, inoculated, and incubated at 30°C as percentage of local lesions induced on bean plants kept at 23°C.

located on different RNA species and that reassortment of RNA species resulted in the formation of wt. However, this test may give erroneous results because complementation of two mutants carrying mutations on the same RNA can take place (De Jager and Breekland, 1979; van Vloten-Doting et al., 1980; Roosien and van Vloten-Doting, 1982).

I. Complementation and Interference between Mutants with ts Defects on the Same RNA

Complementation studies are useful to obtain insight into the number of functions encoded by the viral genome. Mutants carrying ts defects on different RNAs will always fall into different complementation groups because reassortment of the genome parts will result in the formation of wt. New information will only be obtained when the experiments are restricted to mutants carrying ts defects on the same RNA. All complementation tests should always be done with high concentrations of each mutant so as to ensure a high percentage of doubly infected cells.

Systematic complementation studies have only been reported for AMV (van Vloten-Doting et al., 1980: Roosien and van Vloten-Doting, 1982). All mutants used in these studies were ts in virus production on tobacco (expressed as the number of local lesions induced on bean by homogenates of mutant-infected tobacco discs incubated at 23 or 30°C; see Fig. 2). The majority of mutants used in the first study were spontaneous mutants and the mutation was assigned to one of the RNA segments by supplementation tests (see Section IV.H.2). The results were expressed as virus production ratio [= (virus production at 30°C/virus production at 23°C) × 100].

The virus production ratio for the different combinations of mutants varied from experiment to experiment. This variability in the percentage of complementation observed between a pair of mutants (Table IV) might be due to variations in the concentration of the mutants in the leaf homogenate used as inoculum. Attempts to improve the reproducibility by using purified nucleoprotein preparations were unsuccessful, probably because some mutant preparations were inactivated much faster than others. Nevertheless, it was observed that certain combinations consistently produced virus at 30°C while others did not.

1. RNA 1 ts Mutants

Two of the four mutants with a ts defect on RNA 1 [Bts 1(uv)t and Bts 4(s)t,b] were demonstrated to complement each other, while the other two [Bts 2(s)t and Bts 3(s)t] showed no complementation or interference. The simplest explanation for these results is that the product encoded by RNA 1 has two functional domains, one represented by Bts 1(uv)t and one by Bts 4(s)t,b, while both domains are disturbed by the mutations present in Bts 2(s)t and Bts 3(s)t. In later work (Sarachu et al., 1983), it was found that Bts 4(s)t,b carried an additional ts mutation on RNA 2; however, this does not influence the conclusions from the complementation test. This conclusion is corroborated by the results of detailed analysis of cowpea protoplasts infected with AMV RNA 1 ts mutants. One RNA 1 ts mutant was affected in minus-strand RNA synthesis only, while the other was affected in minus-strand RNA synthesis as well as in translatability of viral RNA (see Sections IV.J.5 and IV.J.8).

2. RNA 2 ts Mutants

van Vloten-Doting et al. (1980) studied three AMV RNA 2 ts mutants. One of these mutants could complement the other two. Subsequently, Roosien and van Vloten-Doting (1982) investigated four additional mutants with defects on RNA 2, all of which were induced by UV irradiation of the middle component. Some combinations of mutants consistently produced much less virus than expected at 23°C (Table VI). Because of this anomaly, the virus production ratio could not be used to express the results. Direct comparison of virus produced by a mixture of mutants to that produced by each of the mutants alone, involved comparisons of virus produced in different tobacco leaves and assayed on different beans. A control experiement with wt showed that virus production measured in different plants could vary by a factor of 4 due to differences in susceptibility of the plants. Taking into account this variability, the interactions between the mutants were divided into three classes:

TABLE VI. Interference between AMV RNA 2 ts Mutants

Exp.	Incubation temp.	Virus production[a] in tobacco infected by		
		Mutant X	Mutant Y	Mutant X + mutant Y
		Mts 1(ni)t	Mts 4(uv)t	Mts 1(ni)t + Mts 4(uv)t
1	30	0	0	0	0
	23	105	267	0	(186)
		Mts 1(ni)t	Mts 10(uv)t	Mts 1(ni)t + Mts 10(uv)t
2	30	0	0	0	(0)
	23	105	282	0	(194)
		Mts 4(uv)t	Mts 10(uv)t	Mts 4(uv)t + Mts 10(uv)t
3	30	0	0	0	(0)
	23	267	282	7	(275)
		Mts 4(uv)t	non-ts[b]	Mts 4(uv)t + non-ts[b]
4	30	0	729	0	(360)
	23	267	1430	30	(849)
		Mts 4(uv)t	wt	Mts 4(uv)t + wt
5	30	1	89	62	(45)
	23	315	116	159	(216)

[a] Values represent virus production in tobacco leaf discs measured by local lesion assay on bean (see Section IV.B and Fig. 2). Values in parentheses are those expected if there was no interaction between the mutants [production by mixture = 0.5 (production by mutant X + production by mutant Y)].
[b] Non-ts mutant is Mts 1(uv)b,c (Section IV.J.11).

1. Interference (I): virus produced by the mixture of mutants at 23°C is less than 10% of the expected value (Table VI).
2. Complementation (C): virus produced by the mixture of mutants exceeds the expected value by a factor of 10 (Table VII).
3. No interaction (N) (all intermediate values).

Using these parameters the RNA 2 mutants could be divided into two complementation groups (A and B) (Table VIII). Mixtures of mutants belonging to different groups often complemented each other. Mixtures of mutants belonging to the same group often interfered with each other.

Although the results confirmed those obtained earlier, there is one complication. Mutant Mts 3(s)t,b,c was found to be a very complex mutant (Sarachu et al., 1983). Depending on the host, it expressed ts mutations on different RNA species: in cowpea protoplasts on RNA 1 and RNA 2, in bean on RNA 2 and RNA 3, and in tobacco on RNA 1 and RNA 3 (Table V). Originally the mutation was mapped on RNA 2 by supplementation test in tobacco. It is possible that the assignment of the mutation to RNA 2 was wrong. Alternatively, the mutant character may have changed. Whatever the explanation for the behavior of Mts 3(s)t,b,c may be, it does not negate the conclusion that RNA 2 mutants fall into two complementation groups since for instance Mts 9(uv)t and Mts 7(uv)t,b (both UV-induced mutants) also showed complementation (Table VIII).

TABLE VII. Complementation between AMV RNA 2 ts Mutants

Exp.	Incubation temp. (°C)	Virus production[a] in tobacco infected by		
		Mutant X	Mutant Y	Mutant X + mutant Y
		Mts 2(s)t	Mts 4(uv)t	Mts 2(s)t + Mts 4(uv)t
1	30	0	0	40 (0)
	23	20	267	91 (144)
		Mts 3(s)t,b,c	Mts 4(uv)t	Mts 3(s)t,b,c + Mts 4(uv)t
2	30	0	0	110 (0)
	23	18	267	231 (143)
		Mts 3(s)t,b,c	Mts 10(uv)t	Mts 3(s)t,b,c + Mts 10(uv)t
3	30	0	0	56 (0)
	23	18	282	71 (150)
		Mts 3(s)t,b,c	Mts 9(uv)t,b	Mts 3(s)t,b,c + Mts 9(uv)t
4	30	0	0	36 (0)
	23	17	39	144 (28)

[a] Values represent virus production in tobacco leaf discs measured by local lesion assay on bean (see Section IV.B and Fig. 2). Values in parentheses are those expected if there was no interaction between the mutants [production by mixture = 0.5 (production by mutant X + production by mutant Y)].

There is very little information about the molecular basis of interference between two mutants inoculated simultaneously. The fact that multiplication of both mutants is inhibited even at the permissive temperature suggests that interference takes place during the replication of virus in the primary infected cell and that the interference is not due to exclusion of one mutant by the other during infection of the neighboring cells.

TABLE VIII. Genetic Interaction between AMV Mutants (Mts)

Complementation group		A				B		
		Mts 1	Mts 4	Mts 9	Mts 10	Mts 2	Mts 3	Mts 7
A	Mts 1(ni)t	—	I[a]	I	I	C[b]	C	N[c]
	Mts 4(uv)t		—	?	I	C	C	N
	Mts 9(uv)t				N	N	C	N
	Mts 10(uv)t			—	—	?	C	N
B	Mts 2(s)t					—	N	N
	Mts 3(s)t,b,c						—	I
	Mts 7(uv)t,b							—

[a] I (interference), defined as virus production at 23°C by (mutant X + mutant Y), is less than 10% of virus production expected [(production by mutant X + production by mutant Y)/2].
[b] C (complementation), defined as virus production at 30°C by (mutant X + mutant Y), is at least 10 times higher than virus production expected [(production by mutant X + production by mutant Y)/2].
[c] N (no interaction): virus production at both 23 and 30°C by (mutant X + mutant Y) does not deviate more than a factor of 4 from the virus production expected [(production by mutant X + production by mutant Y)/2].
[d] ?: results not reliable due to lack of infection in some experiments.

TABLE IX. Complementation between AMV RNA 3 ts Mutants[a]

Complementation group		Tbts								
		1	2	3	4	7	8	9	5	6
A	Tbts 1(s)t,b	0	0	0	0	0	0	0	0	16
	Tbts 2(s)t	—	0	0	0	0	0	0	0	45
	Tbts 3(s)t	—	—	0	0	0	0	0	0	44
	Tbts 4 (uv)t	—	—	—	0	0	0	0	0	27
	Tbts 7(uv)t	—	—	—	—	0	0	0	0	100
	Tbts 8(s)t	—	—	—	—	—	0	0	0	19
	Tbts 9(uv)t,b	—	—	—	—	—	—	0	0	13
	Tbts 5(uv)t	—	—	—	—	—	—	—	0	2
B	{Tbts 6(s)t	—	—	—	—	—	—	—	—	0

[a] Values represent (virus production at 30°C/virus production at 23°C) × 100.

The data obtained with the RNA 2 mutants may indicate that RNA 2 codes for one protein which: (1) is active in a multimeric form because combinations of mutants belonging to the same group often interfere with each other [in heteropolymers of subunits with different mutations in the same domains, allosteric interactions between these subunits may lead to a decrease in the activity of the polymer (compare Honess, 1981)] and (2) has two functional domains.

The conclusion that RNA 2 encodes two different functions is supported by the analysis of cowpea protoplasts infected with RNA 2 mutants. The two RNA 2 mutants analyzed showed a defect in minus-strand synthesis and in a function required for the production of infectious virus. Shift-up experiments showed that in some isolates these two functions behaved differently in time (see Sections IV.J.5 and IV.J.9).

3. RNA 3 ts Mutants

Complementation tests have been performed with nine AMV RNA 3 ts mutants. At the permissive temperature, the virus production of all combinations was well within the expected range. Apparently none of these mutants interfered with the multiplication of other mutants. All combinations containing mutant Tbts 6(s)t produced virus at the non-permissive temperature, with the exception of the combination of Tbts 6(s)t and Tbts 5(uv)t (Table IX). Apparently, mutants 1, 2, 3, 4, 7, 8, 9 represent one complementation group (A), and mutant 6 a second (B) while mutant 5 might be a double mutant.

RNA 3 is known to code for coat protein and the 3a protein (van Vloten-Doting and Jaspars, 1977; Chapter 3). The coat protein of mutants 1, 2, 3, 4, 7 (all belonging to complementation group A in Table IX) failed to activate the wt genome at the nonpermissive temperature. Apparently, complementation group A represents mutants with a ts defect in the early function of the coat protein (van Vloten-Doting *et al.*, 1980). It is difficult to determine whether or not the coat protein from these mutants is also disturbed in its encapsidation function.

The coat protein from Tbts 6(s)t can apparently activate the genome at the nonpermissive temperature, and thus the defect has to be in a later function either in the function of 3a or in the encapsidation function of the coat protein. Unfortunately, Tbts 6(s)t was an unstable RNA 3 mutant, initially showing less than 2% virus production at the nonpermissive temperature as compared to that found at the permissive temperature. Upon transfer of the mutant to new plants, this value increased to 15% or higher. The mutant material kept as nucleoprotein at 4°C lost its thermosensitivity upon storage. Apparently, it was contaminated with wt or a revertant which was more stable than the mutant. When mutant-infected dried leaf material stored at − 20°C (see Section IV.D.2) was used as inoculum, the same rapid loss of the ts character was observed and, finally, the mutant was lost completely. Several attempts to obtain other mutants with the characteristics of Tbts 6(s)t were made. All newly induced or selected RNA 3 ts mutants were screened for complementation with several Tbts mutants. However, none of the newly isolated RNA 3 ts mutants showed any complementation.

In view of the fact that RNA 3 codes for the 3a and coat proteins, one would expect to find two complementation groups with about half of the mutants carrying defective 3a protein. Our results clearly conflict with this idea. From the 19 RNA 3 ts mutants analyzed, only one belonged to a second complementation group. Of the other 18, at least 6 were shown to carry a ts defect in the early function of coat protein and the remaining 12 probably also carried similar defects. These results suggest that the 3a protein is so restrictive that mutations are always lethal. However, this cannot be true because several RNA 3 mutants code for a 3a protein which differs from its wt counterpart in electrophoretic mobility on SDS–PAGE. An alternative explanation is that the 3a protein is not essential. Given the fact that all Tricornaviridae contain an open reading frame coding for a 3a protein of similar size, this seems unlikely. The last alternative is that it is very tolerant. If it is involved in the synthesis of subgenomic RNA 4 one could argue that normally there is such an enormous overproduction of RNA 4 that some viable virus might still be synthesized when only 1% of this RNA is produced (see Section IV.J.7). Furthermore, if the transport of virus (or viral RNA–protein complexes) through the host tissues is controlled by 3a (see Section III), one could argue that this function may be very tolerant. Taliansky *et al.* (1982b) actually showed complementation of transport between completely different viruses.

An alternative explanation for the lack of a clear second complementation group on RNA 3 might be that this RNA could code not only for a 3a protein and the coat protein, but also for a fusion product of these two proteins (Joshi *et al.*, 1985). Any ts mutation present in coat or 3a

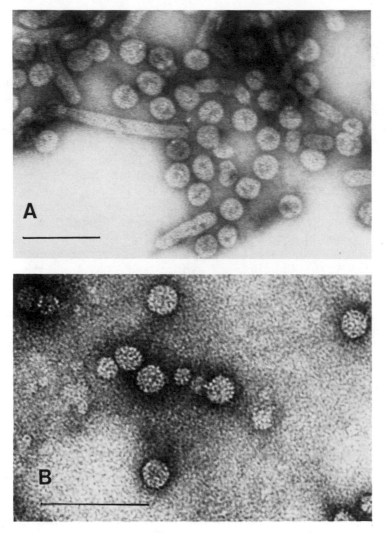

FIGURE 5. Electron micrographs of (A) AMV and (B) BMV mutants with deviating virus architecture. Bars = 100 nm. Courtesy of Drs. J. Roosien and L. Lane, respectively.

protein might also be expressed as a ts mutation in this read-through protein.

J. Viral Functions Affected by Mutations

1. Virus Architecture

Only mutants of AMV and BMV differing in virus architecture from their wt counterparts have been described (Fig. 5).

Part of the nucleoprotein particles of a spontaneous mutant of BMV (BMV-F; Bancroft and Lane, 1973) migrated faster on polyacrylamide gels than nucleoprotein from wt. These particles consisted of RNA 3 encapsidated by approximately 140 coat protein subunits and RNA 4 encapsidated by roughly 110 subunits (Lane, 1974). Alterations in the coat protein are not responsible for this phenomenon because the mutation responsible could be assigned to RNA 2 by the construction of pseudorecombinants (Bancroft and Lane, 1973). Apparently, an RNA 2-coded function is involved in particle assembly *in vivo*; it probably has to do with the combined encapsidation of one RNA 3 plus one RNA 4 molecule. *In vitro*, this function seems to be less important because self-assembly of BMV RNA and protein into particles has been reported (Bancroft *et al.*, 1967; Bancroft, 1970). Lane (1974) also described a nitrous acid-induced BMV mutant (MB 10a) which produced a kind of "middle" component, consisting of RNA 3 encapsidated by 180 coat protein subunits. This mutation, which was not mapped, may affect a structural function. Alternatively, it could affect the synthesis of RNA 4 and the aberrant particles would arise because there is no RNA 4 available for encapsidation.

Bancroft and Lane (personal communication) isolated a CCMV nitrous acid mutant which produced a component with a higher density than the majority of particles. The formation of this dense component was determined by RNA 3. Presumably, the dense component contained several RNA 4 molecules because preparations of mutant virus contained more RNA 4 than RNA 3 molecules.

Roosien and van Vloten-Doting (1983) described a spontaneous AMV mutant [Tbstruct (s)1] which produced mainly spheroidal together with some very long bacilliform particles (Fig. 5A). Both types of particles contained more than one RNA molecule. Analysis of pseudorecombinants showed that information located on RNA 3 was responsible for the altered virus architecture. *In vitro* translation and serology demonstrated that both the coat protein and the 3a protein encoded by RNA 3 contained mutations. Although it is assumed that mutations in the coat protein are responsible for the altered architecture, the evidence is not yet conclusive.

Hull (1969, 1970) described two AMV strains with a tendency to produce significantly longer bacilliform particles. These longer particles contained more than one RNA molecule (Heytink and Jaspars, 1974). For at least one strain (VRU) it was shown that information on RNA 3 is involved in the formation of longer particles. Again it is assumed that it is the coat protein which is responsible (Castel *et al.*, 1979) but involvement of the 3a protein in particle maturation cannot be excluded.

2. Particle Stability

Several virus strains and mutants differing in particle stability have been described. In all cases where this difference has been mapped, it was invariably located on RNA 3, suggesting that alterations in the coat protein are responsible for alterations in particle stability.

The coat protein from an oxidation-sensitive CCMV mutant (Bancroft et al., 1971) contained a cysteine residue in position 25 in place of an arginine in wt (Rees and Short, 1982). The coat protein of the salt-stable CCMV mutant (Bancroft et al., 1973) contained an arginine instead of the lysine in wt in position 105 (Rees and Short, 1982).

The coat protein from a CCMV ts mutant (Dawson et al., 1975) differed by two substitutions: glutamine 21 was changed for lysine and alanine 87 for valine (Rees and Short, 1982). Particles of this ts mutant produced at the permissive temperature were degraded in vivo upon shift-up of the infected protoplasts to the nonpermissive temperature (Dawson et al., 1975).

The coat protein of a CCMV structural mutant (Bancroft et al., 1976) contained at least five amino acid substitutions. Particles from this mutant underwent an abrupt reversible conformational change between pH 5.3 and 5.7, whereas wt particles did so between 6.7 and 7.0. Under mildly acidic conditions, the mutant coat protein formed multilayered "onion skins" but the coat protein of two other CCMV mutants which were unstable at pH 6.0 did not show the formation of such structures.

The amino acid substitutions in six ts coat protein mutants of AMV (van Vloten-Doting et al., 1980) are not yet known. The coat protein from Tbts 1(s)t,b contains at least two amino acid substitutions (Kraal, 1975). Particles from mutant Tbts 1(s)t,b and Tbts 2(s)t were more sensitive to uncoating by AMV RNA than wt particles, while particles from Tbts 7(uv)t,b and particles from an AMV structural mutant [Tbstruct (s)1] were less sensitive in this respect (Smit, 1981; Roosien and van Vloten-Doting, 1983). These results indicate that particles of the first two mutants have less protein–protein interaction than wt, while those of the last-mentioned two mutants have stronger protein–protein interaction.

3. Encapsidation

No mutants of tripartite viruses have been described which at the nonpermissive temperature accumulate free RNA and denatured coat protein. However, accumulation of free RNA at the nonpermissive temperature has been observed in one CCMV RNA 3 ts mutant but no coat protein could be detected (Bancroft et al., 1972). Therefore, it is uncertain whether this mutant has a ts defect leading to a protein which cannot associate with the viral RNA and is enzymatically degraded, or has a ts defect inhibiting the synthesis of coat protein (see Section IV.J.8). The RNA 3 of this mutant probably also carries a mutation in the "nucleation site" for encapsidation. Dawson and co-workers (Dawson et al., 1975; Dawson and Watts, 1979) found that the mutant RNA 3 was dominant over wt RNA 3 both at the permissive and at the nonpermissive temperature. Analysis of phenol extracts of wt- and mutant-infected protoplasts revealed no enhanced synthesis of mutant RNA 3, suggesting that the dominance is due to preferential encapsidation of mutant RNA 3 compared to wt RNA 3.

4. Activation of Genome

In contrast to Cucumo- and Bromoviruses, AMV (Bol et al., 1971) and all other Ilarviruses studied up to now (Gonsalves and Garnsey, 1975a,b,c; van Vloten-Doting, 1975; Huttinga and Mosch, 1976; Gonsalves and Fulton, 1977; Bos et al., 1980) require besides the genomic RNAs, a few coat protein molecules for infectivity. It has been shown that the coat protein activates the genome by associating with each of the genomic RNAs (Smit et al., 1981). In vitro, the coat protein will bind close to the 3'-terminus as well as to some internal sites (Zuidema et al., 1983a, 1984).

The coat protein of six AMV RNA 3 ts mutants (three spontaneous and three UV-induced mutants) was found to be thermosensitive in genome activation. The inactivation was reversible because in leaf discs inoculated at 30°C with wt genomic RNAs plus mutant coat protein, virus was produced upon shift-down to 23°C (van Vloten-Doting et al., 1980).

It is known that the first 25 amino acids are required for binding to the RNAs as well as for genome activation (Zuidema et al., 1983b). In at least one mutant [Tbts 7(uv)t,b], there is no change in this part of the coat protein which can still bind to the genomic RNAs at the nonpermissive temperature but fails to activate wt genomic RNAs (Smit, 1981).

Since there is almost no homology in the primary sequence of coat protein binding sites of AMV and TSV RNAs [both of which can be activated by AMV as well as by TSV coat protein (Gonsalves and Garnsey, 1975c; van Vloten-Doting, 1975)], the conformation of the RNA must be the important factor (Koper-Zwarthoff and Bol, 1980). Therefore, mutants expressing a ts defect in the coat protein binding site of one of the RNA molecules may exist. Since binding of coat protein seems to be a prerequisite for RNA replication (Houwing and Jaspars, 1978; Smit et al., 1981), such a mutation will be expressed by diminished replication of the mutant RNA species. RNA 3 of Tbts 7(uv)t,b may contain such a mutation in its coat protein binding site because in mixed infections of wt and Tbts 7(uv)t,b the mutant RNA 3 is outcompeted by the wt RNA 3 (Roosien et al., 1983b). Alternatively, this phenomenon is due to a diminished affinity of the mutant RNA 3 for the replicase.

5. Synthesis of Minus-Strand RNA

There is ample evidence that plant viruses with plus-strand RNA genomes replicate via a minus-strand RNA template (Matthews, 1981) and there is no evidence for a DNA step in the replication. Minus-strand RNA is present in minute amounts in infected plants or protoplasts. All viral minus-strand RNA present in AMV-infected protoplasts was found to be RNase resistant, suggesting that they are part of a double-stranded structure (Nassuth and Bol, 1983). For studies where plus- and minus-strand RNAs were not probed for separately, the amount of double-stranded RNA can be considered to be indicative of the amount of minus-strand RNA present.

Dawson (1981) used shift-up experiments of cowpea leaves infected with CCMV ts mutants to analyze the effect of the nonpermissive temperature on RNA synthesis determined by incorporation of [³H] uridine. At the nonpermissive temperature, wt CCMV RNA synthesis was about half of that found at the permissive temperature. The same value was found irrespective of the duration of the incubation at the nonpermissive temperature (Fig. 6). Two isolates of an RNA 3 mutant (5f) and two isolates of an RNA 1 mutant (25) were studied. In all cases it was found that incorporation of label into the double-stranded form decreased upon shift to the nonpermissive temperature. One of the isolates continued incorporation of label into the double-stranded form directly after the shift but with the other three isolates there was an immediate decrease of incorporation after the shift-up (Fig. 6). There was no significant difference in the amount of inhibition of synthesis of the three RNA species.

Sarachu et al. (1983) studied the RNA synthesis of two AMV ts mutants in cowpea protoplasts. Minus-strand RNA was detected on "Northern" blots by hybridization with labeled virion RNA. The mutants used behaved as double mutants in cowpea protoplasts, both expressing a ts mutation in RNA 1 as well as in RNA 2. At the nonpermissive temperature both mutants produced little minus-strand RNA. In both cases the synthesis of minus-strand RNA 1 was affected most. Supplementation with wt B or with wt M restored part of the minus-strand synthesis, while supplementation with Tb had no effect (Fig. 7).

In later work (Sarachu et al., 1985), pseudorecombinants containing one ts component and two wt components were studied. The minus strands synthesized by all pseudorecombinants containing mutant RNA 1 or RNA 2 were found to be thermosensitive.

From the CCMV data we have to conclude that at least the products coded for by RNA 1 and RNA 3 are involved in minus-strand synthesis. However, the AMV data indicate that at least the products coded for by RNA 1 and RNA 2 are involved in this process. Analysis of incompletely infected protoplasts showed that although the products of RNA 3 are not required for RNA synthesis, they influence the ratio of plus and minus RNA synthesized (see Section III). Roosien and van Vloten-Doting (1983) observed that in dsRNA preparations from tobacco infected with a non-ts AMV RNA 3 mutant [Tbstruct (s)1], the ratio of dsRNA 1 to dsRNA2 differed from the one found for dsRNA preparations from tobacco infected with wt (Fig. 8). This result suggests that RNA 3-coded product(s) also play a role in the regulation of the synthesis of the different minus RNA species.

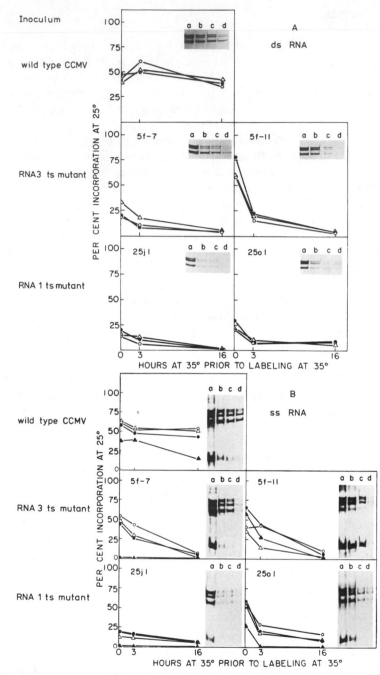

FIGURE 6. Incorporation of [³H]uridine into dsRNAs (A) and ssRNAs (B) of CCMV ts mutants at 35°C after different periods of preincubation at 35°C. Mutant or wild-type virus-infected cowpea plants were incubated for 4 days at 25°C prior to a shift to 35°C for the designated intervals followed by labeling for 4 hr at 35°C. Controls (taken as 100%) consisted of the corresponding isolate labeled for 4 hr at 25°C after 4 days of incubation at 25°C ○, ●, and △ indicate dsRNA 1, 2, and 3 in (A) and ssRNA 1, 2, and 3 in (B); ▲ indicates RNA 4 in (B). Insets show fluorographs of RNA labeled at 25°C (a), or at 35°C following preincubation at 35°C for 0 hr (b), 3 hr (c), or 16 hr (d). Courtesy of Dr. W. O. Dawson.

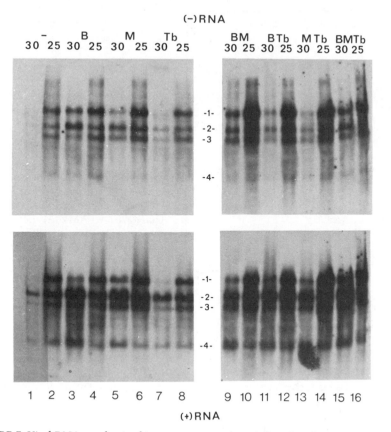

FIGURE 7. Viral RNA synthesized in cowpea protoplasts infected with an AMV RNA 1 and RNA 2 ts double mutant supplemented with wt components and incubated at 25 or 30°C, as indicated. End-labeled virion RNA was used for minus-strand detection and ³²P-labeled cDNA for plus-strand detection. Courtesy of Dr. A. Sarachu.

The RNA sequence or structure of different RNA species may be an important factor in the regulation of the synthesis of these RNA species. Indications for a mutation expressed at the RNA level were obtained with an AMV RNA 3 mutant [Tbts 7(uv)t,b; see Section IV.J.4].

6. Synthesis of Plus-Strand RNA

Plus-strand RNA synthesis was analyzed for the same mutants which were used for the analysis of minus-strand synthesis (Section IV.J.5). The shift-up experiments with the CCMV mutants showed that in most cases incorporation of [³H] uridine in ssRNA followed the same pattern as that of the dsRNA (Fig. 6). There were no significant differences in the effect on the three genomic RNAs.

For all AMV mutants analyzed there was some effect on the amount of plus strand detected; however, the effect was often less than the effect

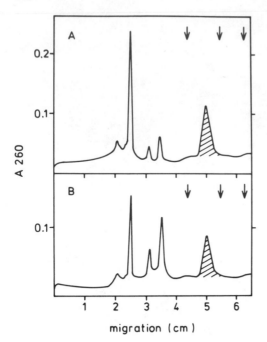

FIGURE 8. Densitograms of polyacry-laminde gels in which dsRNA preparation isolated from tobacco leaves infected with wt AMV (A) and an RNA 3 mutant [Tb struct(s)1] (B) were electrophoresed. Cross-hatched areas represent material also present in preparations from mock-inoculated plants. Arrows indicate the positions of viral RNAs 1, 2, and 3 determined from a sister gel. From Roosien and van Vloten-Doting (1982).

on the corresponding minus strand. Apparently, there is no direct relationship between the amount of plus- and minus-strand RNA synthesized.

None of the mutants studied up to now have shown a decrease of plus-strand synthesis in the presence of normal amount of minus-strand synthesis. This might suggest that synthesis of plus-strand RNA is catalyzed by host-coded proteins, but the number of mutants analyzed up to now is too small to warrant such a conclusion.

One AMV RNA 2 mutant [Myst 1(uv)b,c] produced more virus in cowpea protoplasts than the wt. Furthermore, the mutant virus contained an unusually high amount of RNA 2. Addition of wt RNA 2 to the mutant resulted in virus with a wt phenotype both in yield and in component composition. Apparently, the product encoded by wt RNA 2 limits directly or indirectly (by association with a host protein) the synthesis of RNA 2 itself and to a lesser extent that of the other AMV RNAs. The decrease of RNAs 1 and 3 may be a secondary effect due to difference in the amount of RNA 2 (product?) present (Roosien et al., 1983a). Unfortunately, the amount of minus strand present in these protoplasts was not analyzed and it is uncertain whether the mutation present in Msyst 1(uv)b,c affects minus-strand synthesis, plus-strand synthesis, or both.

7. Synthesis of Subgenomic mRNAs

The only subgenomic RNA with a proven function is RNA 4. This RNA which corresponds to the 3'-terminal part of RNA 3 can *in vivo* and *in vitro* function as mRNA for coat protein (van Vloten-Doting and Jaspars, 1977). There are reports about several other smaller RNAs the functions of which, if any, are unknown. Therefore, only the synthesis of RNA 4 will be discussed.

Theoretically, there are two possible routes for the production of RNA 4: degradation of RNA 3 or partial transcription of RNA 3 minus strands. There are insufficient data to distinguish between these two possibilities. Whatever the mechanism may be, the results of Dawson (1981) indicate that the synthesis of CCMV RNA 4 may differ from that of the genomic RNAs (Fig. 6). With one isolate (5f-7), synthesis of RNA 4 stopped immediately after shift-up, while the genomic RNAs were still being synthesized (Fig. 6). With the other isolate (5f-11), inhibition of RNA 4 synthesis paralleled that of the genomic RNAs. A similar but less pronounced effect was seen when two isolates of an RNA 1 mutant were compared (25jl versus 25ol). One should bear in mind that one out of two isolates of each mutant showed such a difference between the synthesis of genomic and subgenomic RNAs. Therefore, this difference probably reflects additional mutations which do not necessarily map on RNA 3 or RNA 1. The CCMV results show that there is a virus-coded function involved in synthesis of RNA 4 but the location of this function is uncertain.

8. Translatability of Viral RNAs

In cowpea protoplasts infected with an AMV RNA 1 ts mutant (Bts 03,c) or an AMV RNA 2 ts mutant (Mts 0.4,c), synthesis of all three genomic RNAs and of RNA 4 was observed at the nonpermissive temperature, but surprisingly, no coat protein could be detected. Shift-up experiments indicate that the RNA 4 produced at the nonpermissive temperature was not translated. Apparently, the RNA 1 and RNA 2-coded products are involved in the translatability of at least RNA 4 and the product may be involved in a processing event of RNA 4 such as capping. Alternatively, it may be a special factor which enhances the translation of the viral RNA(s) (Huisman *et al.*, 1985).

A CCMV RNA 3 ts mutant (Dawson *et al.*, 1975) produced all three genomic RNAs as well as RNA 4 at the nonpermissive temperature but no coat protein could be detected. According to the authors, the apparent lack of coat protein could have been due to a preferential coat protein turnover at the nonpermissive temperature or to a thermal conditional nonsense codon in the coat protein cistron. The latter possibility seems less likely. If the RNA 4 is indeed not translated at the nonpermissive temperature, the conformation of the RNA may be unfavorable or, more likely, the RNA may not be fully processed or a virus-specific translation factor is missing. No supplementation tests were done with this CCMV mutant which may have been also carrying additional ts mutations on other RNAs. The situation may be analogous to that described for the AMV RNA 1 mutant mentioned above.

9. Maturase?

Shift-up experiments with some isolates of AMV RNA 2 ts mutants (Mts 03,c and Mts 04,c) revealed an early function coded for by RNA 2 which is essential for the production of infectious virus (Huisman *et al.*, 1985). At present we have no indication about this function, but it might be related to the maturation function located on BMV RNA 2 (see Section IV.J.1).

10. Regulation of Component Composition

Each virus strain or mutant has a typical component composition which is a very complex trait. It will be influenced by the following factors: (1) relative synthesis of each of the minus-strand RNA species; (2) relative synthesis of each of the plus-strand species; (3) relative affinity of each of the plus-strand RNA species for the coat protein; (4) relative stability of each of the different nucleoprotein particles.

Although the recognition site for the replicase as well as that for the coat protein of the three genomic RNAs is similar, they are not identical (Koper-Zwarthoff *et al.*, 1979; Barker *et al.*, 1983; Cornelissen *et al.*, 1983a,b). This means that theoretically changes in the replicase (subunit) or in the coat protein may affect the interaction with one RNA more than the other. Thus, the synthesis or the encapsidation of the three genomic RNAs can be influenced to a different extent.

The position with the subgenomic RNA 4 is slightly different. Its sequences are derived from RNA 3 and the nucleation site for encapsidation is probably identical to that of RNA 3, but it cannot be excluded that sequences located at the 5' proximal half of the RNA 3 also play a role in encapsidation. Its mode of synthesis is unknown (see Section IV.J.7) but the involvement of special virus-coded product(s) seems very likely. Furthermore, Nassuth *et al.* (1983), working with AMV-infected cowpea protoplasts, showed that only a small fraction of RNA 4 synthesized became encapsidated, while nearly all the genomic RNA produced was accounted for in particles.

When ts mutants are mapped by the supplementation test, the presence of additional (non-ts) mutations on other RNA components cannot be excluded. This means that there is not necessarily a correlation between alterations in the component composition and the mutation mapped. To avoid this complication, only results obtained with pseudorecombinants or with mutants made by treating one purified component with a mutagen will be discussed here.

Hartmann *et al.* (1976) studied the nucleoprotein component composition of three AMV mutants. One of these mutants lacked top component *a* (particle consisting of 132 coat protein subunits and two RNA 4 molecules). All possible pseudorecombinants of this mutant and wt were constructed and analyzed. The results indicate that RNA 3 determines the presence or absence of top component *a*. In AMV strains or mutants producing top component *a*, the relative amount of top component *a* increases with increasing temperature (Jaspars and Moed, 1966; Hartmann *et al.*, 1976). Assuming that the amount of top component *a* is a direct reflection of the amount of RNA 4 synthesized, Hartmann *et al.* (1976) concluded that the production of RNA 4 from RNA 3 is a temperature-sensitive process. However, other influences (see above) cannot be excluded.

Of the 15 AMV mutants whose nucleoprotein particle ratios were examined (Roosien and van Vloten-Doting, 1982; Roosien *et al.*, 1983a,b, and unpublished results), only 3 showed significant deviations in the component composition. One of these mutants has a very unusual structure [mutant Tbstruct (s)1; see Section IV.J.1]. Its particles contain the normal set of RNAs, but the ratio of the RNA species differs greatly from that found in wt. The mutation(s) responsible for this phenomenon mapped on RNA 3 and both products of RNA 3 were found to contain mutations. Mutant preparations contained about equal amounts of RNAs 1, 2, and 3 together with a very large amount of RNA 4. Another AMV RNA 3 mutant [Tbts 7(uv)t,b] showed a striking increase in the relative amount of encapsidated RNA 4. More research is needed to understand the molecular processes leading to these differences.

11. Transport through Host

Transport of virus through the host is a complex event in which both virus-coded functions as well as host-coded functions (host defense) play a role (Matthews, 1981). Cell-to-cell movement can be distinguished from the more rapid movement of virus through the vascular tissue.

It has been observed that from strains which induce a systemic infection, isolates inducing only local lesions are frequently obtained. However, the reverse is a rare event which supports the suggestion that virus transport is dependent on a virus function (van Vloten-Doting *et al.*, 1983). Until now, no ts mutants in this function have been described (see also Section IV.I.3). Protoplasts infected with BMV RNA 1 and RNA 2 produce substantial amounts of RNAs 1 and 2 (see Section III), but in plants infected with these two RNAs no progeny could be detected. [Similar results have been obtained with AMV (see Section III), but in this case the interpretation is complicated by the coat protein dependency of viral RNA synthesis.] This suggests that the replication of RNAs 1 and 2 is confined to the primarily infected cell. A reason for this confinement could be that only particles can be transported. This is unlikely since for other viruses (see Matthews, 1981) transport of naked RNA has been observed. Taken together, these results suggest that the 3a protein plays a role in virus transport.

One AMV RNA 2 mutant [Msyst 1(uv)b,c] differs from wt in its ability to be transported over long distances. This mutant can invade cowpea plants systemically, while the parent strain will only induce local lesions. However, it is not known whether the confinement of virus in a local lesion is due to a triggering of a host response which blocks virus transport, or to the loss of a virus-coded transport function. Alternatively, the kinetics of virus multiplication may play a decisive role in the type of host response. Cowpea plants infected with wt AMV will respond with a very fast increase in virus followed by a decrease during the necrosis of the cells. Virus accumulation in plants infected with mutant Msyst 1(uv)b,c is much slower but the total amount of virus produced is higher (Roosien, 1983). It is possible that the fast increase of virus in the wt-infected plants totally upsets the metabolism of the host cell, leading to cell death and thus blocking virus transport. The amount of virus produced in the mutant-infected plants is higher but it takes a larger number of cells and a longer time to accumulate this virus. Thus, the stress on each cell may be lower and below the threshold for triggering the necrotic reaction which blocks further transport.

V. SUMMARY AND PROSPECTS

Twenty years ago our knowledge about plant virus replication was very limited. Little was known other than virus-infected plants accumulated progeny virus and that the genetic information for this resided in the viral nucleic acid.

At the end of the 1960s and the beginning of the 1970s we learned that viral genomes could consist of more than one molecule of nucleic acid. Now we know the complete base sequences of several plant virus genomes. With all viruses whose tripartite genomes have been examined, it has been shown that the three RNAs share some homology at their 3'-termini, while the rest of the information is unique (see Chapter 3).

From the analysis of incomplete infections (see Section III) and infections with conditional lethal mutants (Section IV.J), we have learned that the information present in both RNA 1 and RNA 2 is involved in the initial steps of RNA replication leading to minus-strand synthesis. The two proteins encoded by RNAs 1 and 2 probably act in concert which may explain why in the construction of pseudorecombinants between different viruses of the same taxonomic group, exchange of RNA 3 is permitted, while that of RNA 1 or RNA 2 usually is not (Section II). It appears that a protein complex encoded by RNA 1 plus RNA 2 of one virus can recognize the RNA 3 of another virus. This is possible, probably because related viruses have similar 3'-terminal base sequences. The combination of RNA 1 from one virus with RNA 2 of another may be more restrictive because interaction of the proteins they code must be compatible.*

Presently, it is less clear whether one, both, or none of the RNA 1 and RNA 2-coded proteins play a role in plus-strand synthesis. The fact that RNA 3 is not required to induce viral RNA replication does not mean that the information present in this RNA cannot influence the synthesis of RNAs 1 and 2. It has been shown that ts mutation in RNA 3 of CCMV affected both minus- and (indirectly?) plus-strand synthesis, while a protein encoded by AMV RNA 3 is apparently involved in the regulation of plus- and minus-strand synthesis.

Analysis of base substitutions present in different ts mutants will indicate the protein domains involved in the replication of RNA. However, this will be a major undertaking because each mutant will have to be sequenced completely. Since the high mutation frequency of RNA genomes may lead to the accumulation of mutations (Sections IV.C.1 and IV.D.1), the results of such studies will be complicated if the RNAs carry more than one mutation. As manipulation of DNA [e.g., introducing insertions and deletions; base substitutions are probably less suitable (Section IV.C.1)] is easier than RNA, it may be expected that our knowledge about the required sequence and structure of the replicase recognition site will evolve faster than that of the precise function of the proteins involved in replication.

AMV RNA 1 and AMV RNA 2 also encode a function which is required for the efficient translation of viral (sub)genomic RNA. It is to be expected that this function [a capping enzyme or a special translation (initiation?) factor?] will shortly be analyzed in detail.

It appears that a function encoded by RNA 2 is important for the production of infectious virus. This function could be some kind of assembly or maturation factor. AMV RNA 2-coded product also controls (directly or indirectly) the amount of viral RNA (especially RNA 2 itself) synthesized in cowpea protoplasts. A mutation in this regulation function is apparently coupled to a difference in host response (systemic infection instead of hypersensitive reaction). Careful analysis of the replication process of this type of mutant in protoplasts and in intact plants may help us to gain insight into the virus-coded functions which are involved in plant disease and plant defense mechanisms.

It is well established that RNA 3 encodes two proteins, one of which (probably the 3a protein) is involved in cell-to-cell transport. Unfortunately, to date no mutants with a ts defect in this function have been isolated. It is known that naked viral RNA may be transported from cell to cell (Matthews, 1981). However, the infectious entity which is transported through the vascular system (RNA or particles) is still uncertain.

---

* The idea that the RNA 1 and 2-encoded proteins act in a complex is supported by the observed amino acid homology between these two proteins and the 183K protein encoded by TMV RNA (Haselhoff et al. 1984; Cornelissen and Bol, 1984).

Mutants carrying a ts defect in the assembly function of the coat protein may help to solve this question.

More insight into RNA–protein and protein–protein interactions regulating assembly will be gained from further analysis (probably partly at the mRNA level) of the coat protein from mutants showing difference in particle stability or virus architecture.

In AMV, a large number of ts mutations in the function of genome activation by coat protein mutants are known. These mutants will be helpful in investigating the precise mechanism of genome activation and encapsidation.

DNA recombinant techniques may help to reconstruct tripartite genomes into quadrupartite genomes (information for 3a and coat proteins each on a separate replicon; van Vloten-Doting, 1983). Such experiments may tell us why in nature the information for the coat protein is never present on a separate replicon.

Very powerful techniques are now available for sequencing and modifying nucleic acids, as well as for the analysis of minute amounts of nucleic acids and proteins. Therefore, it might be expected that more knowledge about virus-coded functions involved in virus production will become known in the near future. To achieve a complete understanding of the molecular biology of plant viruses and virus diseases, two more barriers have to be overcome: (1) the identification of the host-coded functions and structures involved in virus replication. This problem may be solved at the protoplast level. (2) The identification of host-coded functions involved in virus spread or confinement, and the effect of virus multiplication and/or special virus-coded products on the host (hormone) metabolism. These problems will have to be solved at the level of the intact plant. To achieve this we will have to study not only the genetics of viruses but also the genetics of the host.

VI. GLOSSARY OF MUTANTS CITED IN THIS CHAPTER[a]

Virus	Name	Origin	Effect	Ref.[a]
		Mutations on RNA 1		
AMV	Bts 1(uv)t	uv(B)	ts vp tobacco	1
	Bts 2(s)t	spon.	ts vp tobacco	1
	Bts 3(s)t	spon.	ts vp tobacco	1
	Bts 4(s)t,b,c	spon.	ts vp tobacco	1
			ts 11 bean	1
			ts vp cowpea pps	2
	Bts 03,c	pseudorec.:	ts vp cowpea pps	3
		B from Mts 3(s)t,b	ts minus RNA synthesis	3
		+ wt M + wt Tb	ts viral RNA translatability	3
	Bts 04	pseudorec.:	ts minus RNA synthesis	3
		B from Bts 4(s)t,b,c		
		+ wt M + wt Tb		

Virus	Name	Origin	Effect	Ref.[a]
	A1 ⎫ E2 ⎬ 4HB ⎭	gradual temp. shift	tr vp tobacco tr 11 *Vigna catjang*	4 4
CCMV	MC$_{2d}$	HNO$_2$ (total RNA)	ts 11 *Chenopodium*	5
	E1b	NNG (total virus)	ts vp cowpea ts 11 soybean	6 6
	25	NNG (total virus)	ts vp cowpea ts 11 soybean ts replication complex	6 6 7

Mutations on RNA 2

Virus	Name	Origin	Effect	Ref.[a]
AMV	Mts 1(ni)t	HNO$_2$ (total RNA)	ts vp tobacco	1
	Mts 2(s)t	spon.	ts vp tobacco	1
	Mts 3(s)t,b,c	spon.	ts vp tobacco ts 11 bean ts vp cowpea ts minus RNA synthesis	1 1 1 2
	Mts 4(uv)t	uv(M)	ts vp tobacco	8
	Mts 7(uv)t,b	uv(M)	ts vp tobacco ts 11 bean	8 8
	Mts 9(uv)t	uv(M)	ts vp tobacco	8
	Mts 10(uv)b	uv(M)	ts vp tobacco	8
	Mts 11(uv)b	uv(M)	ts 11 bean	8
	Msyst 1(uv)b,c	uv(M)	systemic inf. bean systemic inf. cowpea enhanced vp cowpea pps	8 9 9
	Mts 03,c	pseudorec.: M from Mts 3(s)t,b,c + wt B + wt Tb	ts vp cowpea pps ts minus RNA synthesis ts maturase(?)	3 3 3
	Mts 04,c	pseudorec.: M from Bts 4(o)t,b,c + wt B + wt Tb	ts vp cowpea pps ts minus RNA synthesis ts maturase(?) ts viral RNA translatability	3 3 3 3
BMV	F	spon.	aberrant particles	5

Mutations on RNA 3

Virus	Name	Origin	Effect	Ref.[a]
AMV	F8a	NNG (total virus)	hypersens. in tobacco component composition	 10
	24$_6$	NNG (total virus)	hypersens. in tobacco component composition	 10
	A$_2$fi	NNG (total virus)	hypersens. in tobacco component composition	 10
	Tbts 1(s)t,b	spon.	ts vp tobacco ts 11 bean ts coat protein two amino acid sub. more sens. to uncoating	1 1 1 11 12
	Tbts 2(s)t	spon.	ts vp tobacco ts coat protein more sens. to uncoating	1 1 12

Virus	Name	Origin	Effect	Ref.[a]
	Tbts 3(s)t	spon.	ts vp tobacco	1
			ts coat protein	1
	Tbts 4(uv)t	uv(Tb)	ts vp tobacco	1
			ts coat protein	1
	Tbts 5(uv)t	uv(Tb)	ts vp tobacco	1
			ts coat protein	1
	Tbts 6(s)t	spon.	ts vp tobacco	13
	Tbts 7(uv)t,b	uv(Tb)	ts vp tobacco	1
			ts 11 bean	1
			ts coat protein	1
			less sens. to uncoating	12
			RNA synthesis recessive to wt RNA 3 synthesis	14
	Tbts 8(uv)t	uv(Tb)	ts vp tobacco	13
	Tbts 9(uv)t	uv(Tb)	ts vp tobacco	13
	Tbstruct (s)1	spon.	aberrant particles	15
CCMV	—	HNO$_2$ (total RNA)	ts vp *Chenopodium*	16
			ts coat protein	16
			ts vp tobacco pps	17
			RNA 3 dominant over wt RNA 3	18
			coat protein: Lys 21 → Glu	19
			Val 87 → Ala	19
	—	HNO$_2$ (total RNA)	oxidation sens. coat protein	20
			coat protein: Arg 25 → Cys	19
	—	HNO$_2$ (total RNA)	salt-stable particles	21
			coat protein: Lys 105 → Arg	19
	—	HNO$_2$ (total RNA)	perturbed assembly	22
			coat protein five amino acid sub.	22
	5f	NNG (total virus)	ts vp cowpea	6
			ts 11 soybean	6
			ts replication complex	7

<u>Mutation not localized</u>

BMV	10[a]	HNO$_2$ (total RNA)	aberrant virus particles	23

[a] Abbreviations: B, bottom component; ll, local lesion induction; M, middle component; pps, protoplasts; sens., sensitivity; spon., spontaneous; sub., substitutions; Tb, top component *b*; tr, thermoresistant; ts, thermosensitive; uv, ultraviolet irradiation; vp, virus production; wt, wild type.

[b] References: 1, van Vloten-Doting *et al.* (1980); 2, Sarachu *et al.* (1983); 3, Sarachu *et al.* (1985); 4, Franck and Hirth (1976); 5, Bancroft and Lane (1973); 6, Dawson (1978); 7, Dawson (1981); 8, Roosien and van Vloten-Doting (1982); 9, Roosien *et al.* (1983a); 10, Hartmann *et al.* (1976); 11, Kraal (1975); 12, Smit (1981); 13, This chapter (Section IV.I.3); 14, Roosien *et al.* (1983b); 15, Roosien and van Vloten-Doting (1983); 16, Bancroft (1972); 17, Dawson *et al.* (1975); 18, Dawson and Watts (1979); 19, Rees and Short (1982); 20, Bancroft *et al.* (1971); 21, Bancroft *et al.* (1973); 22, Bancroft *et al.* (1976); 23, Lane (1974).

ACKNOWLEDGMENTS.   I am grateful for the helpful comments of Drs. Alberta Sarachu and Jan Roosien.

REFERENCES

Atabekov, J. G., and Morozov, S. Y., 1979, Translation of plant virus messenger RNAs, *Adv. Virus Res.* **25**:1–91.

Bancroft, J. B., 1970, The self-assembly of spherical plant viruses, *Adv. Virus Res.* **16**:99–134.

Bancroft, J. B., 1972, A virus made from parts of the genomes of brome mosaic and cowpea chlorotic mottle virus, *J. Gen. Virol.* **14**:223–228.

Bancroft, J. B., and Lane, L. C., 1973, Genetic analysis of cowpea clorotic mottle and brome mosaic virus, *J. Gen. Virol.* **19**:381–389.

Bancroft, J. B., Hills, G. J., and Markham, R., 1967, A study of the self-assembly process in a small spherical virus: Formation of organized structures from protein subunits *in vitro*, *Virology* **31**:354–379.

Bancroft, J. B., McLean, G. D., Rees, M. W., and Short, M. N., 1971, The effect of an arginyl to a cystinyl replacement on the uncoating behavior of a spherical plant virus, *Virology* **45**:707–715.

Bancroft, J. B., Rees, M. W., Dawson, J. R. O., McLean, G. D., and Short, M. N., 1972, Some properties of a temperature-sensitive mutant of cowpea chlorotic mottle virus, *J. Gen. Virol.* **16**:69–81.

Bancroft, J. B., Rees, M. W., Johnson, M. W., and Dawson, J. R. O., 1973, A salt-stable mutant of cowpea chlorotic mottle virus, *J. Gen. Virol.* **21**:507–513.

Bancroft, J. B., McDonald, J. G., and Rees, M. W., 1976, A mutant of cowpea chlorotic mottle virus with a perturbed assembly mechanism, *Virology* **75**:293–305.

Barker, R. F., Jarvis, N. P., Thompson, D. V., Loesch-Fries, L. S., and Hall, T. C., 1983, Complete nucleotide sequence of alfalfa mosaic virus RNA 3, *Nucleic Acids Res.* **11**:2881–2891.

Bol, J. F., van Vloten-Doting, L., and Jaspars, E. M. J., 1971, A functional equivalence of top component *a* RNA and coat protein in the initiation of infection by alfalfa mosaic virus, *Virology* **46**:73–85.

Bos, L., 1969, Experiences with a collection of plant viruses in leaf material dried and stored over calcium chloride, and a discussion of literature on virus preservation, *Meded. Fac. Landbouwwet. Rijksuniv. Gent* **34**:875–887.

Bos, L., Huttinga, H., and Maat, D. Z., 1980, Spinach latent virus, a new Ilarvirus seed-borne in *Spinacia oleracea*, *Neth. J. Plant Pathol.* **86**:79–98.

Castel, A., Kraal, B., De Graaf, J. M., and Bosch, L., 1979, The primary structure of the coat protein of alfalfa mosaic virus strain VRU, *Eur. J. Biochem.* **102**:125–138.

Cornelissen, B. J. C., and Bol, J. F., 1984, Homology between the proteins encoded by tobacco mosaic virus and two Tricornaviruses, *Plant Mol. Biol.* **3**:379–384.

Cornelissen, B. J. C., Brederode, F. T., Moorman, R. J. M., and Bol, J. F., 1983a, Complete nucleotide sequence of alfalfa mosaic virus RNA 1, *Nucleic Acids Res.* **11**:1253–1265.

Cornelissen, B. J. C., Brederode, F. T., Veeneman, G. H., van Boom, J. H., and Bol, J. F., 1983b, Complete nucleotide sequence of alfalfa mosaic virus RNA 2, *Nucleic Acids Res.* **11**:3019–3025.

Davies, J. W., and Hull, R., 1982, Genome expression of positive strand RNA viruses, *J. Gen. Virol.* **61**:1–19.

Dawson, J. R. O., and Watts, J. W., 1979, Analysis of the products of mixed infection of tobacco protoplasts with two strains of cowpea chlorotic mottle virus, *J. Gen. Virol.* **45**:133–137.

Dawson, J. R. O., Motoyoshi, F., Watts, J. W., and Bancroft, J. B., 1975, Production of RNA and coat protein of a wildtype isolate and a temperature-sensitive mutant of cowpea chlorotic mottle virus in cowpea leaves and tobacco protoplasts, *Virology* **29**:99–107.

Dawson, W. O., 1978, Isolation and mapping of replication-deficient, temperature-sensitive mutants of cowpea chlorotic mottle virus, *Virology* **90**:112–118.

Dawson, W. O., 1981, Effect of temperature-sensitive, replication-defective mutations on RNA synthesis of cowpea chlorotic mottle virus, *Virology* **115:**130–136.

Dawson, W. O., and Jones, G. E., 1976, A procedure for specifically selecting temperature-sensitive mutants of tobacco mosaic virus, *Mol. Gen. Genet.* **145:**307–309.

De Jager, C. P., and Breekland, L., 1979, Evidence for intrastrand complementation in cowpea mosaic virus infection, *Virology* **99:**312–318.

Domingo, E., Sabo, D., Taniguchi, T., and Weissmann, C., 1978, Nucleotide sequence heterogeneity of an RNA phage population, *Cell* **13:**735–744.

Donis-Keller, H., Browning, K. S., and Clark, J. M., Jr., 1981, Sequence heterogeneity in satellite tobacco necrosis virus RNA, *Virology* **110:**43–54.

Franck, A., 1978, Contribution a l'etude du fonctionnement du genome multipartite du virus de la mosaique de la Luzerne, Ph.D. thesis, University of Strasbourg.

Franck, A., and Hirth, L., 1976, Temperature-resistant strains of alfalfa mosaic virus, *Virology* **70:**283–291.

Goelet, P., Lomonossof, G. P., Butler, P. J. G., Akain, M. E., Gait, M. J., and Karn, J., 1982, Nucleotide sequence of tobacco mosaic virus RNA, *Proc. Natl. Acad. Sci. USA* **79:**5818–5822.

Gonsalves, D., and Fulton, R. W., 1977, Activation of *Prunes* necrotic ringspot virus and rose mosaic virus by RNA 4 component of some Ilarviruses, *Virology* **81:**398–407.

Gonsalves, D., and Garnsey, S. M., 1975a, Functional equivalence of an RNA component and coat protein for infectivity of citrus leaf rugose virus, *Virology* **64:**23–31.

Gonsalves, D., and Garnsey, S. M., 1975b, Nucleic acid components of citrus variegation virus and their activation by coat protein, *Virology* **67:**311–318.

Gonsalves, D., and Garnsey, S. M., 1975c, Infectivity of heterologous RNA–protein mixtures from alfalfa mosaic, citrus leaf rugose, citrus variegation, and tobacco streak viruses, *Virology* **67:**319–326.

Hartmann, D., Mohier, E., Leroy, C., and Hirth, L., 1976, Genetic analysis of alfalfa mosaic virus, *Virology* **74:**470–480.

Haseloff, H. J., Goelet, P., Zimmern, D., Ahlquist, P., Dasgupta, R. J., and Kaesberg, P., 1984, Striking similarities in amino acid sequence among nonstructural proteins encoded by RNA viruses that have dissimilar genomic organization, *Proc. Natl. Acad. Sci.* **81:**4358–4362.

Heytink, R. A., and Jaspars, E. M. J., 1974, RNA contents of abnormally long particles of certain strains of alfalfa mosaic virus, *Virology* **59:**371–382.

Holland, J., Spindler, K., Horodyski, F., Grabau, E., Nichol, S., and van der Pol, S., 1982, Rapid evolution of RNA genomes, *Science* **215:**1577–1585.

Honess, R. W., 1981, Complementation between phosphonoacetic acid resistant and sensitive variants of herpes simplex viruses: Evidence for an oligomeric protein with restricted intracellular diffusion as the determination of resistance and sensitivity, *J. Gen. Virol.* **57:**297–306.

Houwing, C. J., and Jaspars, E. M. J., 1978, Coat protein binds to the 3'-terminal part of RNA 4 of alfalfa mosaic virus, *Biochemistry* **17:**2927–2933.

Huisman, M. J., Sarachu, A. N., Alblas, F., and Bol, J. F., 1985, Alfalfa mosaic virus temperature-sensitive mutants II. Early functions encoded by RNA 1 and RNA 2, *Virology* (in press).

Hull, R., 1969, Alfalfa mosaic virus, *Adv. Virus Res.* **15:**365–433.

Hull, R., 1970, Studies on alfalfa mosaic virus. IV. An unusual strain, *Virology* **42:**283–292.

Huttinga, H., and Mosch, W. H. M., 1976, Lilac ring mottle virus: A coat protein-dependent virus with a tripartite genome, *Acta Hortic.* **59:**113–118.

Jaspars, E. M. J., and Moed, J. R., 1966, The complexity of alfalfa mosaic virus, in: *Viruses of Plants* (A. B. R. Beemster and J. Dijkstra, eds.), pp. 188–195, North-Holland, Amsterdam.

Joshi, S., Neeleman, L., Pley, C. W. A., Haenni, A. L., Chapeville, F., Bosch, L., and van Vloten-Doting, L., 1984, Non-structural alfalfa mosaic virus RNA-coded proteins present in tobacco leaf tissue, *Virology* **139:**231–242.

Kiberstis, P., Loesch-Fries, L. S., and Hall, T. C., 1981, Viral protein synthesis in barley protoplasts infected with native and fractionated brome mosaic virus RNA, *Virology* **112**:804–808.

King, A. M. Q., McCahon, D., Slade, W. R., and Newman, J. W. I., 1982, Recombination in RNA, *Cell* **29**:921–928.

Koper-Zwarthoff, E. C., and Bol, J. F., 1980, Nucleotide sequence of the putative recognition site for coat protein in the RNAs of alfalfa mosaic virus and tobacco streak virus, *Nucleic Acids Res.* **8**:3307–3318.

Koper-Zwarthoff, E. C., Brederode, F. T., Walstra, P., and Bol, J. F., 1979, Nucleotide sequence of the 3'-noncoding region of alfalfa mosaic virus RNA 4 and its homology with the genomic RNAs, *Nucleic Acids Res.* **7**:1887–1900.

Kraal, B., 1975, Amino acid analysis of alfalfa mosaic virus coat proteins: An aid for viral strain identification, *Virology* **66**:336–340.

Kuhn, C. W., and Wyatt, S. D., 1979, A variant of cowpea chlorotic mottle virus obtained by passage through beans, *Phytopathology* **69**:621–624.

Kunkel, L. O., 1940, *Publ. Am. Assoc. Adv. Sci.* **12**:22.

Lane, L., 1974, The Bromoviruses, *Adv. Virus Res.* **19**:151–220.

Lane, L. C., 1979, The nucleic acids of multipartite, defective and satellite plant viruses, in: *Nucleic Acids in Plants*, Volume 2 (T. C. Hall and J. W. Davies, eds.), pp. 65–110, CRC Press, Boca Raton, Fla.

Lister, R. M., 1966, Possible relationship of virus-specific products of tobacco rattle virus infections, *Virology* **28**:350–353.

Lister, R. M., 1968, Functional relationship between virus-specific products of infection by viruses of the tobacco rattle type, *J. Gen. Virol.* **2**:43–58.

Matthews, R. E. F., 1981, *Plant Virology*, 2nd ed., Academic Press, New York.

Mossop, D. W., and Francki, R. I. B., 1977, Association of RNA 3 with aphid transmission of cucumber mosaic virus, *Virology* **81**:177–181.

Nassuth, A., and Bol, J. F., 1983, Altered balance of the synthesis of plus- and minus-strand RNAs induced by RNAs 1 and 2 of alfalfa mosaic virus in the absence of RNA 3, *Virology* **124**:75–85.

Nassuth, A., ten Bruggencate, G., and Bol, J. F., 1983, Time course of alfalfa mosaic virus RNA and coat protein synthesis in cowpea protoplasts, *Virology* **125**:75–84.

Nishiguchi, M., Motoyoshi, F., and Oshima, N., 1978, Behaviour of a temperature-sensitive strain of tobacco mosaic virus in tomato leaves and protoplasts, *J. Gen. Virol.* **39**:53–61.

Oswald, J. W., Rozendaal, A., and van der Want, J. P. M., 1955, The alfalfa mosaic virus in the Netherlands, its effect on potato and a comparison with the potato aucuba mosaic virus, Proc. Conf. Potato Virus Dis., 2nd, Lisse-Wageningen, 1954, p. 137.

Prabhakar, B. S., Haspel, M. V., McClintock, P. R., and Notkins, A. L., 1982, High frequency of antigenic variants among naturally occurring human Coxsackie B4 virus isolates identified by monoclonal antibodies, *Nature (London)* **300**:374–376.

Racaniello, V. R., and Baltimore, D., 1981, Cloned polio virus complementary DNA is infectious in mammalian cells, *Science* **214**:916–919.

Rao, A. L. N., and Francki, R. I. B., 1981, Comparative studies on tomato aspermy and cucumber mosaic viruses. VI. Partial compatibility of genome segments from the two viruses, *Virology* **114**:573–575.

Rao, A. L. N., and Francki, R. I. B., 1982, Distribution of determinants for symptom production and host range on the three RNA components of cucumber mosaic virus, *J. Gen. Virol.* **61**:197–205.

Reddy, D. V. R., and Black, L. M., 1977, Isolation and replication of mutant populations of wound tumor virions lacking certain genome segments, *Virology* **80**:336–346.

Rees, M. W., and Short, M. N., 1982, The primary structure of cowpea chlorotic mottle virus coat protein, *Virology* **119**:500–503.

Robinson, D. J., 1973, Inactivation and mutagenesis of tobacco rattle virus by nitrous acid, *J. Gen. Virol.* **18**:215–222.

Roosien, J., 1983, Mutants of alfalfa mosaic virus, Ph.D. thesis, University of Leiden.

Roosien, J., and van Vloten-Doting, L., 1982, Complementation and interference of ultra-violet-induced Mts mutants of alfalfa mosaic virus, *J. Gen. Virol.* **63**:189–198.

Roosien, J., and van Vloten-Doting, L., 1983, A mutant of alfalfa mosaic virus with an unusual structure, *Virology* **126**:155–167.

Roosien, J., Sarachu, A. N., Alblas, F., and van Vloten-Doting, L., 1983a, An alfalfa mosaic virus RNA 2 mutant, which does not induce a hypersensitive reaction in cowpea plants, is multiplied to a high concentration in cowpea protoplasts, *Plant Mol. Biol.* **2**:85–88.

Roosien, J., van Klaveren, P., and van Vloten-Doting, L., 1983b, Competition between the RNA 3 molecules of wild type alfalfa mosaic virus and the temperature-sensitive mutant Tbts 7(uv), *Plant Mol. Biol.* **2**:113–118.

Sarachu, A. N., Nassuth, A., Roosien, J., van Vloten-Doting, L., and Bol, J. F., 1983, Replication of temperature-sensitive mutants of alfalfa mosaic virus in protoplasts, *Virology* **125**:64–74.

Sarachu, A., Huisman, M. J., van Vloten-Doting, L., and Bol, J. F., 1985, Alfalfa mosaic virus temperature-sensitive mutants, I. Mutants defective in viral RNA and protein synthesis, *Virology* (in press).

Smit, C. H., 1981, Multiple activation of the genome of alfalfa mosaic virus, Ph.D. thesis, University of Leiden.

Smit, C. H., Roosien, J., van Vloten-Doting, L., and Jaspars, E. M. J., 1981, Evidence that alfalfa mosaic virus infection starts with three RNA–protein complexes, *Virology* **112**:169–173.

Taliansky, M. E., Kaplan, I. B., Yarvekulg, L. V., Atabakov, T. I., Agronovsky, A. A., and Atabekov, J. G., 1982a, A study of TMV ts mutant Ni 2519. II. Temperature-sensitive behaviour of Ni 2519 RNA upon reassembly, *Virology* **118**:309–316.

Taliansky, M. E., Malyshenko, S. I., Pshennikova, E. S., and Atabekov, J. G., 1982b, Plant virus transport function. II. Factor controlling virus host range, *Virology* **122**:327–331.

Taniguchi, T., Palmieri, M., and Weissmann, C., 1978, Qβ DNA-containing hybrid plasmids giving rise to Qβ phage formation in the bacterial host, *Nature (London)* **274**:223–228.

van Vloten-Doting, L., 1975, Coat protein is required for infectivity of tobacco streak virus: Biological equivalence of the coat proteins of tobacco streak and alfalfa mosaic virus, *Virology* **65**:215–225.

van Vloten-Doting, L., 1983, Advantages of multipartite genomes of single-stranded RNA plant viruses in nature, for research, and genetic engineering, *Plant Mol. Biol. Rep.* **1**:55–60.

van Vloten-Doting, L., and Jaspars, E. M. J., 1977, Plant covirus systems: Three component systems, in: *Comprehensive Virology*, Volume 11 (H. Fraenkel-Conrat and L. R. Wagner, eds.), pp. 1–53, Plenum Press, New York.

van Vloten-Doting, L., and Neeleman, L., 1982, Translation of plant virus RNAs, in: *Encyclopedia of Plant Physiology*, Volume 14B (D. Boulter and B. Parthier, eds.), pp. 337–367, Springer-Verlag, Berlin.

van Vloten-Doting, L., Hasrat, J. A., Oosterwijk, E., van't Sant, P., Schoen, M. A., and Roosien, J., 1980, Description and complementation analysis of 13 temperature-sensitive mutants of alfalfa mosaic virus, *J. Gen. Virol.* **46**:415–426.

van Vloten-Doting, L., Francki, R. I. B., Fulton, R. W., Kaper, J. M., and Lane, L. C., 1981, Tricornaviridae—A proposed family of plant viruses with tripartite, single-stranded RNA genomes, *Intervirology* **15**:198–203.

van Vloten-Doting, L., Bol, J. F., Nassuth, A., Roosien, J., and Sarachu, A. N., 1983, Structure and function of plant virus genomes, in: *NATO ASI Series*, Volume A63 (O. Ciferri and L. Dure, III, eds.), pp. 437–449.

Wu, J. H., Blakely, L. M., and Dimitman, J. E., 1969, Inactivation of a host resistance mechanism as an explanation for heat activation of TMV-infected bean leaves, *Virology* **37**:658–666.

Wyatt, S. D., and Kuhn, C. W., 1979, Replication and properties of cowpea chlorotic mottle virus in resistant cowpeas, *Phytopathology* **69**:125–129.

Yarwood, C. E., 1970, Reversible host adaptation in cucumber mosaic virus, *Phytopathology* **60:**1117–1119.

Yarwood, C. E., 1979, Host passage effects with plant viruses, *Adv. Virus. Res.* **25:**169–187.

Zuidema, D., Bierhuizen, M. F. A., Cornelissen, B. J. C., Bol, J. F., and Jaspars, E. M. J., 1983a, Coat protein binding sites on RNA 1 of alfalfa mosaic virus, *Virology* **125:**361–369.

Zuidema, D., Bierhuizen, M. F. A., and Jaspars, E. M. J., 1983b, Removal of the N-terminal part of alfalfa mosaic virus coat protein interferes with the specific binding to RNA 1 and genome activation, *Virology* **129:**225–260.

Zuidema, D., Cool, R. M., and Jaspars, E. M. J., 1984, Minimum requirements for specific binding of RNA and coat protein of alfalfa mosaic virus, *Virology* **136:**282–292.

CHAPTER 6

Virus–Host Relationships
Symptomatological and Ultrastructural Aspects

GIOVANNI P. MARTELLI AND MARCELLO RUSSO

I. INTRODUCTION

As indicated in the title, this chapter covers only two of the many facets characterizing the complex interactions established by viruses of the Bromo-, Cucumo-, Ilar-, and alfalfa mosaic virus groups with their host plants.

Symptomatological aspects will be dealt with rather briefly owing to the availability of comprehensive reviews on the subject (Hull, 1969; Lane, 1974, 1981; Kaper and Waterworth, 1981; Fulton, 1981) and the information provided by *CMI/AAB Descriptions of Plant Viruses*. Ultrastructural aspects will be treated in a somewhat more detailed manner, reviewing information from the literature and some original observations recently made in our laboratory.

Whereas the fine structure of plants infected by some representatives of the above virus groups such as cucumber mosaic (CMV), alfalfa mosaic (AMV), and cowpea chlorotic mottle viruses (CCMV) has been thoroughly investigated by a number of different workers, only limited studies have been published about other viruses. This is the case with tomato aspermy (TAV) and peanut stunt viruses (PSV) whose ultrastructure is described in two papers, one of which is not readily available (Kraev *et al.*, 1975), and with Ilarviruses, to which no more than than three reports are devoted.

GIOVANNI P. MARTELLI AND MARCELLO RUSSO • Dipartimento di Patologia vegetale, Università degli Studi di Bari and Centro di studio del CNR sui virus e le virosi delle colture mediterranee, Bari, Italy.

In the hope of obtaining a better and more informative insight into the ultrastructure of cells infected by viruses belonging to all four groups discussed in this volume, plants infected by AMV, all definitive members (according to Matthews, 1982) of the Cucumovirus (CMV, TAV, and PSV) and Bromovirus [CCMV, brome grass mosaic (BMV), and broad bean mottle (BBMV)] groups, and the following members of the Ilarvirus group: tobacco streak (TSV, five strains), citrus variegation (CVV, two strains), apple mosaic (ApMV), *Prunus* necrotic ringspot (PNRV), prune dwarf (PDV), and asparagus virus 2 (AV-2), were reinvestigated in our laboratory. All these viruses were studied in foliar tissues of different hosts at various stages of infection by Drs. M. A. Castellano, A. Di Franco, and the authors. The noteworthy results of these studies are incorporated in this presentation.

II. EFFECTS OF VIRUS INFECTION ON HOST MORPHOLOGY AND CELL STRUCTURE

Following penetration into a plant, viruses establish intimate relationships with the host cells, the outcome of which is usually a deranged metabolism and structure. Changes ensuing from viral infection, however minor, may have a bearing on the physiology and/or morphology of the host leading to a variety of abnormalities, the most readily perceptible of which are "symptoms," i.e., the outward signs of the disease. However, sometimes symptomless infections do occur which may be devoid of structural modifications of any consequence.

Questions are often posed as to what causes symptoms to develop in a virus-infected plant and how they correlate with ultrastructural changes. As yet, there is no clear-cut answer to these questions. Most ultrastructural studies take little or no account of the type of symptoms shown by infected tissues in the areas sampled or, if they do, an effort is seldom made to explain the appearance of certain outward manifestations in terms of deranged cytology. Experimental data on this subject are scanty and especially so with reference to two of the groups (the Bromoviruses and Ilarviruses). Therefore, drawing generalized conclusions by extrapolating from available information, although tempting in some cases, may be unsafe and misleading.

In the past, light microscopic investigations have given some insight into the gross histological disturbances accompanying mosaic patterns elicited by different viruses in host leaves (Esau, 1948, 1967). CMV infections in cucumber (*Cucumis sativus* L.) seem to conform to the general rule whereby chlorotic islands of the leaf blade are hypoplastic (i.e., have thinner mesophyll tissues, with their cells and intercellular spaces being smaller than normal) whereas the surrounding green areas are either apparently normal or have a thicker hyperplastic mesophyll (Porter, 1954). No wonder, then, that the profound variations in the histology of green and discolored areas may affect tissue growth locally. As a consequence, mottling may be accompanied by irregular deformations of the leaf blade such as blistering or, if veins are involved, crinkling or curling.

Electron microscopic investigations of tobacco (*Nicotiana tabacum* L.) leaves infected with ordinary strains of CMV have confirmed the above findings (Ehara and Misawa, 1973; Honda and Matsui, 1974). Epidermal and mesophyllic cells of green tissues of infected leaves were normal in size and ultrastructure. Virus particles were not readily detected so that use of treatment for removing ribosomes was required and revealed that about 40% of the cells contained a few particles scattered or in small aggregates. Conversely, in the yellowish areas of diseased tissues, only epidermal cells were apparently normal. Cells of the palisade and spongy tissues were deformed, disorderly arranged, and smaller than normal. Virus particles, either scattered or in large aggregates, were plentiful and readily seen in about 90% of the cells (Honda and Matsui, 1974).

The distribution of CMV particles in cells of dark green and discolored areas of tobacco leaves, is similar to that reported for several AMV strains in basil (*Ocimum basilicum* L.), alfalfa (*Medicago sativa* L.) (Gerola *et al.*, 1969a), and grapevine (*Vitis vinifera* L.) (Bovey and Cazelles, 1978), for BBMV in broad bean (*Vicia faba* L.) (Rubio and Van Slogteren, 1956), and for a few additional viruses in other hosts (Gibbs and Harrison, 1976).

Whether the concentration of intracellular virus is directly related to the severity of cytological alterations and, in turn, with type and intensity of symptoms is not clear. There are cases in which such a relationship has been reported as a likely possibility. In *Nicotiana* plants infected with three different CMV isolates (Gerola *et al.*, 1965; Honda and Matsui, 1974; Ehara and Misawa, 1975) and in basil infected with AMV (Gerola *et al.*, 1969a), cells of discolored (light green to bright yellow) leaf areas exhibited the most pronounced chloroplast abormalities and the highest virus content. Ehara and Misawa (1975) state that: "the larger the virus content in the cell, the higher was the frequency of abnormal chloroplasts and inhibition of chloroplast development."

These reports, however, do not agree with findings concerning other AMV isolates and TAV. In AMV-infected tobacco tissues, little or no modification of chloroplasts was observed even in cells whose cytoplasm was literally filled with particles (de Zoeten and Gaard, 1969; M. Russo, unpublished results). Likewise, *Nicotiana clevelandii* Gray and *N. benthamiana* Domin. infected with TAV, exhibited no apparent difference in chloroplast morphology and structure regardless of age of infection, virus concentration—which often was extremely high—and symptomatology (M. Russo and A. Di Franco, unpublished results).

That in virus-infected plants discolorations of mottled leaves arise primarily from chloroplast disturbances, is a widely accepted idea. However, alterations suffered by plastids may not always be such as to result in morphological or structural changes detectable even with the electron microscope.

To the examples quoted above, a few more may be added. Thus, Francki *et al.* (1985) observed no consistent chloroplast abnormalities in chlorotic tissues of young leaves infected with several CMV and TAV isolates. Similarly, no changes or very minor changes were seen in chloroplasts of French bean (*Phaseolus vulgaris* L.) and cucumber leaves infected with CVV and ApMV, respectively, which showed bright yellow vein banding (M. A. Castellano, unpublished results).

On the other hand, there are studies which consistently report severe plastidial modifications specifically in cells of chlorotic or yellow islands of mottled leaves of different plants infected with Cucumoviruses (Gerola *et al.*, 1965; Honda and Matsui, 1974; Ehara and Misawa, 1975), BMV (Paliwal, 1970), and AMV (Gerola *et al.*, 1969a; Favali and Conti, 1970, 1971; Conti *et al.*, 1972; Wilcoxson *et al.*, 1974; Bovey and Cazelles, 1978). Also in cowpea (*Vigna unguiculata* Walp.) plants infected with CCMV, structural changes of chloroplasts were minor in leaves with light mottling, but became progressively severe with increasing discoloration of the tissues. In intensely yellow leaves, chloroplasts were clumped, individual plastids were misshapen, swollen, and showed disarranged thylakoids and increased numbers of plastoglobuli (Fig. 1).

A very detailed description of chloroplast abnormalities ensuing from CMV infection was given by Ehara and Misawa (1975), who also studied the ontogeny of these modifications in systemically infected tobacco tissues. Deranged chloroplasts were grouped in four categories according to type and severity of structural alterations they showed. These depended on the developmental stages of the organelles at the time of infection. Whereas profound modifications of the lamellar system—i.e., loss of grana, development of myelinlike structures and tubular networks—characterized chloroplasts that were in the proplastidial stage when infection occurred, disintegration of internal structure was more typical of those plastids that were already differentiated or in an advanced stage of differentiation. In chlorotic areas, a rapid decrease of chlorophyllase activity and of chlorophyll content was observed as the infection progressed. Concomitantly, there was a marked increase of carotenoids and pheophytin, possibly derived from chlorophyll breakdown (Kato and Misawa, 1974). According to these authors (Kato and Misawa, 1974; Ehara and Misawa, 1975), none of the above modifications could be ascribed to intraplastidial multiplication of CMV.

Interference with the proplastid–chloroplast transformation process through impairment of thylakoid development and retention of stromacenter, was suggested as the most likely cause of the severe chloroplast alterations found in basil leaves infected with an isolate of AMV (Gerola *et al.*, 1969a). Similar general modification but no presence of stromacenter, was also observed in plastids of cells of the same host infected with a different AMV isolate (Favali and Conti, 1970). These structural

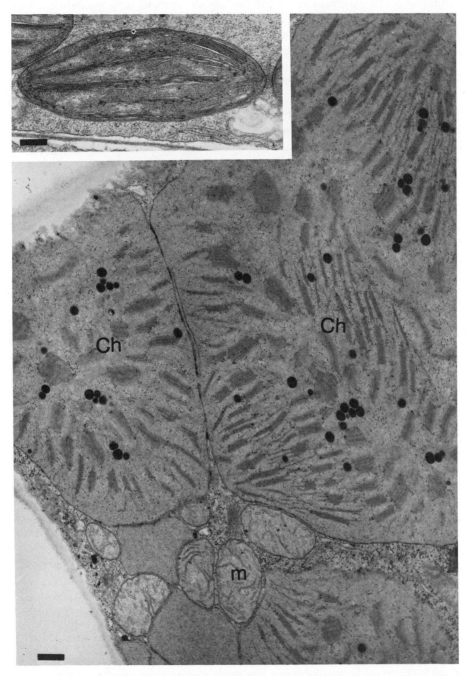

FIGURE 1. Structurally deranged and clumped chloroplasts (Ch) in a cell of a yellow cowpea leaf infected with cowpea chlorotic mottle virus (Bromovirus). Mitochondria (M) are also altered. Inset shows a chloroplast from a healthy cell. Bars = 200 nm.

changes were interpreted by all the above authors as a possible consequence of osmotic imbalance and deranged intraplastidial protein metabolism brought about by virus infection.

Degenerative changes of chloroplasts of basil leaves infected with the WY strain of CMV, consisted essentially in breakdown of lamellae and formation of myelinlike whorls, features attributed to virus-induced variation in the lipid/protein ratio of thylakoid membranes (Favali and Conti, 1970). However, as pointed out by Butler and Simon (1971), most if not all modifications observed in chloroplasts of virus-infected cells are reminiscent of those induced by senescence. Nevertheless, generalizations may be dangerous. For instance, carbendazim, a compound with cytokininlike properties (Skene, 1972), like cytokinins, was found to exert an antisenescent action on lettuce leaves infected with beet western yellows virus by preserving chloroplast integrity and suppressing disease symptoms (Tomlinson and Webb, 1978). The same compound, however, was unable to prevent mosaic symptoms in tobacco leaves infected with CMV and lettuce infected with lettuce mosaic virus (Tomlinson et al., 1976), suggesting that processes leading to chloroplast degeneration may differ according the virus–host combination (Tomlinson and Webb, 1978).

In CMV-infected cucumber cotyledons, enhanced production of ethylene, a senescence-inducing compound (Aharoni and Lieberman, 1979) known to affect chlorophyll content (Burg, 1962), was implicated as a likely inducer of tissue chlorosis consequent to possible chloroplast derangement (Marco and Levy, 1979). However, whether and to what extent this applies also to other CMV–host combinations and types of infection (i.e., local versus systemic), remains to be seen (Balázs et al., 1969; De Laat and Van Loon, 1983).

Mitochondria have been reported to be variously affected by virus infections. For instance, we have observed that in cells infected by any of the Bromoviruses, mitochondria showed low electron density of the stroma, whereas cristae were reduced in size and number or, sometimes, were totally absent (M. Russo, unpublished results). The structural alterations progressed with age of infection but no clear-cut correlation could be established between their occurrence and the symptoms expressed by the host.

Very impressive modifications of mitochondria have also been recorded in tissues infected with a number of viruses other than those with tripartite genomes discussed in this volume (see reviews by Martelli and Russo, 1977; Rubio-Huertos, 1978; Weintraub and Schroeder, 1979; Francki et al., 1985), but in no case has the impact of these changes on host physiology and/or symptoms been established.

Alterations of the Golgi apparatus in infected plants may be more subtle. Enhanced vesiculation has been reported in plants infected with several viruses such as CMV, TAV, BBMV, AMV, and TSV-B (see Section IV). It is not clear if cytoplasmic accumulations of secretory vesicles originate from an excessive vesiculating activity of the Golgi apparatus or from an inhibition of intracellular vesicle transport. Virus-induced vesicle aggregates like those in Fig. 15 are strongly reminiscent of those induced in maize cells by cytochalasin B, a substance that prevents transfer of secretory vesicles from Golgi to the cell surface (Mollenhauer and Morré, 1976). Should virus infections elicit a similar blocking action on the movement of dictyosomal vesicles in developing cells, the ensuing inhibition of cell wall growth would be likely to result in malformations of the infected tissue.

III. SYMPTOMATOLOGY

Viruses belonging in the four taxonomic groups discussed in this volume incite a wide array of symptoms which, as is the case with virus infections, are influenced by several interacting variables the major of which are the virus genome, the host genome, and the environment. Because of these interactions, consistent symptomatological patterns in the natural or experimental hosts of any of these viruses are hardly recognizable.

CMV, a cosmopolitan and one of the least specialized plant viruses, constitutes a primary example of this. Douine *et al.* (1979) have counted about 800 susceptible species in 86 botanical families and Kaper and Waterworth (1981) list some 100 natural hosts. These comprise representatives of both mono- and dicotyledons and range from herbaceous plants to shrubs to woody plants, including unsuspected hosts like olive (Savino and Gallitelli, 1983). In weeds, CMV infections are often asymptomatic. *Portulaca oleracea* L. and *Stellaria media* (L.) Will., just to mention a couple of the many weed hosts, are symptomless carriers of different strains of this virus which can play a primary epidemiological role as natural reservoirs for their perpetuation and spread (see reviews by Quiot, 1981; Martelli and Quacquarelli, 1983).

The array of symptoms induced by CMV in most cultivated plants is tremendous. It varies from mild mottling to severe malformation of aboveground plant organs, dwarfing, fruitlessness, flower color breaking and other chromatic disorders of leaves and fruits. Some strains elicit yellow discolorations of the foliage and some localized or generalized necrosis. In tomato, a strong necrotic reaction that may kill the plant is associated with occurrence of type D CARNA-5, a low-molecular-weight nongenomic RNA molecule replicating only in the presence of the helper virus. This same satellite RNA can also attenuate the symptoms of CMV in several other hosts (e.g., tobacco and corn). The nonnecrogenic type R CARNA-5 attenuates CMV symptoms also in tomato (Jacquemond and Lot, 1981). Hence, the presence of a satellite RNA in a CMV inoculum represents an additional source of symptomatological variation. A comparable effect on disease expression is produced by PARNA-5, the satellite RNA of PSV, a virus with a natural host range virtually limited to legumes in which it causes pronounced stunting, leaf distortion, and mottling (Kaper and Waterworth, 1981).

Equally restricted is the host range of TAV, whose most distinctive pathogenic effect is the induction of seedless fruits in tomato and strongly malformed flowers in chrysanthemum (Hollings and Stone, 1971).

Among Bromoviruses, BMV is the one with the widest natural host range which, however, is largely restricted to the Gramineae (Lane, 1974, 1981). The symptoms induced by BMV are reported to be the mosaic type although, because of the parallel-veined leaves of the hosts, the discolored areas tend to be elongated, resulting in a variegation better defined as striping or streaking. CCMV and BBMV are natural pathogens of Leguminosae in which they elicit chlorotic to bright yellow mottling and deformation of the leaves (Bancroft, 1971; Gibbs, 1972).

In the Ilarvirus group, 9 of the 12 members (Fulton, 1981, 1983) infect only woody plants in nature. In many of the hosts, the symptoms are chlorotic or, more frequently, bright yellow discolorations of the leaves in the form of blotching, stippling, ring spotting, "oak leaf" and line patterns. Necrotic reactions may occur, especially in the early stages of infection ("shock phase"). Shock symptoms of PNRV in cherry consist of chlorotic–necrotic rings and lines on newly formed leaves which develop into holes ("shot-hole" effect) ensuing from detachment of the necrotic tissue. Necrotic patterns of various kinds characterize also the shock phase of TSV infections in inoculated tobacco leaves. Successive leaves are invaded by the virus but show no visible symptoms. This phenomenon, known as "recovery," is typical of herbaceous and woody plants chronically infected by a number of ringspot-type viruses, including several members of the Ilarvirus group and can also occur in plants infected with AMV (Bos, 1970; Fulton, 1981).

AMV is another plant virus with a worldwide distribution and an extended host range. Hull (1969) lists more than 300 plant species in 47 families as hosts of AMV. Van Regenmortel and Pinck (1981) list over 60 in 14 families as hosts infected in nature.

As pointed out by Van Regenmortel and Pinck (1981) the AMV group is a large conglomerate of strains with different biological properties. This fact and the high number of susceptible hosts account for the tremendous range of symptoms displayed by AMV-infected plants. Most field plants affected by this virus show distinct patterns of chrome-yellow discoloration.

IV. CYTOPATHOLOGY

A. Cucumovirus Group

1. General Cytology of Infected Cells

In artificially inoculated plants, Cucumovirus infections often result in severe cytological modifications in tissues underlying symptomatic areas. These changes involve the ground cytoplasm, whose structural organization may be altered by heavy secondary vacuolation (Gerola *et al.*, 1965; Russo and Martelli, 1973). Fragmentation of the protoplast results in many small cytoplasmic islands connected to one another by thin bridges (Fig. 2). In TAV- and PSV-infected cells, where the same phenomenon occurs (M. Russo and A. Di Franco, unpublished results) secondary vacuoles arise from swellings of the endoplasmic reticulum (ER) cisternae which, as the infection progresses, may merge with one another and the central vacuole.

Increased development of cytoplasmic membranes is another feature common to cells invaded by all three definitive members of the group (Gerola *et al.*, 1965; Honda and Matsui, 1968, 1974; M. Russo and A. Di Franco, unpublished results). Membrane accumulations may develop from: (1) proliferation of ER and gathering of ER strands in defined cytoplasmic areas (Figs. 3 and 4); (2) enhanced presence of plasmalemmasomes; (3) development of large bodies made up of membranes of tubular or convoluted form or concentrically arranged so as to resemble myelinic structures. These membranous bodies (MB) are located in the cytoplasm or, more commonly, next to secondary or central vacuoles into which they often protrude (Fig. 2, inset b).

MB are not a specific feature of infected cells for they are also present in healthy tissues (Gerola *et al.*, 1965; Ehara, 1979). However, in Cucumovirus-invaded tissues, MB are larger, structurally more complex, and occur with a much higher frequency. In fact, Honda and Matsui (1968) concluded that CMV may be a specific elicitor of these structures for MB appeared to be absent in tobacco cells infected with tobacco mosaic virus (TMV), but were plentiful in comparable samples doubly infected by TMV and CMV or by CMV alone.

MB are thought to originate from the plasma membrane (Ehara, 1979), but their function, if any, is controversial. Gerola *et al.* (1965) consider membrane proliferations as a by-product of a deranged cellular condition nonspecifically induced by different agents, including viruses. Ehara (1979) envisages an active role for MB in detoxifying cells through disposal to the vacuole of unwanted metabolites derived from virus infection. Whatever their significance, MB are structures less intriguing and probably less important in the economy of virus replication than the tonoplast-associated vesicles first detected in CMV- and TAV-infected cells by Hatta and Francki (1981a) and more recently found also in PSV infections (Fig. 3, inset a).

These vesicles are round to ovoid MB bound by a unit membrane, measuring up to about 100 nm in diameter and connected with the tonoplast from which they protrude into the vacuoles (Fig. 3). Some of the vesicles contain darkly staining amorphous material but many have fine fibrils which proved sensitive to RNase digestion under low-salt incubation conditions. Based on this evidence, Hatta and Francki (1981a) concluded that the fibrils consisted of dsRNA and that the vesicles were likely sites of viral RNA synthesis. These authors did not indicate how

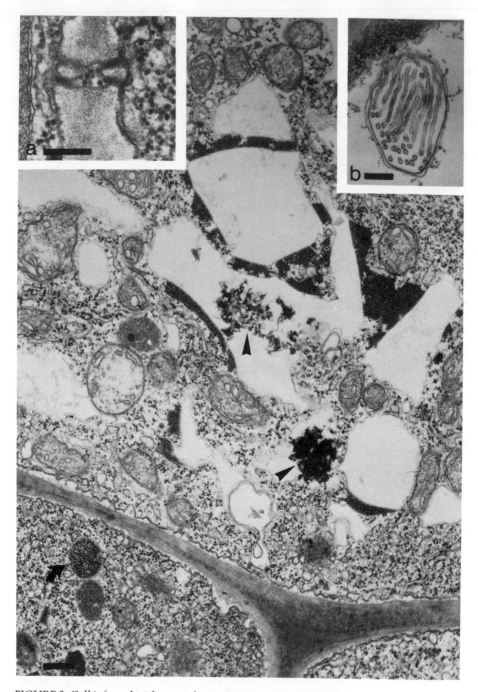

FIGURE 2. Cell infected with cucumber mosaic virus (Cucumovirus) showing fragmentation of ground cytoplasm due to secondary vacuolation and the appearance of crystalline arrays of virus particles. Clumps of particles embedded in a darkly staining material are also present in the vacuoles (arrowheads). In the cell on the lower left, a spheroidal viral aggregate is visible (arrow). Inset (a) shows CMV virus particles in a plasmodesma; inset (b) shows a membranous body protruding into the vacuole. CW, cell wall. Bars = 200 nm.

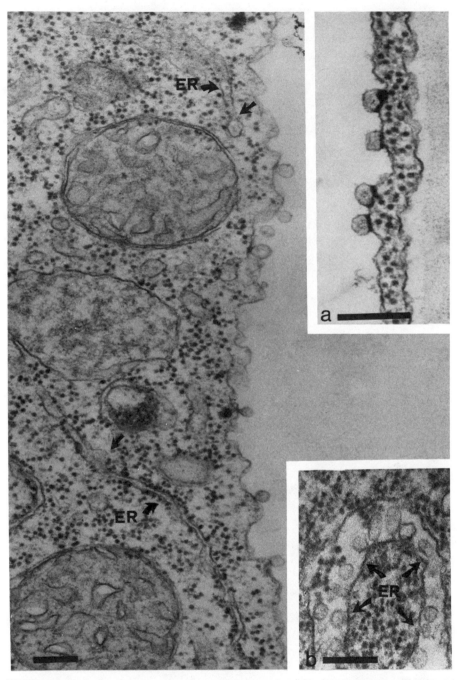

FIGURE 3. Vesicular structures in the cytoplasm and on the tonoplast of a cell infected with tomato aspermy virus (Cucumovirus). Some of the vesicles appear to develop from the endoplasmic reticulum (ER) (arrows and inset b). Inset (a) shows tonoplast-associated vesicles in a cell infected with peanut stunt virus (Cucumovirus). Many virus particles are visible in the cytoplasm. Bars = 200 nm.

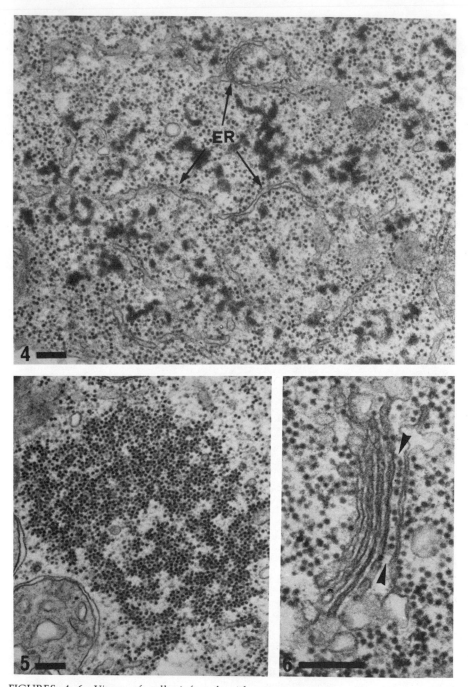

FIGURES 4–6. Views of cells infected with tomato aspermy virus (Cucumovirus).
FIGURE 4. Endoplasmic reticulum (ER) strands, vesicles, clumps of darkly staining mater-
ial, and scattered virus particles constituting a loosely textured cytoplasmic inclusion.
FIGURE 5. A cytoplasmic aggregate of virions associated with electron-dense material.
FIGURE 6. Virus particles (arrowheads) associated with dictyosomal cisternae. Bars = 200
nm.

the vesicles develop nor was it possible to provide a straightforward explanation for their origin in subsequent studies carried out in our laboratory.

In TAV-infected *N. benthamiana* and *N. clevelandii* plants, vesicular structures were equally well represented and distributed in inoculated leaves and in leaves systemically infected for 1, 2, and 5 weeks. In some samples, infection was in a very advanced stage, suggesting that the vesicles are not transient structures associated only with active virus replication.

In tissues infected with all three definitive Cucumoviruses, vesicles were found lining the tonoplast of central and secondary vacuoles—hence, they keep developing together with other cytological alterations—but were also present inside vacuoles away from the tonoplast and in the cytoplasm (Fig. 3), sometimes being connected with the ER as though they were budding from it (Fig. 3 and inset b). It seems possible that RNA-containing vesicles originate in the cytoplasm from the ER, migrate toward the vacuole, fuse with the tonoplast from which they bulge, and may ultimately be released into the vacuole.

Other major cytoplasmic structures are affected by Cucumoviruses, though to an extent that varies depending on the virus or its strain, host, and age of infection. Chloroplasts are indeed those affected the most. As mentioned in Section II, extensive structural and functional modifications are evoked by CMV, TAV, and PSV through impairment of differentiation of proplastids or disruption of the internal structure of mature chloroplasts. Detailed accounts of the striking variety of alterations shown by chloroplasts have been described by a number of authors working in different laboratories with diverse virus–host combinations (Gerola *et al.*, 1965; Misawa and Ehara, 1966; Honda and Matsui, 1968, 1974; Favali and Conti, 1970; Russo and Martelli, 1973; Ehara and Misawa, 1975; Kraev *et al.*, 1975).

Mitochondria are not spared by Cucumovirus infections and seem to follow the fate of other cellular components in relation to severity of protoplast damaging. Thus, regardless of the infecting virus, modification of their shape and architecture is detected with a greater frequency in seriously altered cells. In CMV-infected cells, Gerola *et al.* (1965) reported the occurrence of peripherally vesiculated mitochondria somewhat resembling those induced by viruses belonging to different taxonomic groups (see discussion in Russo and Martelli, 1982). However, this finding was not confirmed by any of the subsequent studies found in the literature or by recent observations made in our laboratory.

Enhanced vesiculation of dictyosomes, the significance of which is unknown, was found associated with CMV infections in *N. glutinosa* (Gerola *et al.*, 1965) and TAV in *Chrysanthemum murifolium* Ramat (Lawson and Hearon, 1970).

Nuclei seem to be little affected by Cucumoviruses as judged from published information. Only in two instances, nuclear modifications induced by CMV were reported in tobacco mesophyll cells (Honda and Matsui, 1974) and protoplasts (Honda et al., 1974) consisting, in both cases, of depletion of heterochromatin.

2. Appearance and Intracellular Distribution of Virus Particles

It has been reported that CMV particles, unless arranged in ordered aggregates, are not readily discernible in cells where they may be confused with ribosomes (Matsui and Yamaguchi, 1966; Honda and Matsui, 1974). Efforts have therefore been made to facilitate their visualization by destruction of cytoplasmic ribosomes. This was achieved by Honda and Matsui (1974) by floating leaf disks from infected leaves on phosphate buffer for 24 hr at 25°C under continuous illumination prior to fixation and embedding. Similar results were obtained by Hatta and Francki (1979) with a more elegant technique based on pancreatic ribonuclease treatment of glutaraldehyde-fixed tissue samples. Ribosomes were digested to a great extent whereas CMV particles were unaffected. This method had the advantage of preserving cellular structures which were damaged by the more drastic treatment devised by Honda and Matsui (1974). However, pretreatment of Cucumovirus-infected tissues does not seem to be essential for visualization of intracellular virus particles because individual particles are often readily detected by their size, electron density, and general outward appearance, even when they are randomly dispersed in the cytoplasm (Lawson and Hearon, 1970; Russo and Martelli, 1973; Tolin, 1977) (see also Figs. 3, 4, and 6).

In thin-sectioned cells, Cucumovirus particles appear as isometric, solid, or more frequently doughnut-shaped bodies (i.e,. having a dark-staining outer shell and a lighter center) with a rounded contour. Estimates of particle diameter vary from 16–19 nm for TAV (Lawson and Hearon, 1970) to 20–29 nm for CMV (Gerola et al., 1965; Honda and Matsui, 1968, 1974; Russo and Martelli, 1973; Honda et al., 1974; Tolin, 1977; Hatta and Francki, 1979). Comparative observations of CMV, TAV, and PSV made in our laboratory (M. Russo and A. Di Franco, unpublished results) have confirmed that the outward appearance of intracellular particles of the three viruses is virtually the same and their mean diameter approximates to 22 nm, a figure in agreement with Hatta and Francki's (1981b) recent estimate for CMV.

Cucumovirus particles are present in parenchyma and conducting tissues of systemically invaded hosts (Lyons and Allen, 1969). Intracellular particles are distributed at random in the cytoplasm (Figs. 3 and 4) but, as discussed in the next section, they often aggregate in various forms. Virus aggregates are also found in the vacuoles (Figs. 7–9). Some of these aggregates are made up of virus particles interspersed with a darkly staining amorphous material of unknown nature (Figs. 8 and 9) resembling structures seen in cells infected with Como- and Nepoviruses (Russo et al., 1982; Di Franco et al., 1983).

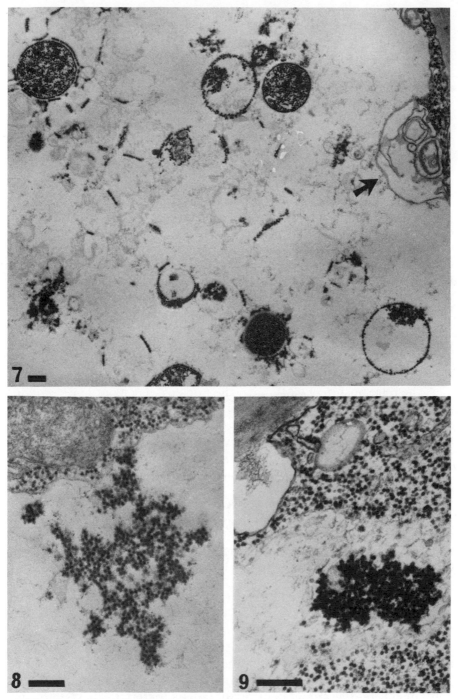

FIGURES 7–9. Views of intravacuolar Cucumovirus particles.
FIGURE 7. Spheroidal aggregates of TAV particles. Arrow indicates a membranous body.
FIGURES 8 and 9. Irregular clumps of CMV (Fig. 8) and PSV (Fig. 9) particles intermingled
with electron-dense amorphous material. Bars = 200 nm.

Association of virions with plasmodesmata was found in TAV (Lawson and Hearon, 1970) and CMV infections (Fig. 2, inset a). In both instances, discrete accumulations of virus particles were seen within otherwise normal cell walls. However, Francki *et al.* (1984) observed cell wall outgrowths and tubules of viruslike particles passing through plasmodesmata in cells infected with a TAV strain, and like those induced by a number of viruses of other groups (reviewed by Martelli, 1980).

Cucumoviruses have been reported to establish topological relationships with cell organelles other than microbodies and mitochondria. A most peculiar association is that of TAV with dictyosomes. Lawson and Hearon (1970) were the first to observe rows of TAV particles packed between dictyosomal cisternae, a finding which was recently confirmed in our laboratory (Fig. 6). These authors, however, as we, were unable to establish whether the particles were assembled in the dictyosomes or became trapped there during membrane development. Trapping is also a likely explanation for the presence of CMV and TAV particles betwen the whorled membranes of myelinlike cytoplasmic bodies (Honda and Matsui, 1968; M. Russo and A. Di Franco, unpublished results).

Association of CMV particles with chloroplasts has been reported once, by Ehara and Misawa (1975). The organelles containing particles were necrotic and disrupted, suggesting that they may have come from the surrounding cytoplasm.

The relationship of CMV with the nucleus may have a different and greater significance. Honda and Matsui (1974) found that up to 70% of the nuclei of CMV-infected tobacco cells contained virus particles, and particles have also been detected in nuclei of *N. clevelandii* infected with the T strain of CMV (Francki *et al.*, 1985). Francki *et al.* (1985) point out that RNA synthesis and particle assembly by Cucumoviruses may take place in the cytoplasm; hence, nuclei are more likely to be sites of particle accumulation than production. However, it has been observed that visibly modified nuclei of tobacco protoplasts infected with CMV, 24 hr after inoculation consistently contained aggregates of virus particles, mostly connected with the nucleolus, indicating that CMV particles were produced in the nucleus (Honda *et al.*, 1974).

3. Inclusion Bodies

When viewed with the light microscope after appropriate staining, mesophyllic and epidermal cells of plants infected with Cucumoviruses may reveal the presence of amorphous and/or crystalline inclusion bodies. These are resistant to detergent (Triton X-100) treatment and thus do not contain substantial amounts of membranous material. They stain positively for protein and RNA (Christie and Edwardson, 1977; Edwardson and Christie, 1979). Such cytoplasmic and intravacuolar inclusions have been observed in thin-sectioned tissues invaded by some isolates of all three definitive Cucumoviruses.

In TAV-infected *N. benthamiana* and *N. clevelandii* cells, unusual cytoplasmic inclusions were seen which consisted of rather loose aggregates of virus particles, ER strands and flakes of electron-opaque amorphous material of unknown nature (Fig. 4). It may be that a closer association of this dense material with virus particles gives rise to the rather compact viral clusters, like that shown in Fig. 5, which are a rather typical feature of TAV infections (Lawson and Hearon, 1970; Francki *et al.*, 1984; M. Russo, unpublished results).

Christie and Edwardson (1977) found that the amorphous inclusions seen in the light microscope, correspond to cytoplasmic or, sometimes, vacuolar aggregates of virus particles. These aggregates in TAV-infected (Fig. 10) and, to a lesser extent, in CMV-infected cells (Gerola *et al.*, 1965; see also Fig. 2) consist of peculiar spheroidal structures, surrounded by a thin membrane (Fig. 10, inset), which are essentially made up of virus particles (Fig. 10). The bounding membrane confers stability on these bodies which retain their shape also when present in the vacuole (Fig. 7).

Angular inclusions seen with the light microscope correspond to virus crystals. All three Cucumoviruses are known to produce crystalline aggregates (Christie and Edwardson, 1977; Edwardson and Christie, 1979) but these are less frequently encountered in TAV and PSV than in CMV infections.

Crystalline CMV is seldom found in the cytoplasm (Russo and Martelli, 1973; Honda and Matsui, 1974; see also Fig. 2). Crystals are more often located in the vacuoles (Gerola *et al.*, 1965; Honda and Matsui, 1968, 1974; Russo and Martelli, 1973; Honda *et al.*, 1974; Ehara and Misawa, 1974; Ehara *et al.*, 1976; Tolin, 1977; Hatta and Francki, 1979). It has been suggested that this may depend on the interfering action exerted by ribosomes on CMV crystallization. Ehara *et al.* (1976) were in fact unable to obtain CMV particles arrayed into a regular crystalline lattice when these were mixed with ribosomes *in vitro*.

CMV crystals are extremely variable in size and shape (rhombic, rectangular, hexagonal, octagonal) and the particles are arranged in a square or hexagonal pattern, thus indicating a possible cubic-packed structure. Seldom are these crystals compact solid bodies like those of other isometric viruses; they exhibit gaps in the lattice which may be so large so as to give rise to hollow structures rimmed by a few rows of orderly arranged virions (Fig. 11) (Honda and Matsui, 1968; Edwardson and Christie, 1979).

B. Bromovirus Group

1. General Cytology of Infected Cells

All three definitive members in the bromovirus group have been studied ultrastructurally in infected host tissues (de Zoeten and Schlegel,

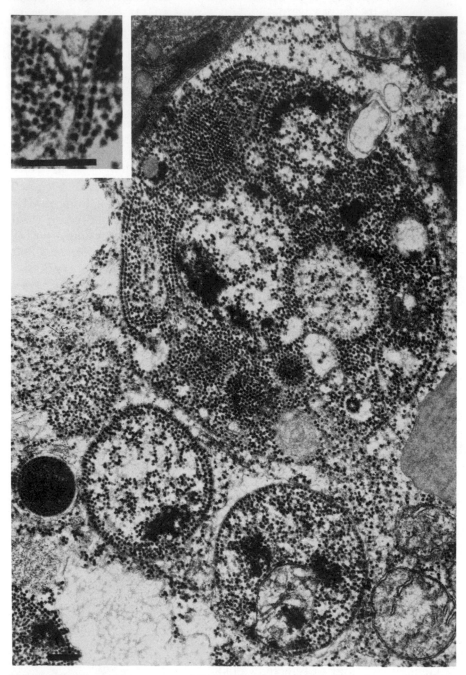

FIGURE 10. Spheroidal aggregates of TAV particles (Cucumovirus) in the cytoplasm of an infected cell. Inset shows the thin membranes surrounding viral aggregates. Bars = 200 nm.

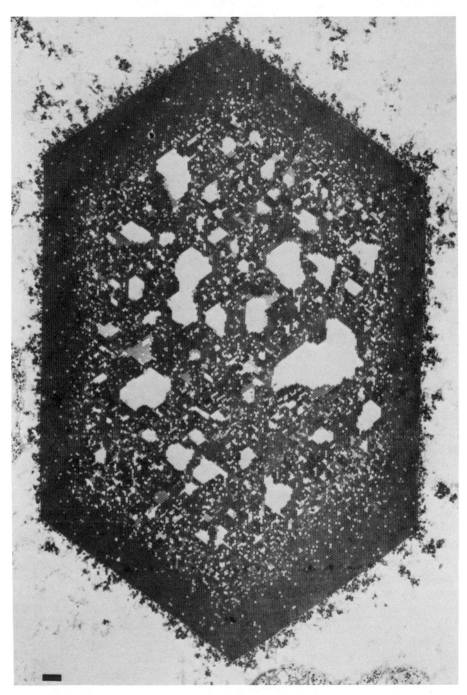

FIGURE 11. An intravacuolar CMV (Cucumovirus) crystal. Bar = 200 nm.

1967a,b; Paliwal, 1970; Moline and Ford, 1974; Lastra and Schlegel, 1975; Kim, 1977; M. Russo, unpublished results). CCMV and BMV have also been examined in protoplasts (Motoyoshi et al., 1973; Burgess et al., 1974).

Some of the cytological modifications suffered by infected cells resemble those elicited by Cucumoviruses. Thus, for instance, in N. benthamiana infected with BBMV, heavy secondary vacuolation of the protoplast was observed together with the presence of many intracytoplasmic and intravacuolar MB (M. Russo, unpublished observations).

Disturbances of the cytoplasmic membrane system consisting of proliferation and accumulation of ER strands and production of vesicles are a characteristic feature of Bromovirus infections. Increased amounts of ER adjacent to the nucleus were detected as soon as 6 and 7 hr after inoculation in BMV- and CCMV-infected protoplasts, respectively (Burgess et al., 1974) and 3 days after inoculation in primary cowpea leaves infected with CCMV (Kim, 1977). The vesicles are round to ovoid membrane-bound structures of variable size and diameter up to 120 nm. Internally, they contain dots of electron-dense material or a network of fine fibrils resembling nucleic acid (Figs. 13, 14, and 16). It appears that these vesicular structures originate from blebbing of the ER cisternae (Motoyoshi et al., 1973; Burgess et al., 1974; Kim, 1977; see also Figs. 14 and 16) and accumulate in the cytoplasm, usually in clusters of several elements bounded by a peripheral membrane (Figs. 14 and 16), where they take part in the formation of inclusion bodies.

Cytoplasmic aggregates of membranous vesicles may also originate from enhanced vesiculating activity of the dictyosomes, as in the case of BBMV (Fig. 15). Distyosomal vesicles are electron-lucent and gather into rather compact clusters in well-defined cytoplasmic areas.

Fibril-containing vesicles, resembling those derived from the ER, are also present inside localized dilations of the nuclear envelope in CCMV (Burgess et al., 1974; Kim, 1977), BMV (Burgess et al., 1974) and BBMV infections (Fig. 14, inset b). It appears that these vesicles originate from the inner nuclear membrane and while being released into the cytoplasm, acquire an additional peripheral membrane derived from the outer lamella of the nuclear envelope. Hence, nuclear vesicles of Bromoviruses may arise through a budding process comparable to that evoked by pea enation mosaic virus (de Zoeten et al., 1972). Nuclei may not be otherwise affected, except for a depletion of chromatin as observed in the yellow foliar tissues of cowpea infected with CCMV (M. Russo, unpublished observations), or the occurrence of filamentous inclusions as those elicited by the yellow stipple strain of the same virus (Kim, 1977).

As reported in Section II, severely altered chloroplasts were present in chlorotic cowpea leaves infected with CCMV. Those shown in Fig. 1 are the most pronounced modifications seen in the course of investigations carried out in our laboratory on a number of hosts infected with

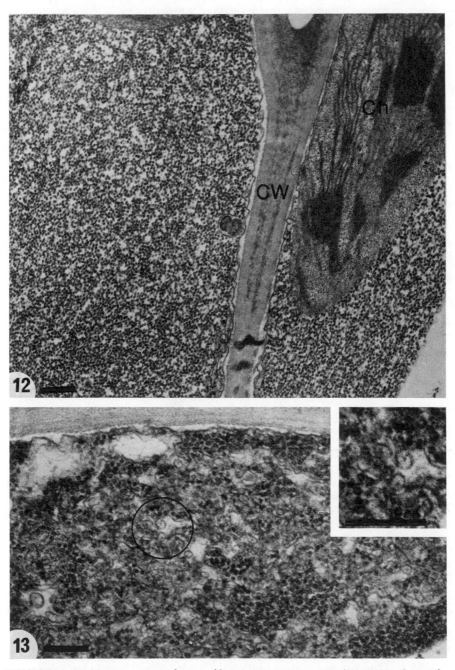

FIGURES 12, 13. Massive accumulation of bromegrass mosaic virus (Bromovirus) particles in the cytoplasm of infected cells (Fig. 12). A cytoplasmic inclusion body in a BMV-infected cell made up of membranous material, virus particles, and vesicles (Fig. 13). Inset shows a close-up of the vesicular structures in the encircled area. CW, cell wall; Ch, chloroplast. Bars = 200 nm.

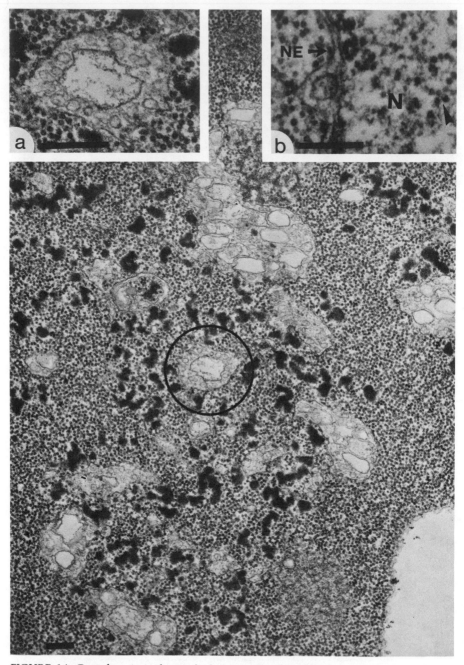

FIGURE 14. Cytoplasmic inclusion body in a cell infected with broad bean mottle virus (Bromovirus) made up of clusters of membranous vesicles, flakes of darkly staining amorphous material, and virus particles. Inset (a) shows at higher magnification the encircled area with fibril-containing vesicles associated with ER strands. Inset (b) shows a vesicle in the perinuclear space. Arrowhead points to intranuclear virus particles. N, nucleus; NE, nuclear envelope. Bars = 200 nm.

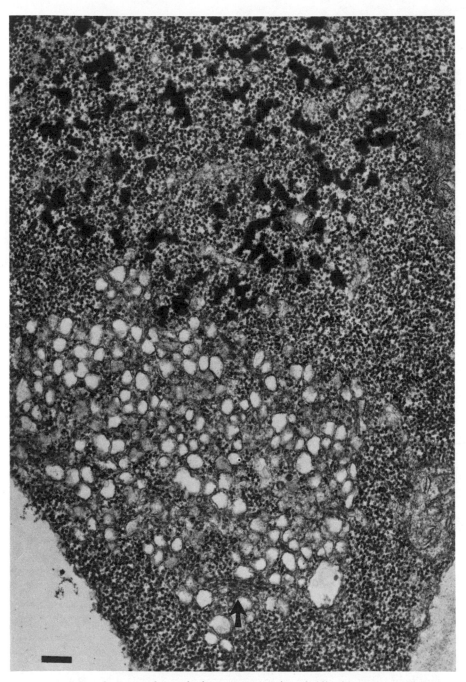

FIGURE 15. Cytoplasmic inclusion body in a BBMV-infected cell. The aggregate of electron-lucent vesicles of dictyosomal origin is separated from the electron-dense material. Arrow points to a dictyosome. Bar = 200 nm.

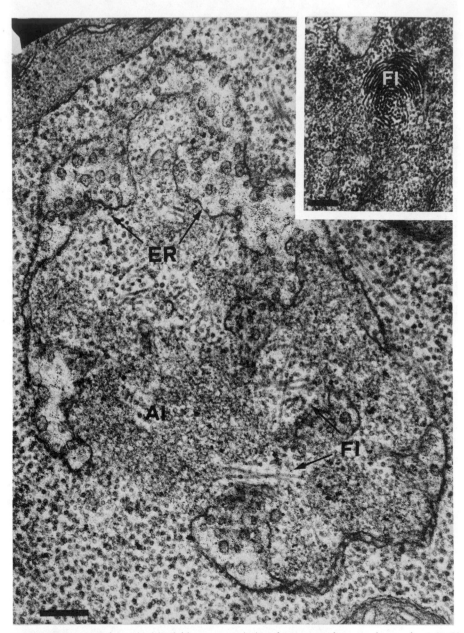

FIGURE 16. Amorphous (AI) and filamentous (FI) inclusions in the cytoplasm of a CCMV (Bromovirus)-infected cell. The inclusions are surrounded by endoplasmic reticulum (ER) strands from which vesicles appear to be developing. Inset shows a bundle of fibrils (FI). Bars = 200 nm. From Kim (1977).

each of the three definitive Bromoviruses. However, a variety of less spectacular changes, such as modifications of size and shape of the organelles, increased size of starch granules, or slight reduction of the thylakoid system, were seen in chloroplasts of most infected samples.

Unusually large mitochondria having a somewhat ameboid appearance were reported to occur in CCMV-infected protoplasts (Burgess *et al.*, 1974) whereas very small mitochondria were observed in roots of several Gramineae infected with BMV (Moline and Ford, 1974). In our studies, regardless of the infecting virus, the shape and size of the mitochondria did not appear to be affected as much as their internal organization. In general, and especially as infections grew older, an increasingly high number of mitochondria exhibited a marked reduction of the stroma density and in the size and numbers of cristae. The cristae became progressively smaller and fewer in number so as to leave a double-membraned, hollow electron-lucent structure. Similarly deranged mitochondria were observed by Paliwal (1970) in BMV-infected plants. In no case were microbodies reported to be affected by Bromovirus infections.

2. Appearance and Intracellular Distribution of Virus Particles

Bromoviruses are very invasive. In infected plants their particles have been observed in the epidermis, foliar, root, and phloem parenchyma, differentiating and mature sieve tubes, tracheary elements, and sclerenchymatous cells (Paliwal, 1970; Moline and Ford, 1974; Kim, 1977; M. Russo, unpublished results). The consensus is that virus multiplication or assembly takes place in these different tissues and, in the case of BMV, it was suggested that it occurs also in meristematic root cells (Moline and Ford, 1974).

Intracellularly, Bromovirus particles have an outward appearance comparable to that of Cucumoviruses (i.e., doughnutlike bodies with a rounded contour) and a diameter ranging from 20 to 26 nm (de Zoeten and Schlegel, 1967b; Hills and Plaskitt, 1968; Paliwal, 1970; Kim, 1977). Our measurements indicate a diameter of about 23 nm.

Owing to their abundance, intracellular virus particles are readily identified even when they are not orderly arrayed. Nevertheless, staining procedures like the uranyl soak method (Hills and Plaskitt, 1968) have been devised which help in recognizing virus particles especially when they are located within organelles (e.g., nuclei) or cytoplasmic areas where they may be confused with ribosomes.

In CCMV-infected cells, treatment with pronase and subtilisin improved considerably the identification of virus particles, which were relatively unaffected by the enzymes as compared with the surrounding cytoplasmic constituents (Kim, 1977).

In tobacco protoplasts, progeny BMV and CCMV particles were detected in the cytoplasm as soon as 10 and 15 hr postinfection, respectively

(Burgess *et al.*, 1974). Their number increased rapidly so that discrete viral aggregates were discernible 24 hr after inoculation (Motoyoshi *et al.*, 1973; Burgess *et al.*, 1974). In tissues of intact plants, intracellular virus particles were first seen 3 days after inoculation (Kim, 1977).

In tissues systemically infected for 7 days or longer, regardless of the host or the virus, very large amounts of virus particles are present. These accumulate preferentially in the cytoplasm (Figs. 13–16) often in such quantities so as to completely fill the cell. Cytoplasmic virus particles are usually scattered, but crystalline aggregates have also been observed in cells infected with all members of the group (de Zoeten and Schlegel, 1967b; Hills and Plaskitt, 1968; Moline and Ford, 1974; Christie and Edwardson, 1977; M. Russo, unpublished observations). Unusual forms of particle aggregation were encountered in tobacco protoplasts infected with BMV. These consisted of: (1) short rows of particles stacked between electron-dense lamellalike structures; (2) particles arranged in a roughly helical manner on the surface of tubular structures (Burgess *et al.*, 1974). Comparable aggregates have not been detected in any subsequent ultrastructural investigations on Bromoviruses.

Besides the ground cytoplasm, Bromovirus particles can accumulate in the central vacuole (Paliwal, 1970; Moline and Ford, 1974; Christie and Edwardson, 1977) and BMV has also been detected in the intercellular spaces of barley mesophyll and vascular bundle sheath (Paliwal, 1970). This seems to be one of the very few records of a plant virus having such a peculiar localization. As an explanation, Paliwal (1970) suggested that either virus particle assembly takes place within intercellular spaces from protein and nucleic acid migrating from adjoining cells, or whole particles move into these areas through plasmodesmata. However, unlike the Cucumoviruses, in no instance have Bromovirus particles been observed within plasmodesmata of infected tissues.

Association of Bromoviruses with cell organelles has also been reported. Intranuclear occurrence of virions either randomly scattered or clustered in discrete aggregates was repeatedly recorded in foliar (Hills and Plaskitt, 1968; Kim, 1977; M. Russo, unpublished observations) and root (Moline and Ford, 1977) tissues of different hosts infected with all definitive members of the group, but not in tobacco protoplasts infected with CCMV or BMV (Motoyoshi *et al.*, 1973; Burgess *et al.*, 1974). This latter finding and the observation that virus particles appear in the karyoplasm later than in the cytoplasm, have led some authors to suggest that virus particles may enter into nuclei through nuclear pores rather than being assembled there (Motoyoshi *et al.*, 1973; Moline and Ford, 1974). Contrary to these views, Kim (1977) suggests that replication of CCMV may take place in the nucleus in connection with the presence of certain inclusions.

Occurrence of Bromovirus particles inside chloroplasts is rather common. However, they always appear to be derived from invagination of virus-containing ground cytoplasm (de Zoeten and Schlegel, 1967b; Paliwal, 1970; M. Russo, unpublished observations).

3. Inclusion Bodies

Rubio and Van Slogteren (1956) were the first to describe inclusion bodies in BBMV-infected broad bean cells. The inclusions were only present in discolored areas of symptomatic leaves and, under the light microscope, they appeared as granular bodies which became vacuolated as infection progressed. In the electron microscope, the inclusions appeared to be made up mostly of virus particles. More recent investigations have confirmed that all definitive Bromoviruses elicit the formation of amorphous inclusions detectable with the light microscope in the cytoplasm of infected cells (Christie and Edwardson, 1977). Such inclusions are different in size, are round to elongate in shape, and stain positively for protein, RNA, and lipids. They correspond to the amorphous cytoplasmic inclusions seen with the electron microscope (Figs. 14 and 15).

Inclusion bodies elicited by BMV and BBMV have a similar outward appearance and gross composition. These are rather loose to compact structures made up of ER strands, accumulations of membranous vesicles, virus particles, and flakes of very intensely electron-opaque material with amorphous texture. Micrographs published by various authors do not differ much from Figs. 13–15 presented here (de Zoeten and Schlegel, 1967b; Lastra and Schlegel, 1975; Christie and Edwardson, 1977). In BBMV-induced inclusions, however, there seems to be a different distribution of the vesicles in relation to other components according to the vesicle origin. Whereas vesicles derived from the ER and probably also from the nuclear envelope, are intermingled with the electron-dense material shown in Fig. 14, vesicles of dictyosomal origin tend to remain in compact aggregates next to, but not mixed with, the electron-dense matter (Fig. 15). Such differences were consistently observed in all samples of *N. clevelandii* and *N. benthamiana* infected with BBMV (M. Russo, unpublished observations). They may arise from the suggestion previously mentioned, that virus infection impairs intracellular transfer of dictyosomal secretory vesicles which accumulate close to their place of origin. Vesicles, on the other hand, develop in different places owing to the wide distribution of the ER membranes from which they appear to develop. In any case, dictyosomal and ER or nuclear vesicles may be distinguished from each other by the lack of fibrillar material in the former (Figs. 14 and 15).

Two types of inclusion bodies are also associated with CCMV infections. These were described in detail by Kim (1977) and referred to as amorphous (AI) and filamentous (FI) inclusions, respectively. AI consist of accumulations of finely granular, rather electron-dense material which occurs in the cytoplasm, sometimes near the nucleus, whereas FI are long, flexuous, rodlike structures about 17 nm in diameter loosely aggregated in bundles and located in the cytoplasm and nucleus. These flexuous structures are similar to the filamentous form of viral protein found sometimes *in vitro* (Bancroft *et al.*, 1969).

Both AI and FI, together with the ER strands, fibril-containing vesicles, and virus particles, give rise to complex cytopathic structures as shown in Fig. 16. According to Kim (1977), AI and FI have the same gross chemical composition and are both sensitive to proteolytic enzymes; they may represent different developmental stages of similar materials. The function of these structures is unknown although they are suspected of being involved in virus multiplication. As indicated by radioautographic (de Zoeten and Schlegel, 1967a) and immunoautoradiographic (Lastra and Schlegel, 1975) studies, cytoplasmic inclusions associated with BBMV infections may be the sites of viral RNA replication and contain viral antigen. The latter is also present in the nuclei, in high amounts in the initial stages of infection and later in similar quantities to that in the cytoplasm, suggesting that virus coat protein synthesis may take place in the nuclei and assemblage of virions in the cytoplasm (Lastra and Schlegel, 1975). These views, however, contrast with the observation that, as previously mentioned, whole virus particles can be found in the nuclei, and that there is biochemical evidence that BBMV coat protein is synthesized on cytoplasmic ribosomes (Gibbs and McDonald, 1974).

Crystalline arrays of virus particles constitute another type of intracellular inclusion which is relatively rare, except with BMV; in the latter, crystals were frequently seen in both plant (Paliwal, 1970; Moline and Ford, 1974; Christie and Edwardson, 1977) and insect cells in which they accumulated following ingestion (Paliwal, 1972).

C. Ilarvirus Group

Published information on the cytology of Ilarvirus infections consists only of one study concerning CVV in French bean tissues (Gerola *et al.*, 1969b) and two reports on TSV (Edwardson and Purcifull, 1974; Christie and Edwardson, 1977). It has been suggested by Francki *et al.* (1985) that the scarcity of ultrastructural data on this group of viruses is because investigations "have turned up nothing dramatic or specific enough to be published." Our experience largely confirms this view. The investigations in our laboratory found minor cytological changes in most of the virus–host combinations studied. Only in the case of Fulton's B strain of TSV, a Brazilian virus serologically different from the type culture and other North American isolates (Fulton, 1972), were distinct cytological modifications observed in infected cells (M. A. Castellano and G. P. Martelli, unpublished observations). The ultrastructural observations, which constitute the essence of the description that follows, were made on foliar tissues of *Chenopodium quinoa* Willd. Infected by this virus sampled 24, 48, and 96 hr after inoculation.

1. General Cytology of Infected Cells

In *C. quinoa* leaves, symptoms and ultrastructural changes appeared as early as 48 hr postinoculation. The most remarkable of these changes was an enhanced development of cytoplasmic membranes. Proliferation of ER (Fig. 17), production of vesicles, plasmalemmasomes, and membranous myelinlike bodies akin to those associated with Cucumovirus infections, were consistent features of infected cells.

Vesicles were single-membraned round structures up to 200 nm in diameter, either electron-lucent and apparently empty or containing dots of amorphous electron-dense material or a network of fine fibrils (Figs. 17 and 18). They appeared to originate from dictyosomes or from blebbing of ER cisternae. Vesicular structures lay scattered in the ground cytoplasm or, more often, were in clusters of several elements (Fig. 18) that gave rise, together with virus particles and other cell constituents, to cytoplasmic inclusion bodies. Occasionally, tonoplast-associated vesicles were found protruding into the central vacuole (Fig. 18 inset).

Mitochondria were the only organelles to show clear signs of degeneration. The stroma had a much reduced electron density and cristae were disarranged, smaller and fewer than normal (Fig. 18 inset). No appreciable alterations of nuclei or chloroplasts were detected.

Localized modifications of the cell wall–plasmalemma interface were plentiful. These consisted of most peculiar funnellike membranous structures associated with plasmalemmasomes. These "funnels" appeared to develop from a distension of the plasma membrane at the level of plasmodesmata, through a process that produced a sort of tapering membrane-walled channel, an extremity of which was connected with the plasmodesmata while the other end was free and open toward the cytoplasm (Figs. 19 and 20). These membranous structures always contained virus particles. In no case were they surrounded by cell wall material, thus differing substantially from the virus-containing protrusions induced by Como-, Nepo-, and viruses of other groups (for review see Martelli, 1980). The significance of these alterations is not known. However, since in 48-hr infections virus-containing "funnels" were seen protruding from the cytoplasm into plasmodesmata (Figs. 19 and 20), whereas in 96-hr infections they were outside the protoplast trapped between plasma membrane and cell wall (Fig. 21), it appears likely that these structures are part of a transient and, perhaps, rather unsuccessful mechanism for intercellular transport of virus particles.

2. Appearance and Intracellular Distribution of Virus Particles

CVV particles have been observed in mature sieve elements of French bean (Gerola *et al.*, 1969b) whereas, regardless of the strain, TSV particles were only seen in the cytoplasm or, more rarely, nuclei of meristematic

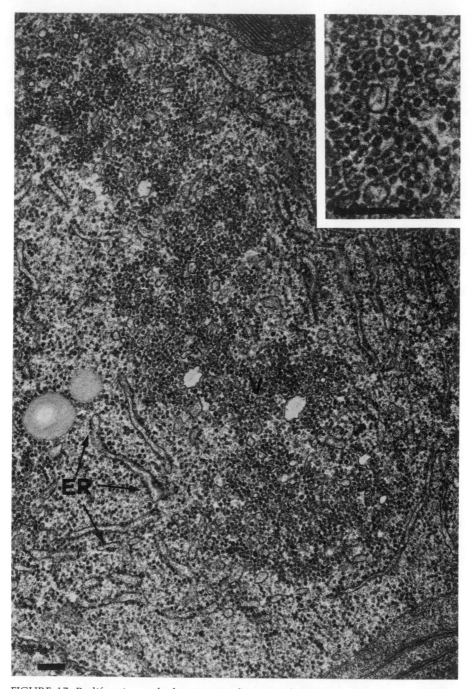

FIGURE 17. Proliferating endoplasmic reticulum strands (ER), membranous vesicles, and virus particles (V) making up a cytoplasmic inclusion in a cell infected with tobacco streak virus strain B (Ilarvirus). Inset shows detail of vesicles and virions. Bars = 200 nm.

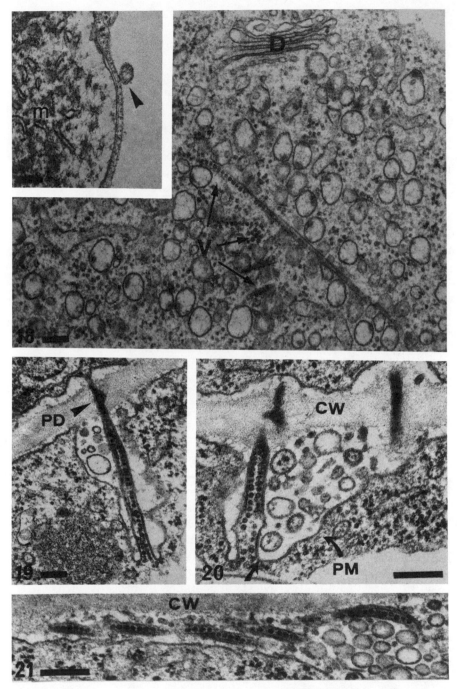

FIGURES 18–21. Accumulation of vesicles possibly of dictyosomal (D) origin and virus particles (V), some in rows contained within tubules in a cell infected with TSV-B (Fig. 18). Inset shows part of a deranged mitochondrion (m) and a tonoplast-associated vesicle (arrowhead). Funnellike extensions of the plasma membrane (PM and arrow in Fig. 20) associated with plasmodesmata (PD) and containing rows of virus particles. In Fig. 21 the same structures are flattened against the cell wall (CW). Bars = 200 nm.

and parenchymal cells of their hosts (Edwardson and Purcifull, 1974; M. A. Castellano and G. P. Martelli, unpublished observations).

Intracellular virions appear as solid or doughnut-shaped round bodies varying in size. Gerola *et al.* (1969b) reported a diameter of about 30 nm for CVV but our estimates for TSV indicate a diameter ranging from 18 to 23 nm, a figure in agreement with measurements made by Edwardson and Purcifull (1974).

In infected cells, particles of Ilarviruses may be: (1) scattered at random throughout the cytoplasm; (2) packed in large but disordered aggregates (Fig. 17); (3) in paracrystalline arrays (Edwardson and Purcifull, 1974; Christie and Edwardson, 1977); (4) in single rows within incomplete tubular structures (Gerola *et al.*, 1969b; see also Fig. 18). The virus particles can be identified with confidence only when they gather in one of the above aggregation forms.

Intracellular arrays of TSV particles do not show the cubic or pseudocubic crystalline packing that is customary for Cucumo- and Bromoviruses. Rather, these aggregates appear as stacked rows of virus particles arranged as if derived from stratification of individual rosaries of virions like those contained within cytoplasmic tubules. Thus, TSV aggregates resemble the paracrystals of Nepoviruses (Martelli and Russo, 1977).

3. Inclusion Bodies

In epidermal strips of *D. stramonium*, Christie and Edwardson (1977) detected with the light microscope, amorphous inclusions that corresponded to cytoplasmic accumulations of virus particles and aggregates of fibers of undetermined nature. Inclusion bodies of TSV-B as seen with the electron microscope, consist of areas of dense cytoplasm surrounded by mitochondria, where aggregates of virus particles are intermingled with ER strands and membranous vesicles (M. A. Castellano and G. P. Martelli, unpublished observations).

Nothing is known of the intracellular replication sites or the morphogenesis of Ilarviruses. However, for TSV-B, there are indications that virus production and/or assembly may take place in the cytoplasm. This possibility is substantiated by: (1) appearance of progeny virus in the cytoplasm of infected cells very early after inoculation; (2) some similar cytological modifications, such as proliferation of membranes and production of vesicles with RNA-like fibrils, with those elicited by Bromoviruses and Cucumoviruses known to multiply in the cytoplasm.

D. Alfalfa Mosaic Virus

AMV, the sole representative of this monotypic group, has been extensively studied in both natural (Gerola *et al.*, 1969a; Wilcoxson *et al.*, 1974, 1975) and experimentally infected hosts (Desjardin, 1966; de Zoeten and Gaard, 1969; Hull *et al.*, 1969, 1970; Hull and Plaskitt, 1970; Conti *et al.*, 1972; Diaz-Ruiz and Moreno, 1972; Bovey and Cazelles, 1978; Hatta and Francki, 1981b; M. Russo, unpublished studies). Altogether, the literature lists some 20 papers from different laboratories on the ultrastructure of AMV infections, which makes this virus one of the best known among those with tripartite genomes.

1. General Cytology of Infected Cells

Several authors (Gerola *et al.*, 1969a; Wilcoxson *et al.*, 1974; Bovey and Cazelles, 1978) have reported that in AMV-infected plants, cytological modifications of some consequence occur only in cells of organs showing symptoms. Symptomless tissues may contain high concentrations of virus particles but their fine structure is not visibly altered (Wilcoxson *et al.*, 1974).

Fragmentation of the ground cytoplasm appears to be one of the AMV-induced cytopathic effects, consequent to formation of many secondary vacuoles (Gerola *et al.*, 1969a; de Zoeten and Gaard, 1969; Hull, 1969; Wilcoxson *et al.*, 1974). Furthermore, an increased number of membrane-bound vesicles are sometimes present in the cytoplasm (Gerola *et al.*, 1969a) or on the tonoplast, protruding into the central vacuole (Rubio-Huertos, 1978; see also Fig. 22, inset a). These vesicles, whose origin has not been ascertained, often contain dots of electron-dense material, or fine fibrils resembling nucleic acid.

The effect of AMV on major organelles like chloroplasts and nuclei is erratic. Sometimes, chloroplasts suffer severe modifications of the lamellar system, whose development is impaired, thus resulting in highly abnormal structures (Gerola *et al.*, 1969a; Favali and Conti, 1970; Conti *et al.*, 1972; Bovey and Cazelles, 1978), but in other instances they appear undamaged. Likewise, except for a single report (Conti *et al.*, 1972), extensive degeneration of nuclei has not been observed, although these organelles may show deep invaginations of the boundary membrane (Fig. 23) or contain virus-related inclusions (Hull *et al.*, 1969, 1970; Conti *et al.*, 1972).

In the literature, no mention is made of specific modifications suffered by mitochondria, microbodies, and dictyosomes nor did these organelles appear to be affected in tobacco tissues recently investigated (M. Russo, unpublished observations; see also Fig. 22). Evidently, in necrosing cells, all constituents, including major organelles, undergo progressive degeneration and disruption (Wilcoxson *et al.*, 1974).

2. Appearance and Intracellular Distribution of Virus Particles

In a detailed study on the distribution of AMV in different organs of systemically infected alfalfa plants, Wilcoxson *et al.* (1975) detected virus particles in the epidermis, mesophyll, and vascular parenchyma and

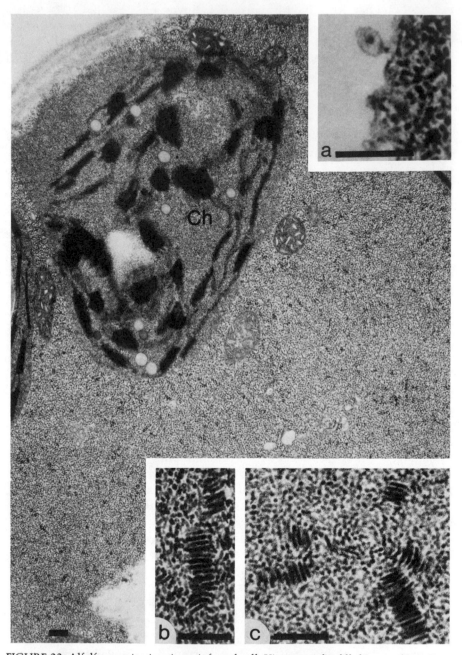

FIGURE 22. Alfalfa mosaic virus in an infected cell. Virus particles fill the cytoplasm. Inset (a) shows tonoplast-associated vesicles whereas insets (b) and (c) show different aggregation forms of virus particles. Ch, chloroplast. Bars = 200 nm.

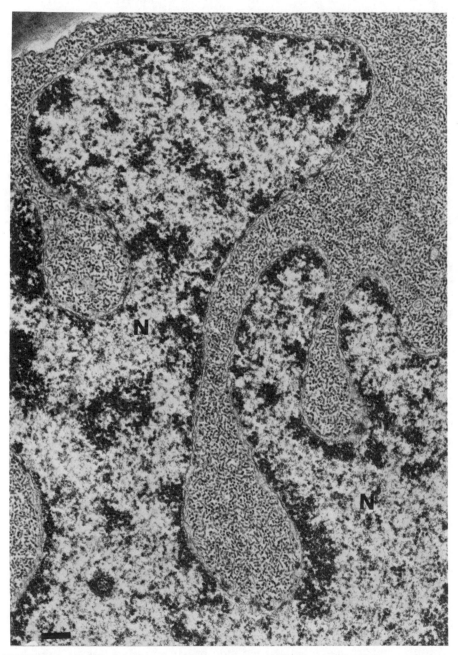

FIGURE 23. A deeply lobed nucleus (N) in a tobacco cell infected with alfalfa mosaic virus. The cytoplasm is packed with virus particles. Bar = 200 nm.

transfer cells of leaves, ovary wall, bud receptacle, anthers, embryonic cotyledons, as well as in pollen grains. Occasionally, virions were also observed in phloem elements (de Zoeten and Gaard, 1969). This is in line with the alleged invasiveness of AMV, which is known to spread systemically in the great majority of its natural and experimental hosts, in some of which it is also transmitted through seed (Hull, 1969; Frosheiser, 1974). Because of the particle pleomorphism shown also in thin sections, AMV particles are readily identified in infected cells where they accumulate in the cell sometimes in such huge amounts so as to completely fill the cytoplasm and also accumulate in the vacuoles (Figs. 22 and 23).

In thin sections, AMV particles appear as electron-dense solid bodies with round or bacilliform shapes with a diameter of 16–18 nm (Fig. 22). From their shape it is virtually impossible to identify with certainty the four morphological types of AMV particles. However, based on their relative length (about 40 and about 54 nm), de Zoeten and Gaard (1969) were able to recognize with confidence Tb and M particles in the cytoplasm of cowpea and tobacco cells, whereas other workers (Wilcoxson et al., 1974; Bovey and Cazelles, 1978) found alfalfa and grapevine cells to contain bacilliform particles up to 60 nm long, possibly representing longitudinally sectioned B particles.

No clear-cut association of virus particles with cell organelles has been reported in any of the ultrastructural studies on AMV-infected cells, except for a couple of records of particles in nuclei (Hull et al., 1970; Conti et al., 1972).

Whether virus assembly may take place in the nucleus is unknown. However, there are indications from actinomycin D treatments and autoradiographic studies that the nucleus is involved in viral RNA synthesis (Bassi et al., 1970). However, these authors reported that substantial amounts of viral RNA were also present in the cytoplasm.

3. Inclusion Bodies

Amorphous–granular or vacuolate and, more rarely, hexagonal crystalline inclusion bodies have been observed by several authors with the light microscope in epidermal and parenchymal cells of different host plants (Hull et al., 1969; Diaz-Ruiz and Moreno, 1972; Christie and Edwardson, 1977). The consensus is that these inclusions may correspond to the cytoplasmic or vacuolar aggregates of virus particles seen in thin sections (see review by Christie and Edwardson, 1977). Indeed, many AMV strains form intracellular aggregates of virus particles, which were grouped in four distinct types by Hull and co-workers (Hull, 1969; Hull et al., 1969, 1970): (1) short rafts of particles arranged in a hexagonal array; (2) long bands of particles aligned side by side, sometimes in a stacked layer configuration (Fig. 22 inset b); (3) aggregates made up of a series of whorllike structures (centers of aggregation) connected to one another by bands of particles to form a three-dimensional irregular network throughout the cell (Fig. 22, inset c); (4) complex aggregates made up predominantly of long particles and exhibiting a different morphology when seen in longitudinal (featherlike structures), transverse (clusters of interlocking rings), or tangential (series of alternating bands of longitudinally and cross-sectioned particles) section. Aggregates of the same kind, especially of type 3 and 4, were also seen within nuclei (Hull *et al.*, 1970; Conti *et al.*, 1972).

In most studies on the ultrastructure of AMV infections, cytoplasmic accumulations of particles were detected which could be reconciled with one or more of the above aggregation forms (de Zoeten and Gaard, 1969; Conti *et al.*, 1972; Diaz-Ruiz and Moreno, 1972; Wilcoxson *et al.*, 1974, 1975; Rubio-Huertos, 1978; M. Russo, unpublished observations). However, the type of aggregation was found to vary with the host (de Zoeten and Gaard, 1969), the plant organ examined (Wilcoxson *et al.*, 1975) or, within the same organ, with the area sampled (Wilcoxson *et al.*, 1974). On this account, the reliability of intracellular particle aggregates as diagnostic markers for strain identification (Hull and Plaskitt, 1970) becomes questionable; perhaps even when following the suggestion by Wilcoxson *et al.* (1975) that ultrastructural observations should be done on standard plant organs.

The ontogeny of viral aggregates has been tentatively explained in terms of: (1) ionic conditions present in the vacuole or restricted cytoplasmic areas, which, by reducing the net surface charge of the particles, would propitiate their aggregation in certain configurations (i.e., side by side) (Hull *et al.*, 1970); (2) loss of selective permeability of cell membrane and of compartmentalization of some basic proteins resulting from degeneration of virus-infected cells which would trigger aggregation of virus particles (de Zoeten and Gaard, 1969). Contrary to these views, Wilcoxson *et al.* (1974) maintain that aggregation of AMV does not depend on cell damage because with many of the AMV strains they investigated, less particle aggregation was seen as cell damage increased whereas viral aggregates were also present in apparently intact cells.

V. CONCLUDING REMARKS

The comparative studies of intracellular behavior of the Bromoviruses, Cucumoviruses, Ilarviruses, and alfalfa mosaic virus reveal that there are no common ultrastructural features specific enough to characterize these viruses at the "supergroup" or "family" level. If it is true that the increased production of cytoplasmic membranes is a remarkably consistent characteristic of cells infected with all viruses considered in the present review, it should be kept in mind that this is a common phenomenon also of many other virus groups. Various types of membranous structures derived either from proliferation of the ER or sloughing off of the nuclear envelope are known to develop in cells invaded by most plant viruses. In fact, these changes often constitute the earliest signs of infection. The membranes may accumulate to form cytoplasmic inclusions which, similarly to other membranous bodies originating from peripheral vesiculation of cell organelles (i.e., mitochondria, chloroplasts, microbodies), are likely sites of viral nucleic acid replication (see reviews by Martelli and Russo, 1977; Francki et al., 1985).

Membranes, especially those in vesicular form, have been implicated in the transport of viral RNA from cell to cell or, intracellularly, from the site of RNA synthesis to the site of particle assembly (de Zoeten, 1981; de Zoeten and Gaard, 1983). The latter may also be the case of the tonoplast-associated vesicles thought to contain double-stranded RF or RI forms of viral RNA which are abundant in Cucumovirus-infected cells and which are also present to a lesser extent in TSV-B and AMV infections. These vesicles occur in cells infected with a number of viruses belonging to quite distinct taxonomic groups (Francki et al., 1985).

Most viruses discussed in this volume induce the formation of cytoplasmic inclusions of two types; (1) those made up primarily of modified cell constituents, which are (or may be) relevant to virus multiplication, and (2) virus particle aggregates. Unfortunately, inclusions of the former type are so inconsistent in outward appearance and gross constitution that they are of little or no use for identification purposes at the group level, contrary to what often happens with other virus groups (Martelli and Russo, 1977, 1984; Edwardson and Christie, 1978; Francki et al., 1985).

Crystalline viral inclusions were suggested by Edwardson and Christie (1978) as being "a main characteristic of the Bromovirus group." However, in this and other taxonomic groups, intracellular virus crystals are too inconsistently found to constitute a reliable diagnostic character at the group level. Furthermore, Bromovirus crystals may be sufficiently labile so as to be disrupted by faulty fixation techniques (Langenberg, 1979). Crystalline viral aggregates, however, may not be totally useless for diagnostic purposes at the "species" level. For instance, CMV crystals have such a distinctive outward appearance that, when present, they constitute a powerful hint to virus identification. The snag is that intracellular crystallization is not a consistent and widespread phenomenon with CMV infection, except perhaps when it occurs in the cells together with other viruses like TMV (Honda and Matsui, 1968), potato virus Y (Russo and Martelli, 1973), or artichoke latent virus (M. Russo, unpublished observations). Mixed infections seem to accentuate the tendency of CMV particles to crystallize.

Noncrystalline particle aggregates may also be virus-specific, or nearly so. For instance, membrane-bound spheroidal aggregates of virions seem to be much more frequently found in cells infected with TAV than with any other member of the Cucumovirus group.

Despite the above examples, there is very little in the ultrastructural changes in cells infected with the viruses discussed here to be relied upon for identification. In fact, with these, more than with other plant viruses, there is a wide gap between knowledge of the structure and significance of intracellular abnormalities.

Conventional electron microscopy in the last couple of decades has played a primary role in unraveling the fine structure of virus-infected cells. Indeed, this task has not yet been totally accomplished for much ground still remains to be covered. Nevertheless, there is an increasing need for wider use of newly developed complementary biochemical and cytochemical techniques, without which deeper insight cannot be obtained of the meaning of ultrastructural modifications and their impact on the plant's physiology and morphology.

ACKNOWLEDGMENTS.   Grateful thanks are expressed to Drs. A. A. Brunt, R. W. Fulton, R. Hull, V. Lisa, F. Marani, and G. I. Mink for supplying cultures of some viruses used in this work; Drs. R. G. Milne and R. I. B. Francki for providing manuscripts prior to publication; Drs. K. S. Kim for supplying micrographs; and to Drs. M. A. Castellano and A. Di Franco for collaboration in the ultrastructural study of several viruses.

REFERENCES

Aharoni, N., and Lieberman, M., 1979, Ethylene as a regulator of senescence in tobacco leaf discs, *Plant Physiol*, **64:**801–804.

Balázs, E., Gáborjányi, R., Tóth, A., and Király, Z., 1969, Ethylene production in Xanthi tobacco after systemic and local virus infections, *Acta Phytopathol. Acad. Sci. Hung.* **4:**355–358.

Bancroft, J. B., 1971, Cowpea chlorotic mottle virus, *CMI/AAB Descriptions of Plant Viruses* No. 49.

Bancroft, J. B., Bracker, C. R., and Wagner, G. W., 1969, Structures derived from cowpea chlorotic mottle and brome mosaic virus protein, *Virology* **38:**324–335.

Bassi, M., Favali, M. A., Conti, G. G., and Betto, E., 1970, Uridine-^{3}H incorporation in leaf cells infected with lucerne mosaic virus: A quantitative electron microscopic autoradiographic study, *Phytopathol. Z.* **69:**247–255.

Bos, L., 1970, *Symptoms of Virus Diseases in Plants*, PUDOC, Wageningen.

Bovey, R., and Cazelles, O., 1978, Alfalfa mosaic virus on grapevine, in: *Proc. VI Int. Conf. ICVG, Cordova, Spain, 1976, Monogr. INIA* **18:**131–134.

Burg, S. P., 1962, The physiology of ethylene formation, *Annu. Rev. Plant Physiol.* **13:**265–302.

Burgess, J., Motoyoshi, F., and Fleming, E. N., 1974, Structural changes accompanying infection of tobacco protoplasts with two spherical viruses, *Planta* **117:**133–144.

Butler, R. D., and Simon, E. W., 1971, Ultrastructural aspects of senescence in plants, *Adv. Gerontol. Res.* **3:**73–129.

Christie, R. G., and Edwardson, J. R., 1977, Light and electron microscopy of plant virus inclusions, *Fla. Agric. Exp. Stn. Monogr.* No. 9.

Conti, G. G., Favali, M. A., and Vegetti, G., 1972, The behaviour of yellow spot mosaic virus, a strain of alfalfa mosaic virus , in different host plants, symptomatology and ultrastructural observations, *Riv. Pat. Veg.* **8**(Ser. IV):323–340.

De Laat, A. M. M., and Van Loon, L. C., 1983, The relationship between stimulated ethylene production and symptom expression in virus-infected tobacco leaves, *Physiol. Plant Pathol.* **22:**261–273.

Desjardins, P. R., 1966, Inclusion bodies in *Nicotiana tabacum* produced by two strains of alfalfa mosaic virus, *Phytopathology* **56**:875.

de Zoeten, G. A., 1981, Early events in plant virus infections, in: *Plant Diseases and Vectors* (K. Maramorosch and K. F. Harris, eds.), pp. 221–239, Academic Press, New York.

de Zoeten, G. A., and Gaard, G., 1969, Possibilities for inter- and intracellular translocation of some icosahedral plant viruses, *J. Cell Biol.* **40**:814–823.

de Zoeten, G. A., and Gaard, G., 1983, Mechanisms underlying systemic invasion of pea plants by pea enation mosaic virus, *Intervirology* **19**:85–94.

de Zoeten, G. A., and Schlegel, D. E., 1967a, Nuclear and cytoplasmic uridine-^{3}H incorporation in virus-infected plants, *Virology* **32**:416–427.

de Zoeten, G. A., and Schlegel, D. E., 1967b, Broad bean mottle virus in leaf tissue, *Virology* **31**:173–176.

de Zoeten, G. A., Gaard, G., and Diez, R. F., 1972, Nuclear vesiculation associated with pea enation mosaic virus-infected plant tissue, *Virology* **48**:638–647.

Diaz-Ruiz, J. R., and Moreno, R., 1972, Caracteristicas sintomatologicas, morfologicas y ultrastructurales de una raza del virus del mosaico de la alfalfa encontrada en España, *Microbiol. Esp.* **25**:1–16.

Di Franco, A., Martelli, G. P., and Russo, M., 1983, An ultrastructural study of olive ringspot virus in *Gomphrena globosa*, *J. Submicrosc. Cytol.* **15**:539–548.

Douine, L., Quiot, J. B., Marchoux, G., and Archange, P., 1979, Recensement des espèces vegetales sensibles au virus de la mosaïque du concombre (CMV): Etude bibliographique, *Ann. Phytopathol.* **11**:439–475.

Edwardson, J. R., and Christie, R. G., 1978, Use of virus-induced inclusions in classification and diagnosis, *Annu. Rev. Phytopathol.* **16**:31–55.

Edwardson, J. R., and Christie, R. G., 1979, Light microscopy of inclusions induced by viruses infecting pepper, *Fitopatol. Bras.* **4**:341–373.

Edwardson, J. R., and Purcifull, D. E., 1974, Relationship of *Datura quercina* and tobacco streak viruses, *Phytopathology* **64**:1322–1324.

Ehara, Y., 1979, Complex membranous organelles in the cells of the local lesion area on cowpea leaves infected with cucumber mosaic virus, *Ann. Phytopathol. Soc. Jpn.* **45**:17–24.

Ehara, Y., and Misawa, T., 1973, Studies on the infection of cucumber mosaic virus—Leaf growth and appearance of mosaic symptoms, *Shokubutsu Byogai Kenkyu* **8**:245–260.

Ehara, Y., and Misawa, T., 1975, Occurrence of abnormal chloroplasts in tobacco leaves infected systemically with the oridinary strain of cucumber mosaic virus, *Phytopathol. Z.* **84**:233–252.

Ehara, Y., Misawa, T., and Nagayama, H., 1976, Arrangement of cucumber mosaic virus particles in the virus aggregates, *Phytopathol. Z.* **87**:28–39.

Esau, K., 1948, Some anatomical aspects of plant virus disease problems, *Bot. Rev.* **14**:413–449.

Esau, K., 1967, Anatomy of plant virus infections, *Annu. Rev. Phytopathol.* **5**:45–76.

Favali, M. A., and Conti, G. G., 1970, Ultrastructural observations on the chloroplasts of basil plants either infected with different viruses or treated with 3-amino-1,2,4-triazole, *Protoplasma* **70**:153–166.

Favali, M. A., and Conti, G. G., 1971, The fine structure of healthy and virus-infected leaves following ultracentrifugation, *Cytobiologie* **3**:153–161.

Francki, R. I. B., Milne, R. G., and Hatta, T., 1985, *An Atlas of Plant Viruses*, Volume II, CRC Press, Boca Raton, Fla. (in press).

Frosheiser, F. I., 1974, Alfalfa mosaic virus transmission to seed through alfalfa gametes and longevity in alfalfa seed, *Phytopathology* **64**:102–105.

Fulton, R. W., 1972, Inheritance and recombination of strain-specific characters in tobacco streak virus, *Virology* **50**:810–820.

Fulton, R. W., 1981, Ilarviruses, in: *Handbook of Plant Virus Infections and Comparative Diagnosis* (E. Kurstak, ed.), pp. 377–413, Elsevier/North-Holland, Amsterdam.

Fulton, R. W., 1983, Ilarvirus group, *CMI/AAB Descriptions of Plant Viruses* No. 275.

Gerola, F. M., Bassi, M., and Belli, G., 1965, Some observations on the shape and localization of different viruses in experimentally infected plants, and on the fine structure of the host cells. II. *Nicotiana glutinosa* systemically infected with cucumber mosaic virus, strain Y, *Caryologia* **18**:567–597.

Gerola, F. M., Bassi, M., and Betto, E., 1969a, A submicroscopical study of leaves of alfalfa, basil, and tobacco experimentally infected with lucerne mosaic virus, *Protoplasma* **67**:307–318.

Gerola, F. M., Lombardo, G., and Catara, A., 1969b, Histological localization of citrus variegation virus (CVV) in *Phaseolus vulgaris*, *Protoplasma* **67**:319–326.

Gibbs, A. J., 1972, Broad bean mottle virus, *CMI/AAB Descriptions of Plant Viruses* No. 101.

Gibbs, A. J., and Harrison, R. D., 1976, *Plant Virology: The Principles*, Arnold, London.

Gibbs, A. J., and McDonald, P. W., 1974, RNAs labeled *in vitro* by polymerase from leaves infected with broadbean mottle virus, *Intervirology* **4**:45–51.

Hatta, T., and Francki, R. I. B., 1979, Enzyme cytochemical method for identification of cucumber mosaic virus particles in infected cells, *Virology* **93**:265–268.

Hatta, T., and Francki, R. I. B., 1981a, Cytopathic structures associated with tonoplasts of plant cells infected with cucumber mosaic and tomato aspermy viruses, *J. Gen. Virol.* **53**:343–346.

Hatta, T., and Francki, R. I. B., 1981b, Identification of small polyhedral virus particles in thin sections of plant cells by an enzyme cytochemical technique, *J. Ultrastruct. Res.* **74**:116–129.

Hills, G. J., and Plaskitt, A., 1968, A protein stain for the electron microsocpy of small isometric plant virus particles, *J. Ultrastruct. Res.* **25**:323–329.

Hollings, M., and Stone, O. M., 1971, Tomato aspermy virus, *CMI/AAB Descriptions of Plant Viruses* No. 79.

Honda, Y., and Matsui, C., 1968, Electron microscopy of intracellular modifications of tobacco by mixed infections with cucumber mosaic and tobacco mosaic viruses, *Phytopathology* **58**:1230–1235.

Honda, Y., and Matsui, C., 1974, Electron microscopy of cucumber mosaic virus-infected tobacco leaves showing mosaic symptoms, *Phytopathology* **64**:534–539.

Honda, Y., Matsui, C., Otsuki, Y., and Takebe, I., 1974, Ultrastructure of tobacco mesophyll protoplast inoculated with cucumber mosaic virus, *Phytopathology* **64**:30–34.

Hull, R., 1969, Alfalfa mosaic virus, *Adv. Virus Res.* **15**:365–433.

Hull, R., and Plaskitt, A., 1970, Electron microscopy on the behavior of two strains of alfalfa mosaic virus in mixed infections, *Virology* **42**:773–776.

Hull, R., Hills, G. J., and Plaskitt, A., 1969, Electron microscopy of *in vivo* aggregation forms of a strain of alfalfa mosaic virus, *J. Ultrastruct. Res.* **26**:465–479.

Hull, R., Hills, G. J., and Plaskitt, A., 1970, The *in vivo* behaviour of twenty-four strains of alfalfa mosaic virus, *Virology* **42**:753–772.

Jacquemond, M., and Lot, H., 1981, L'ARN satellite du virus de la mosaïque du concombre. I - Comparaison de l'aptitude à induire la nécrose de la tomate d'ARN satellites isolés de plusierus souches du virus, *Agronomie* **1**:927–932.

Kaper, J. M., and Waterworth, H. E., 1981, Cucumoviruses, in: *Handbook of Plant Virus Infections and Comparative Diagnosis*, (E. Kurstak, ed.), pp. 258–332, Elsevier/North-Holland, Amsterdam.

Kato, S., and Misawa, T., 1974, Studies on the infection and the multiplication of plant viruses. VII. The breakdown of chlorophyll in tobacco leaves systemically infected with cucumber mosaic virus, *Annu. Phytopathol. Soc. Jpn.* **40**:14–21.

Kim, K. S., 1977, An ultrastructural study of inclusions and disease development in plant cells infected by cowpea chlorotic mottle virus, *J. Gen. Virol.* **35**:535–543.

Kraev, V. G., Semernikova, L. I., Vutenko, S. I., and Parembskaya, N. V., 1975, Certain structural changes induced by peanut stunt virus in cells of host plants, *Mikrobiol. Zh. (Kiev)* **37**:261–264.

Lane, L. C., 1974, The Bromoviruses, *Adv. Virus Res.* **19**:151–220.

Lane, L. C., 1981, Bromoviruses, in: *Handbook of Plant Virus Infections and Comparative Diagnosis* (E. Kurstak, ed.), pp. 334–376, Elsevier/North-Holland, Amsterdam.

Langenberg, W. G., 1979, Chilling of tissue before glutaraldehyde fixation preserves fragile inclusions of several plant viruses, *J. Ultrastruct. Res.* **66:**120–131.

Lastra, J. R., and Schlegel, D. E., 1975, Viral protein synthesis in plants infected with broad-bean mottle virus, *Virology* **65:**16–26.

Lawson, R. H., and Hearon, S., 1970, Subcellular localization of chrysanthemum aspermy virus in tobacco and chrysanthemum leaf tissue, *Virology* **41:**30–37.

Lyons, H. R., and Allen, T. C., 1969, Electron microscopy of viruslike particles associated with necrotic fleck of *Lilium longiflorum*, *J. Ultrastruct. Res.* **27:**198–204.

Marco, S., and Levy, D., 1979, Involvement of ethylene in the development of cucumber mosaic virus-induced chlorotic lesions in cucumber cotyledons, *Physiol. Plant Pathol.* **14:**235–244.

Martelli, G. P., 1980, Ultrastructural aspects of possible defence reactions in virus-infected plant cells, *Microbiologica* **3:**369–391.

Martelli, G. P., and Quacquarelli, A., 1983, The present status of tomato and pepper viruses, *Acta Hortic.* **127:**39–64.

Martelli, G. P., and Russo, M., 1977, Plant virus inclusion bodies, *Adv. Virus Res.* **21:**175–266.

Martelli, G. P., and Russo, M., 1985, The use of thin sectioning for visualization and identification of plant viruses, in: *Methods in Virology*, Volume 8 (K. Maramorosch and H. Koprowski, eds.), Academic Press, New York (in press).

Matsui, C., and Yamaguchi, A., 1966, Some aspects of plant viruses *in situ*, *Adv. Virus Res.* **12:**127–174.

Matthews, E. R. F., 1982, Fourth report of the International Committee on Taxonomy of Viruses: Classification and nomenclature of viruses, *Intervirology* **17:**4–199.

Misawa, T., and Ehara, Y., 1966, Electron microscopic observation of host cells infected with cucumber mosaic virus, *Tohoku J. Agric. Res.* **16:**159–173.

Moline, H. E., and Ford, R. E., 1974, Bromegrass mosaic virus infection of seedling roots of *Zea mays*, *Triticum aestivum*, *Avena sativa* and *Hordeum vulgare*, *Physiol. Plant Pathol.* **4:**209–217.

Mollenhauer, H. H., and Morré, D. J., 1976, Cytochalasin B, but not colchicine, inhibits migration of secretory vesicles in roots tips of maize, *Protoplasma* **62:**44–52.

Motoyoshi, F., Bancroft, J. B., Watts, J. M., and Burgess, J., 1973, The infection of tobacco protoplasts with cowpea chlorotic mottle virus and its RNA, *J. Gen. Virol.* **20:**177–193.

Paliwal, Y. C., 1970, Electron microscopy of bromegrass mosaic virus in infected leaves, *J. Ultrastruct. Res.* **30:**491–502.

Paliwal, Y. C., 1972, Brome mosaic virus infection in wheat curl mite *Aceria tulipae*, a non vector of the virus, *J. Invertebr. Pathol.* **20:**288–302.

Porter, C. A., 1954, Histological and cytological changes induced in plants by cucumber mosaic virus (*Marmor cucumeris* H.), *Contrib. Boyce Thompson Inst.* **17:**453–471.

Quiot, J. B., 1981, Ecology of cucumber mosaic virus in the Rhone valley of France, *Acta Hortic.* **88:**9–21.

Rubio, M., and Van Slogteren, D. H. M., 1956, Light and electron microscopy of X-bodies associated with broad-bean mottle virus and *Phaseolus* virus 2, *Phytopathology* **46:**401–402.

Rubio-Huertos, M., 1978, *Atlas on Ultrastructure of Plant Tissues Infected with Viruses*, Consejo Superior de Investigaciones Cientificas, Madrid.

Russo, M., and Martelli, G. P., 1973, The fine structure of *Nicotiana glutinosa* L. cells infected with two viruses, *Phytopathol. Mediterr.* **12:**54–60.

Russo, M., and Martelli, G. P., 1982, Ultrastructure of turnip crinkle and saguaro cactus virus-infected tissues, *Virology* **118:**109–116.

Russo, M., Castellano, M. A., and Martelli, G. P., 1982, The ultrastructure of broad bean stain and broad bean true mosaic virus infections, *J. Submicrosc. Cytol.* **14:**149–160.

Savino, V., and Gallitelli, D., 1983, Isolation of cucumber mosaic virus from olive in Italy, *Phytopathol. Mediterr.* **22:**76–77.

Skene, K. G. M., 1972, Cytokinin-like properties of the systemic fungicide benomyl, *J. Hortic. Sci.* **47:**179–182.

Tolin, S. A., 1977, Cucumovirus (cucumber mosaic virus) group, in: *The Atlas of Insect and Plant Viruses* (K. Maramorosch, ed.), pp. 303–309, Academic Press, New York.

Tomlinson, J. A., and Webb, M. J. W., 1978, Ultrastructural changes in chloroplasts of lettuce infected with beet western yellows virus, *Physiol. Plant Pathol.* **12:**13–18.

Tomlinson, J. A., Faithfull, E. M., and Ward, C. M., 1976, Chemical suppression of the symptoms of two virus diseases, *Ann. Appl. Biol.*, **84:**31–41.

Van Regenmortel, M. H. V., and Pinck, L., 1981, Alfalfa mosaic virus, in: *Handbook of Plant Virus Infections and Comparative Diagnosis* (E. Kurstak, ed.), pp. 415–421, Elsevier/North-Holland, Amsterdam.

Weintraub, M., and Schroeder, B., 1979, Cytochrome oxidase activity in hypertrophied mitochondria of virus-infected leaf cells, *Phytomorphology* **29:**273–285.

Wilcoxson, R. D., Frosheiser, F. I., and Johnson, L. E. B., 1974, The appearance of eight strains of alfalfa mosaic virus in alfalfa leaves and their effect on the ultrastructure of infected cells, *Can. J. Bot.* **52:**979–985.

Wilcoxson, R. D., Johnson, L. E. B., and Frosheiser, F. I., 1975, Variation in the aggregation forms of alfalfa mosaic virus strains in different alfalfa organs, *Phytopathology* **65:**1249–1254.

CHAPTER 7

Serology and Immunochemistry

E. P. RYBICKI AND M. B. VON WECHMAR

I. INTRODUCTION

The Bromo-, Cucumo-, Ilar-, and alfalfa mosaic virus groups share a number of properties in addition to their tripartite ssRNA genomes. This has prompted the recent proposal that they be grouped in one family for which the name "Tricornaviridae" was suggested (van Vloten-Doting et al., 1981). An important similarity between the viruses that directly affects serological studies on them, is their lability at neutral pH. Viruses of all four groups are sensitive to relatively low concentrations of neutral salts and anionic detergents at neutral pH (see Johnson and Argos, this volume; Lane, 1981; Kaper and Waterworth, 1981; Fulton, 1982). Although no serological relationships between viruses from the different groups have yet been demonstrated, serological studies on each of the groups are likely to be complicated by the same problems. These would include low antivirus serum titers, the presence in antisera of a high proportion of antibodies to dissociated capsid subunits, and virion dissociation in the course of serological testing. It seems obvious that a thorough understanding of the physical properties of each of the viruses is necessary in order to enable comprehensive and reproducible serological studies on them. Equally important is a thorough knowledge of the advantages and limitations of the various serological techniques commonly used in the study of viruses.

Recent reviews on the "Tricornaviridae" have dealt at some length with the serology of the viruses: Lane (1981) has reviewed the Bromoviruses, Kaper and Waterworth (1981) and Francki et al. (1984) the Cuc-

E. P. RYBICKI AND M. B. VON WECHMAR • Microbiology Department, University of Cape Town, Rondebosch 7700, South Africa.

umoviruses, Fulton (1981) and Francki *et al.* (1984) the Ilarviruses; and van
Regenmortel (1982) has reviewed plant virus serology in general. Thus,
we shall concentrate more on updating general conclusions reached elsewhere, rather than preparing an exhaustive review of all the literature.
Because of the considerable impact that the advent of sophisticated en-
zyme- and other immunoassay techniques has had on plant virology in
recent years, we shall attempt to include suggestions wherever relevant,
on the most appropriate use of the techniques in the comparative serology
of the viruses.

II. IMMUNOLOGICAL TECHNIQUES

The following section consists of a review of various techniques of
antiserum preparation and antigen stabilization, and of the serological
techniques most often used to study plant virus serology. Our intention
is to propose the use of "standard methods" for studying the serology of
the "Tricornaviridae," both as guidelines for new workers in the field,
and to help make direct comparisons of data from different laboratories.

A. Problems with Immunization

There is an extensive literature on plant virus antigen preparation,
on immunization schedules, and on the preparation of virus-specific antibodies. Much of this is not applicable to the viruses discussed in this
chapter, as they are usually rather labile under physiological conditions,
and are in some cases notoriously weak immunogens. The protocols detailed below were largely evolved in our laboratory for use with Bromo-
and Cucumoviruses, and have been in use for several years.

1. Immunization Schedules

Antisera to plant viruses are most commonly raised in rabbits, both
because of their suitability as laboratory animals and because relatively
large volumes of serum may be obtained. For example, we have collected
up to 70 bleedings of 30 ml each of high-titer antisera from single rabbits,
over a 3-year period.

Chickens are also a convenient source of antibodies to plant viruses,
from egg yolks rather than from serum (Polson *et al.*, 1980a,b). Chickens
appear to require fewer booster injections than rabbits to maintain a high
antibody titer. Large amounts of antibodies may be prepared as mature
hens can lay almost every day for up to 2 years.

Different laboratories favor different immunization protocols for the
production of high-titered antisera. An immunization schedule developed
in our laboratory for tobacco mosaic virus (TMV) strains (van Regenmortel and von Wechmar, 1970) was found to work equally well for a

variety of filamentous, rodlike, and spherical viruses including CMV, AMV, and the Bromoviruses. The schedule entails three "priming" injections into thigh muscles of chickens or rabbits, spaced at weekly intervals, of 1 ml of a 1:1 (v/v) mixture of Freund's incomplete adjuvant and 0.5–3 mg/ml virus preparation. Booster injections are given 6 weeks later, and at 9- to 12-week intervals for the lifetime of the animal. Rabbits are bled from marginal ear veins at weekly intervals, starting 3 weeks after the first priming injection. Between 20 and 40 ml of blood is collected on each occasion. The serum fraction is collected and stored at −20°C. Bleeding schedules are interrupted at 3-month intervals for 3 weeks, to rest the animals. No booster injections are given during this time. Chickens are primed when 20 weeks old, and eggs collected from 10 days after the first primer until the hens molt. Eggs are stored at 4°C, and processed as batches of 30 pooled yolks to produce purified IgY antibody fraction as described by Polson *et al.* (1980a). Yield of IgY is approximately 300 mg, for a 30-ml yolk (Polson *et al*, 1980b). IgY preparations concentrated 5× from native yolks had precipitin reaction titers against Bromo-, Tymo-, and Tobamoviruses equivalent to those of unconcentrated homologous rabbit antisera.

Antisera specific for capsid protein subunits—formaldehyde-treated or merely dissociated—are best obtained from rabbits, as chickens do not mount a strong immune response to low-molecular-weight antigens (Polson *et al.*, 1980b; von Wechmar, unpublished results). Intravenous rather than intramuscular injections are also better for priming rabbits against small antigens. Booster injections are normally intramuscular, and should be given weekly because of the rapid falloff in antiserum titer after the secondary response (M. B. von Wechmar, unpublished results). Intramuscular priming gives a slower response, but this reaches the same levels eventually (M. B. von Wechmar, unpublished results). Assay of antiprotein titers is a tricky procedure in precipitin tests; it is perhaps better and more sensitively done by indirect ELISA (Rybicki and von Wechmar, 1981). Rabbit antisera to intact viruses obtained after a prolonged immunization schedule often contain a high proportion of antibodies reacting with dissociated capsid protein (von Wechmar and Rybicki, unpublished results). For this reason it may be sufficient merely to immunize rabbits with unstabilized virus preparations in order to obtain antisubunit sera.

2. Virus Stabilization

Viruses from all four groups have been described as relatively poor immunogens. This is presumably due to breakdown of virus particles in the bloodstream of immunized animals (for reviews see Lane, 1981; Kaper and Waterworth, 1981). Serological reactions of Bromo- and Cucumoviruses in particular are complicated by instability of the particles in certain buffers (Matthews, 1982). Consequently, attempts have been made to stabilize virus particles by chemical cross-linking in order to increase serum titers to the intact capsids. For instance, BMV has been stabilized with dimethyl adipimidate (Bancroft and Smith, 1975), glutaraldehyde (GA) (Richter et al., 1973; Rybicki, 1979), and formaldehyde (FA) (von Wechmar, 1967; von Wechmar and van Regenmortel, 1968; Richter et al., 1972; Rybicki and von Wechmar, 1981). CMV has been successfully stabilized with FA (Francki and Habili, 1972), although some groups report no need for stabilization of some strains, and FA appears to have no stabilizing effect on CMV-Y (Kaper and Waterworth, 1981). In our laboratory, only two out of six CMV isolates remained unstable after overnight dialysis against 2% FA (P. Lupuwana, personal communication). Such chemical treatment has been supposed not to affect the antigenic reactivity of particles (van Regenmortel, 1982); however, it is known that the serological relationships among the Bromoviruses are less easily demonstrated after FA treatment (Rybicki and von Wechmar, 1981), and that formalinized BMV is serologically distinguishable from the native virus particle (Rybicki, 1979; Rybicki and Coyne, 1983; see also Section IV.A). Similar observations have been made for Cucumovirus isolates (Musil and Richter, 1983).

Although antigenic alterations caused by chemical stabilization of particles may complicate investigations of the comparative serology of the virus groups, if high-titer capsid-specific antisera are required for labile viruses such as CMV or AMV, then such stabilization appears to be the only answer. FA is probably the simplest agent to use, and perhaps the most effective in preservation of essentially native virus structure. It is routinely used at 0.2–2% (w/v) (von Wechmar and van Regenmortel, 1968; Francki and Habili, 1972; Rybicki and von Wechmar, 1981; Devergne et al., 1981). In our laboratory, viruses are dialyzed for 12 hr at 4°C against 2% FA. This concentration is reduced to 0.2% by subsequent dialysis in order to prevent adverse effects on experimental animals on injection (von Wechmar and van Regenmortel, 1968). With BMV, GA is effective at concentrations as low as 0.08% (w/v), but causes pronounced interparticle cross-linking above this concentration, especially at pH 7.0 or higher (Richter et al., 1973; Rybicki, 1979). Preparations stabilized with 0.2% GA at pH 6.0 were stable in gel precipitin reactions in 2.0 N NaCl—containing agar at pH 7.5 (Rybicki, 1979), despite containing largely dimerized and trimerized virions. It is not known whether GA affects particle antigenicity in an analogous manner to FA.

The procedure followed in Devergne's laboratory—treating Cucumoviruses with 0.4% FA before injection (Devergne et al., 1981)—appears to be a good general rule for any of the "Tricornaviridae." In conjunction with an immunization schedule as outlined earlier (II.A.1), such stabilization should result in relatively high-titered antisera for most of the viruses. This would help ensure the reproducibility of precipitin-based serological comparisons done in different laboratories.

3. Immunoglobulin Preparation

The normal practice in producing virus-specific antisera is to inject an animal with highly purified virus preparation. However, increasingly stringent criteria of purity have shown that even highly purified virus is often contaminated with host-derived antigens (Rybicki and von Wechmar, 1982; O'Donnell *et al.*, 1982; E. P. Rybicki, unpublished results). Some of these contaminants are pigment derivatives of M_r about 10,000 which are highly immunogenic and often tightly bound to the virus particles (M. B. von Wechmar and E. P. Rybicki, unpublished results). Consequently, production of virus-specific antibodies entails both the laborious procedure of virus purification, and the removal of host-reactive antibodies from the resulting antiserum. We have found it easier to immunize animals with concentrated semipurified virus antigen preparations, and to exhaustively host-absorb the resulting antisera, than to attempt to extensively purify the virus. Our absorption procedure utilizes ultracentrifuge-concentrated whole plant extracts. These are made by homogenizing plants 1:1 (w/v) with 0.05 M phosphate buffer, pH 7.0, clarifying the extracts by brief low-speed centrifugation, and centrifuging the supernatant at 300,000g for 3 hr at 4°C. Pellets are resuspended in small volumes of antiserum which are incubated at 37°C for 1 hr, or 4°C for 12 hr, clarified by centrifugation, absorbed a second time, and centrifuged at high speed to remove small immune complexes. This absorbed serum can be used for precipitin assays, or the IgG may be isolated as described by Clark and Adams (1977).

An alternative, easier procedure for absorption of large volumes of pooled antisera is to mix antiserum 2:1 (v:v) with freshly homogenized plant material [1:1 (w/v) in 0.05 M PO_4 pH 7.0), incubate the mixture at 37°C for 30 min, clarify by centrifugation, and repeat the absorption and clarification and precipitate the immunoglobulin fraction by mixing 1:1 (v/v) with 4 M $(NH_4)_2SO_4$. This precipitation is repeated, and the ammonium sulfate dialyzed out of the final, concentrated product. Pigments are easily removed by Whatman DE-52 anion-exchange chromatography. Antibodies prepared in this way are routinely used in our laboratory for enzyme conjugation for ELISA, and do not react detectably with plant sap even in indirect ELISA (M. B. von Wechmar, Kaufmann, and E. P. Rybicki, unpublished results). O'Donnell *et al.* (1982) have described a simple procedure using plant sap extracts covalently linked to CNBr-activated Sepharose, which yields antibody preparations sufficiently pure to be used in their electroblot radioimmunoassay technique (see Section II.B.3).

We feel that undue attention has hitherto been given to purification of antigens prior to immunization: more critical requirements for purity necessitate evermore stringent purification schedules, which not only involve much labor and much loss of labile virions, but are also costly. This effort is often to no avail as the antigens may still be contaminated with host components. Our procedures entail concentration of viruses by centrifugation and/or polyethylene glycol (PEG) followed by relatively cheap and easy absorption procedures, for production of pure antibody.

4. Monoclonal Antibodies

Several laboratories are now making use of hybridoma-derived mouse monoclonal antibodies in serological assays with plant viruses. Antibodies specific for Ilarviruses have been prepared (Halk *et al.*, 1982; Halk and Franke, 1983; Hsu *et al.*, 1984) with a view to aiding serotype identification (see Section III.B.3). At first sight, monoclonal antibodies appear to be the plant virologist's dream come true: antibodies of rigidly defined specificity, available in relatively large amounts, which require neither pure virus for their production nor any absorption procedures prior to use. However, Al-Moudallal *et al.* (1982) have pointed out that monoclonal antibodies obtained after antigenic stimulation by a specific virus strain may in fact react better with a related strain, may not react at all with closely related strains, or may be unable to distinguish between strains that are distantly related by all other criteria. These phenomena are due to the peculiar specificity of monoclonals: a given preparation will react well with a particular, short sequence of amino acid residues or a particular conformation; any change in sequence or conformation could result in reduced affinity of binding or no binding at all. Consequently, recognition of viruses related to the one which provided the immunogenic stimulation for the production of a given monoclonal antibody depends on whether or not any coat protein sequence differences are reflected either in the sequence or in the structure of the particular epitope. The lesson from the studies with Tobamoviruses (Al-Moudallal *et al.*, 1982) is that gross serological differences or similarities as defined by polyclonal antisera are more likely to be usable for taxonomic purposes, than are the overprecise comparisons made using monoclonals. Modern methods for the purification of monospecific antibodies from polyclonal sera (Olmsted, 1981; see also Section II.B.2) have also lessened the need for monoclonals.

In conclusion, it is obvious that monoclonal antibodies have an important place in plant virus serology; whether they are of any more use than polyclonal antisera in comparative studies remains to be seen. We feel their use will be in the detection of specific serotypes of viruses in diagnostic tests, or in the detailed immunochemical analysis of capsids and proteins, rather than in the elucidation of the comparative serology of the groups.

B. Serological Techniques

Viruses in the four groups have been subjected to study by a large number of serological techniques over the years, though most often by the double-diffusion gel precipitin technique of Ouchterlony (1962). We will briefly review the applicability of the most commonly used modern immune techniques to the comparative serology of the "Tricornaviridae," so that perhaps some new comparisons can be made in various centers in the light of results obtained by different methods, or using different viruses.

1. Precipitin Techniques

The Ouchterlony gel double-diffusion technique has been used extensively to detect and determine differences in strains of CMV (Devergne and Cardin, 1973; Musil and Richter, 1983), but was found to be inadequate for relationship studies among the Bromoviruses due to insufficient sensitivity (see Section II.B.2). The advent of the more sensitive amplified immunoassays should result in less sensitive methods like the ring test, tube precipitin test, and microprecipitin tests being reserved for crude diagnostic work only.

The two factors which most influence the success of gel double-diffusion precipitin reactions are the antigen/antibody ratio, and the stability of the antigen before and during the test. For example, BMV produces a single precipitin band in 0.7% agar buffered with 0.05 M sodium acetate/0.075 M phosphate pH 6.0, but above pH 6.0 multiple bands appear (von Wechmar, 1967). This is due to swelling and partial dissociation of the capsids (Incardona and Kaesberg, 1964), and the subsequent exposure of new antigenic sites recognized by a different subset of antibodies (von Wechmar and van Regenmortel, 1968). Cucumovirus stability in agar gel is also greatly influenced by the type, concentration, and pH of the gel buffer. This in turn determines what type of immunoprecipitate forms in the gel (Kaper and Waterworth, 1981). The endpoint sensitivity of the gel double-diffusion tests is about 8 µg/ml, for intact particles of both CMV and BMV (Devergne, 1975; von Wechmar, 1967; M. B. von Wechmar, unpublished results). Results in our laboratory indicate that the antibody/antigen ratio giving precipitates with BMV at low pH (4.0) is far wider than for CMV at pH 7 in phosphate buffer, possibly because of the greater instability of CMV (M. B. von Wechmar, unpublished results). Sensitivity with dissociated capsid proteins is in the range 10–15 µg/ml; however, the range of antibody/antigen ratios that give a visible precipitate is far narrower than for an unstable or partly degraded virion. This may result in false-negative results. In addition, virus proteins usually require great care in preparation, and very specific buffers in order to minimize aggregation. This has perhaps adversely affected serological work with them (von Wechmar, 1967; Rybicki, 1979).

Given that different strains of the same virus have their own peculiar buffer requirements for optimum stability, no one standard gel buffer formulation will be ideal for direct comparisons between virus strains, let alone intra- and intergroup comparisons. For example, Ilarviruses are

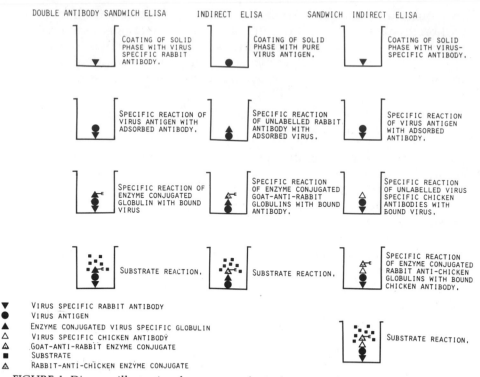

FIGURE 1. Diagram illustrating three commonly used ELISA techniques for plant viruses.

usually stabilized by 30 mM EDTA at pH 6–8 in ammonium phosphate, but can be labile in potassium phosphate (Fulton, 1982). CMV precipitates in saline and at low pH whereas BMV is stable at pH 4–6 in 0.01–1.5 M NaCl, but salt-sensitive above pH 7.0 (Kaper and Waterworth, 1981; Lane, 1981).

2. Enzyme Immunoassay (EIA) Techiques

There are basically three enzyme-linked immunosorbent assay (ELISA) techniques that have become established in plant virology. These are the popular double-antibody "sandwich" (DAS-) ELISA; the simple indirect ELISA (Rybicki and von Wechmar, 1981; Koenig, 1981); and compound or sandwich indirect ELISAs (Bar-Joseph and Malkinson, 1980; van Regenmortel and Burckard, 1980; Barbara and Clark, 1982) (see Fig. 1). DAS-ELISA has been proved to be an extremely sensitive means of virus detection, but is also very strain-specific. This limits its usefulness for detection of related strains of viruses, but makes it the method of choice for demonstrating subtle serological differences between virus isolates (Clark and Adams, 1977; Koenig, 1978; Rochow and Carmichael, 1979; Rybicki and von Wechmar, 1981). The specificity of the technique is largely due to the conjugated antibody, whose affinity is attenuated by chemical cross-linking to the enzyme (Koenig, 1978; Ford *et al.*, 1978). Rao *et al.* (1982) have shown that heterologous Cucumoviruses may be detected more efficiently using a "heterologous" coating antibody and a "homologous" conjugate, which would tend to affirm this. However, we have shown that "direct" ELISA—where antigen is adsorbed directly to a surface—is less specific than DAS-ELISA, even when the same antivirus conjugate is used in both tests (Rybicki and von Wechmar, 1981).

The simple indirect ELISA is extremely useful for the detection of distant serological relationships which may not be demonstrable by any precipitin test. It has to date only found limited application in plant virus serology (Rybicki and von Wechmar, 1981; Koenig, 1981; Lommel *et al.*, 1982), but few objections exist to its being used far more widely. One disadvantage is that relatively pure virus or virus protein must be used for coating; another is that not all proteins adsorb equally well to microtiter trays (E. P. Rybicki, unpublished results). Most antigens, however, do adsorb well, and may be adsorbed in buffers of a wide variety of ionic strength and pH (E. P. Rybicki, unpublished results).

Sandwich or compound indirect techniques rely on trapping virus with adsorbed antibody or F(ab')$_2$ fragment, and subsequent probing with an unlabeled virus-specific antibody. This is detected with enzyme-labeled antiglobulin antibody or staphylococcal protein A (van Regenmortel and Burckard, 1980; Barbara and Clark, 1982). A variant which utilizes adsorbed Clq complement component to trap immune complexes (Torrance and Jones, 1981) has not become popular. The original sandwich indirect technique utilizes virus-specific antibodies from two animal species: this could be difficult to apply if animal house facilities are limited. The F(ab')$_2$ technique appears more generally useful, in that only one rabbit antiserum need be used for each virus to provide both coating F(ab')$_2$ and intact probing antibody molecules.

Sandwich indirect techniques are similar to DAS-ELISA in that impure virus or sap samples may be tested; they share a potential disadvantage in that both techniques rely on two separate antigen–antibody reactions (two-site immunoassay). This is known to increase the specificity of antigen recognition (Koenig, 1978; Rybicki and von Wechmar, 1981). An advantage of "compound" over "simple" indirect ELISA is that whereas different antibodies adsorb relatively uniformly to microtiter trays, different viruses and proteins differ in adsorption ability. A disadvantage of sandwich indirect ELISA using chicken IgY for coating, rabbit serum for probing, and goat anti-rabbit IgG for detection, is that it may not be reliably used for demonstration of Bromovirus serological relationships. If BMV antibodies are used for coating, BBMV is not detectable even if BBMV-specific rabbit serum is used for probing. Increasing the coating antibody concentration may result in weak detection of BBMV, but also leads to unacceptably high "background" nonspecific color development (E. P. Rybicki, unpublished results).

An EIA technique that has recently been applied to plant virus serology is enzyme-assisted immunoelectroblotting (IEB) (Towbin *et al.*, 1979; Rybicki and von Wechmar, 1982). It entails SDS–PAGE of virus or sap preparations, electrophoretic transfer of the resolved polypeptides to a nitrocellulose membrane, and probing of the membrane with antivirus serum. Binding of antibodies to specific polypeptides is detected by a peroxidase-labeled antiglobulin preparation. The technique allows characterization of antigens both by their molecular weights and by their antigenic reactivity. This makes it a useful analytical as well as diagnostic method. In addition, because antigen detection is by an indirect enzyme-immune reaction, distant serological relationships can be detected (Rybicki and von Wechmar, 1982). A possible advantage of the technique in comparative serological investigations is that the antigens are the dissociated and partially denatured coat protein subunits, which may be expected to be more antigenically similar than the intact virus particles (van Regenmortel, 1982). The technique has been used successfully with the Bromoviruses (Rybicki and von Wechmar, 1982) and with South African AMV (G. Pietersen, personal communication), and South African and Israeli CMV isolates (P. Lupuwana, personal communication).

Contrary to what may be expected, it does not appear to be necessary to use antisera to dissociated or SDS-treated capsid protein in order to detect polypeptides bound to nitrocellulose or other solid support (Rybicki and von Wechmar, 1982; Shukla *et al.*, 1983). As with indirect ELISA, antisera to intact viruses are found to bind strongly to dissociated subunit proteins. It appears that the sensitivity of enzyme-immune techniques makes the use of even low-titer antisera quite feasible. Another use of electroblotting lies in the preparation of monospecific antibodies to virus capsid proteins. SDS-PAGE of a virus preparation followed by electroblotting onto a suitable medium allows the excision of particular immobilized polypeptides, and their use as micropreparative immunosorbents (Olmsted, 1981; Rybicki, 1984). We have prepared monospecific antibodies to BMV from a mixture of anti-BMV and anti-"Fraction 1" barley protein sera in one cycle of adsorption and elution (Rybicki, 1984).

3. Radioimmunoassay (RIA) Techniques

Various serological tests for the detection of plant viruses by radioimmunosorbent or solid-phase assays have been described (reviewed by van Regenmortel, 1982). The techniques most applicable to plant virus serology appear to be those of Ball (1973), Ghabrial and Shepherd (1980), and O'Donnell *et al.* (1982). The first-mentioned paper describes a solid-phase, ^{125}I-labeled antigen competition assay. The assay system of Ghabrial and Shepherd (1980) is a double-antibody "sandwich" assay like DAS-ELISA except that the detecting antibody is labeled with ^{125}I. The method appears to be a little more sensitive than DAS-ELISA for quantitation of low antigen concentrations, but undoubtedly suffers the same specificity constraints. The swapping of enzyme for a radioisotope as a labeling agent could be done for any of the EIA techniques described earlier, to give assays with the same advantages and limitations, but with the added disadvantage of having to handle high-activity radioisotopes.

The RIA of O'Donnell *et al.* (1982) is an electroblotting technique, utilizing diazophenylthioether paper for covalent binding of the blotted proteins, and ^{125}I-labeled protein A for autoradiographic and/or scintillation counter detection, of specific virus–antibody complexes. The method is not much more sensitive than the enzyme-assisted technique described earlier (see Section II.B.2); however, it is more amenable to quantification, and blots can be reprobed many times with different antisera (O'Donnell *et al.*, 1982; Shukla *et al.*, 1983). The ease of detection of distant serological relationships, and the detection of small amounts of virus in sap, make the RIA electroblot technique as attractive as the enzyme-immune variant for the study of comparative virus serology. However, its use can only be recommended in laboratories where handling of radioisotopes is routine.

4. Immune Electron Microscopic (IEM) Techniques

IEM has been used for the detection of Cucumoviruses in plant sap (Kaper and Waterworth, 1981; Cohen *et al.*, 1982), and of BMV eluted from stem rust spores and contaminated wheat seeds (Erasmus, 1982; Erasmus *et al.*, 1983; von Wechmar *et al.*, 1984). The technique has been used mainly for diagnostic detection, rather than for differentiation of strains or for demonstrating serological relationships. It is possible that EM techniques are not particularly well suited to the "Tricornaviridae" because of the generally low stability of the viruses.

The simplest of the serological EM techniques is "clumping," or visualization of Ab–Ag precipitates after mixing of antigen with antibody (Milne and Luisoni, 1975). The technique has been used in various contexts, such as in virus detection, establishing strain relationships, and identifying particle surface features with specific antigenicity (van Regenmortel, 1981). One drawback of the technique is that an optimal antiserum/virus ratio needs to be determined for each Ab–Ag combination, just as in ordinary precipitin tests. Another is that antibody–coat protein aggregates may not be distinguishable from background debris.

The most popular IEM technique is the "trapping" of virus particles onto antibody-precoated grids (Derrick, 1973; Milne and Luisoni, 1977). This procedure is now termed "immunosorbent electron microscopy" (ISEM) (Roberts *et al.*, 1982). Although apparently simple and highly sensitive, a specific testing procedure is necessary for each virus–host–antiserum combination, before optimum "trapping" conditions can be defined (Cohen *et al.*, 1982). The precoating of grids with staphylococcal protein A (Shukla and Gough, 1979; Gough and Shukla, 1980) may or may not increase test efficiency, depending on grid preparation and other factors (Milne, 1980; van Balen, 1982). A major disadvantage inherent in ISEM is that small subunit proteins or degraded capsids are not detectable, due to both their size and their lack of a characteristic shape: this has caused complications in our research on CMV (M. B. von Wechmar and P. Lupuwana, unpublished results). Another drawback is that excess free coat protein in sap or purified preparations may inhibit the trapping and/ or "decoration" of virus particles on an EM grid (R. G. Milne, personal communication). That ISEM may not be suitable for investigating the serology of the Cucumoviruses was shown by Rao et al. (1982), who found that a combination of "trapping" and "decoration" was rather unsatisfactory for the demonstration of serological relatedness between CMV and TAV strains.

Two IEM techniques that have been used for immunohistological studies may find use as ISEM "decoration" techniques. These are gold-labeled antibody decoration (GLAD) (Pares and Whitecross, 1982) and immunoferritin decoration (Singer and Schick, 1961; Shepard et al., 1974b). If heavy-metal-labeled antibodies are used for either direct or in-direct "probing" of virus adsorbed onto EM grids, it should be possible to detect both capsid fragments and intact particles, unlike normal decoration/trapping experiments.

Given that ISEM requires both an electron microscope and a skilled operator, and that ISEM and IEM techniques in general are neither particularly reproducible nor particularly easy, it seems that EM serology on most plant viruses will remain a hobby of the more fortunate and better-equipped laboratories. It is always pleasant to be able to see the particles one is working with; however, enzyme immunoassays and the newer electrophoretic immune techniques appear more reliable, easier to apply, and require far less expensive facilities.

III. SEROLOGICAL RELATIONSHIPS

A. Relationships between Viruses

1. Bromoviruses

The published data on the comparative serology of the Bromoviruses BMV, CCMV, and BBMV provide an object lesson on how choice of serological technique may influence the apparent relatedness of viruses in a taxonomic group. For instance, Scott and Slack (1971) used Ouchterlony and intra-gel absorption Ouchterlony tests (van Regenmortel, 1967) to investigate serological relationships between both the virus particles, and their purified capsid subunit proteins. They showed that BMV and CCMV particles and their respective coat proteins were distantly related in tests done at pH 6; however, no relationship could be demonstrated between either BBMV particles or BBMV protein, and either of the other two viruses or their proteins.

Later studies in this laboratory (Rybicki, 1979; Rybicki and von Wechmar, 1981) made use of the same techniques as well as "rocket" immunoelectrophoresis, direct-, indirect-, and DAS-ELISA tests to investigate relationships between intact and FA-treated virus particles, and FA-treated coat protein. An important departure from earlier work on the viruses was the duplication of tests at pH 6 and at pH 7 or above, and the preparation of the viruses at pH 5 or below: BMV, CCMV, and BBMV all swell above neutral pH (see Chapter 2), and the effect of this on the serology of the viruses had not previously been investigated. Ouchterlony gel precipitin tests showed that BMV and CCMV proteins were related in tests at pH 6 and pH 7, using antisera to each of the proteins and high-titer antisera to intact particles. BBMV protein did not react with sera to the other proteins or viruses, and neither did BBMV-specific sera react with the other virus antigens. Tests done at pH 6.0 with intact BMV and CCMV—extracted at pH 4 and 5, respectively—showed no reactions between the viruses and heterologous antisera, and only very weak cross-reactions between the viruses and their coat proteins when antiprotein sera were used. At pH 7 and 7.4, however, the viruses cross-reacted weakly. In addition, there was a stronger cross-reaction between the virus particles and their coat proteins at pH 7 and at pH 6, with both virus- and protein-specific antisera.

The conclusions from this study were that BMV and CCMV proteins had antigenic determinants (epitopes) in common, some of which were also present on intact swollen particles; but that nonswollen particles had no epitopes in common, and only few in common with their respective proteins. The last point was in agreement with an earlier study on BMV serology (von Wechmar and van Regenmortel, 1968). Further evidence for these conclusions came from experiments performed with the sensitive passive hemagglutination system (E. P. Rybicki, unpublished). These showed that antivirus sera absorbed with compact BMV particles at pH 6 did not react with compact virus, but still recognized swollen particles at pH 7.4. Sera absorbed with swollen virus at pH 7.4 did not recognize virus at all.

Direct, indirect, and DAS-ELISA tests on the Bromovirus (Rybicki and von Wechmar, 1981) gave results that initially appeared contradictory, until the differing specificities and areas of suitable application of the different techniques were realized. For example, DAS-ELISA tests at pH 6, using BMV-specific coating and conjugated antibodies, showed that neither CCMV nor BBMV were related to BMV. In direct ELISA tests at pH 6 and 7.4—where microtiter trays were coated with virus and probed directly with BMV-specific conjugate—CCMV appeared closely related to BMV, and BBMV appeared distantly related. There appeared to be little difference in the degree of relationship with change in pH. Indirect ELISA tests (see Fig. 1) were performed with either viruses or their proteins

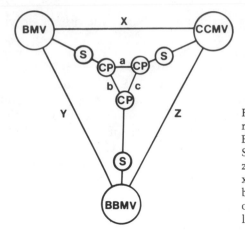

FIGURE 2. A serological map illustrating relationships among BMV, CCMV, and BBMV. CP, coat protein (dissociated virus); S, swollen virus (above pH 7); a, b, c, x, y, z, serological distances, not drawn to scale. x < y and z. y, z not necessarily equal. a, b, c approximately equal. Redrawn from original data (Rybicki, 1979, and unpublished).

adsorbed to microtiter trays, reacting with antisera to the three viruses and their proteins. All Bromovirus antisera tested—whether protein- or virus-specific, whether collected early or late in an immunization schedule—reacted specifically with the three viruses and their protein subunit preparations at pH 6. Chicken IgY antivirus preparations reacted similarly to the rabbit antisera. The indirect assays showed that the virus proteins were more closely related than the intact viruses. These and earlier observations are summarized in Fig. 2. The antigenic relationships between Bromovirus proteins have been confirmed using IEB tests (Section II.B.2), when BMV and BBMV proteins reacted specifically with anti-BMV, -CCMV, and -BBMV sera (Rybicki and von Wechmar, 1982).

The antigenic changes that occur with the pH-induced swelling of the three Bromoviruses are presumed to be caused by physical alterations in capsid organization and morphology. It is known that swelling of both BMV and CCMV particles resulted in "relaxed" structures in which the subunits are structurally similar to unpolymerized coat proteins (Chauvin et al., 1978; Krüse et al., 1981). In addition, RNA sequencing has shown that CCMV and BMV proteins share highly conserved amino acid sequences that are probably present on the surfaces required for particle assembly (Moosic, 1978; Dasgupta and Kaesberg, 1982). It is probably these sequences that are responsible for much of the antigenic similarity between the virus proteins, and perhaps for the similarity between swollen viruses.

Two new Bromoviruses have recently been described. Melandrium yellow fleck virus (MYFV) (Hollings and Horvath, 1982) was proposed to be a member of the taxonomic group on the basis of physical properties such as sedimentation coefficient, particle diameter, and the presence of four ssRNA molecules and one coat protein molecule in virus preparations, among other in vitro properties. The virus is apparently a relatively poor immunogen. No evidence of a serological relationship between MYFV and BMV, CCMV, or BBMV was detected in gel precipitin tests using antisera to all of the viruses (Hollings and Horvath, 1978, 1981).

Another new Bromovirus—cassia yellow blotch virus (CYBV)—was isolated in Australia (Dale *et al.*, 1984). The virus appears to share more physical properties with the "established" Bromoviruses than with any other group. The presence of four virion (and three genomic) RNAs and the decrease in sedimentation coefficient at pH 7 appear to be especially convincing evidence for its inclusion in the group. CYBV is evidently a poor immunogen: antiserum to pure virus had a low titer, and the authors had to resort to FA treatment to presumably increase Ouchterlony titers. No reaction was detected between CYBV and antisera to BMV, CCMV, or BBMV; the CYBV antiserum also did not react with the other antigens.

In light of the results obtained using EIA techniques with BMV, BBMV, and CCMV, it is unfortunate these tests have not been utilized more often in serological comparisons of the new putative Bromoviruses and the older members of the group. Indeed, the serological relationship between BBMV, and BMV and CCMV, is one of the best reasons for regarding these viruses as related in view of the many differences between BBMV and the others (Lane, 1981).

To end this section on a cautionary note, mention should be made of the effects of chemical stabilization on the serology of the Bromoviruses (see also Section II.A.2). Stabilization of BMV and CCMV by FA treatment decreases the degree of apparent serological relatedness between the viruses in both gel precipitin and indirect ELISA tests (Rybicki and von Wechmar, 1981). Stabilized viruses did not cross-react in Ouchterlony tests at any pH, and reactions of FA-treated viruses with each other's antisera were reduced by 50—60% relative to the equivalent reactions between untreated virus and the same antisera. This observation may be found to be important in assisting the evaluation of results of comparative serological studies on other virus groups.

2. Cucumoviruses

The serology of the Cucumovirus group has been reviewed twice recently (Kaper and Waterworth, 1981; van Regenmortel, 1982). Both reviews highlighted the contradictions in the literature on serological relationships within the group, with various groups of authors reporting success or failure in demonstrating serological relationships among TAV, PSV, and CMV. The consensus of both reviewers is that these three Cucumoviruses are serologically related: this conclusion appears to be based mainly on the data of Devergne and Cardin (1975). More recent evidence is provided by Devergne *et al.* (1981) and Rao *et al.* (1982). The former compared and contrasted DAS-ELISA and sandwich indirect ELISA techniques (IgY coating, rabbit serum for probing) for the detection and for the demonstration of antigenic relationships between CMV, TAV, and PSV. In DAS-ELISA tests using CMV-specific antibodies, no relationship

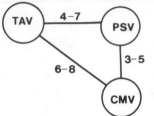

FIGURE 3. A serological map illustrating relationships among the major Cucumovirus serogroups. Diagram drawn to scale. Lines represent "antigenic distance" between serogroups, in SDI units. CMV serogroup includes serotypes ToRS, DTL, and Co. Drawn from the data of Devergne and Cardin (1975) and Devergne et al. (1981).

was detectable between CMV and PSV, or CMV and TAV. In one experiment, the CMV antibodies did not recognize another CMV strain. In sandwich indirect tests, however, all available strains of all three viruses were detected. These included even those with serological differentiation indices [$SDI_i = -\log_2$ (homologous titer/heterologous titer)] as great as 7. These and earlier data from Devergne's laboratory were used to construct Fig. 3.

The report of Rao et al. (1982) concerned the serological relationship between TAV and CMV, which these workers had failed to demonstrate on a number of other occasions (Habili and Francki, 1974, 1975; Mossop et al., 1976). Using a large number of antisera to virus stabilized with 0.25% GA, and stabilized viruses as test antigens, they found a definite—although weak—serological relationship between Australian CMV and TAV strains. The SDI's of five CMV strains and two TAV strains ranged between 4 and 8. This is in good agreement with the results of Devergne et al. (1981). Thus, any controversy over the relatedness of these, the most distantly related Cucumoviruses, has been resolved. It is interesting that Rao et al. (1982) were not able to use the antibody decoration technique of IEM to prove the relationship. They did, however, manage to use a variant of DAS-ELISA to show relationships between a TAV strain and a CMV strain which had reciprocal SDI's of 6–8. The tests utilized heterologous coating and homologous conjugated antibodies, which is known to lessen the specificity of the technique (Koenig, 1978; Bar-Joesph and Salomon, 1980). The result is surprising when one considers the distant relationships shown by Devergne et al. (1981) using a supposedly less specific technique (see Fig. 3).

It seems that too much emphasis has been placed—in Cucumovirus serology as in the serology of other labile viruses—on the necessity of obtaining high-titer antisera against intact virus particles. That this is difficult with the Cucumoviruses is amply clear from the literature, which abounds with references to the chemical stabilization of virus particles in order to (1) improve their immunogenicity, and (2) improve the sensitivity of virus antigen detection in tube- or gel-precipitin tests. It is important to realize, when evaluating results from serological tests using FA-treated virus particles and antisera raised against such particles, that FA treatment can alter the antigenicity of viruses (Rybicki and von Wechmar, 1981; Rybicki and Coyne, 1983; see also Section III.A.1). Musil and Richter (1983) have shown that the parallel use of FA-stabilized and un-stabilized CMV strains can lead to anomalous results in Ouchterlony tests. Other agents probably have similar effects.

Particle stabilization by FA- or other treatments may in fact be un-necessary. Strong reactions occur in DAS-ELISA tests between disso-ciated BMV protein and virus-specific antibodies; in indirect ELISA tests, sera with very low precipitin titers against protein reacted well with both intact viruses and free protein (Rybicki, 1979; Rybicki and von Wechmar, 1981). It would be interesting to test some of the low-titer Cucumovirus antisera in similar assays with free protein, unstabilized virions, and chemically stabilized virions. It is possible that epitopes on subunit sur-faces that are not exposed in intact viruses are in fact more closely related (i.e., the polypeptide sequence is more conserved) than metatopes, or sub-unit epitopes on the virus capsid surface (Shepard et al., 1974a; Purcifull et al., 1981; Rybicki and von Wechmar, 1981; van Regenmortel, 1982). DAS-ELISA detection of virus subunit proteins may also be more sen-sitive than detection of intact virus (Rybicki and Coyne, 1983), so that care need not be taken to preserve virus particle integrity in sap extracts.

3. Ilarviruses

Any dicussion of the serology of the Ilarviruses is complicated by the problem that viruses with different names may in fact by closely related strains. This has been recognized in the literature (Matthews, 1982; Uyeda and Mink, 1983), where, for example, black raspberry latent virus is now classified as a strain of tobacco streak virus (TSV; type member).

Distinct serological subgroups of the Ilarviruses have recently been proposed (Mink and Uyeda, 1982; Uyeda and Mink, 1983). Another very similar classification has also been proposed elsewhere (Francki et al., 1984). The latter proposal differs from the former in the grouping of Dan-ish plum line pattern virus with ApMV rather than with PNRSV, and the grouping of asparagus stunt virus with the TSV serotype. An updated grouping of the Ilarviruses is presented in Chapter 1 (Table III) of this volume. Evidence that lends support to at least a part of the proposed grouping are data concerning serological relationships among viruses in Subgroups 1 and 2 (Uyeda and Mink, 1983). There appears to be a clear-cut dividing line between TSV, and the viruses in Subgroup 2. Both sets of viruses were formerly included in Subgroup A of a two-subgroup di-vision of the Ilarviruses (Matthews, 1979). A serological map of the pres-ent Subgroup 2 is presented in Fig. 4: the map was drawn from a table of reciprocal SDI's calculated from the data published by Uyeda and Mink (1983). Further justification of the classification was provided by Mc-Morran and Cameron (1983). Using DAS-ELISA with infected leaf ho-mogenates, they detected 41 Ilarvirus isolates that represented the symp-tomatic and serological extremes of PNRSV, ApMV, and PDV, using only

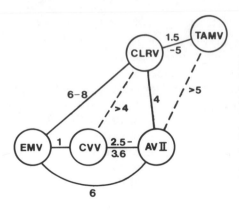

FIGURE 4. A serological map illustrating relationships among Ilarviruses in Subgroup 2. AV II, asparagus virus II; CVV, citrus variegation virus; EMV, elm mottle virus; CLRV, citrus leaf rugose virus; TAMV, tulare apple mosaic virus. Lines represent antigenic distance between serotypes, in SDI units. Broken lines indicate very weak relationships, or nonreciprocal cross-reactions. Drawn from the data of Uyeda and Mink (1983).

antibodies to PNRSV-G, rose mosaic virus (RMV; an ApMV strain), and PDV-B. The antibody preparations recognized only their "own" serotype and reacted with all of the viruses in that serotypic group. No evidence was found for a serological relationship between any of the PNRSV strains and the ApMV serotype, even though viruses in each serotype cross-reacted in precipitin tests (Fulton, 1968). This may be seen as further evidence of (1) the suitability of DAS-ELISA as a discriminatory tool; and (2) its unsuitability as a means of determining distant relationships.

Thomas *et al.* (1983) described a new Ilarvirus that does not react—in micro-Ouchterlony tests—with antisera to ApMV, PNRSV (two rose and one cherry isolates), AVII, spinach latent virus, two TSV isolates, and a black raspberry latent strain of TSV. Thus, the virus is apparently unrelated to Ilarviruses from Subgroups 1, 2, or 3 (no data for Subgroups 4 and 5), and neither is it related to spinach latent virus. Much more data are required to clarify the often confusing serological relationships among Ilarviruses.

The techniques most often used to study Ilarvirus serology have been precipitin-based, such as Ouchterlony's test (see McMorran and Cameron, 1983), and the ring interface test used by Uyeda and Mink (1983) to determine relationships among the viruses of Subgroups 1 and 2. However, DAS-ELISA has recently been extensively used (e.g., Clark and Adams, 1977; Barbara *et al.*, 1978; McMorran and Cameron, 1983), and monoclonal antibodies specific for various Ilarviruses have been tested by indirect ELISA (Halk *et al.*, 1982; Halk and Franke, 1983). The use of DAS-ELISA has been to the detriment of proving interserotype or intersubgroup relationships (McMorran and Cameron, 1983). It was recently stated that its use would create a hopeless situation in determining subgroup relationships because of its pronounced specificity (Mink and Uyeda, 1982). Indirect ELISA has only recently begun to be used with the Ilarviruses, and then apparently only in connection with monoclonal antibodies (see also Section IV.A.3). We suggest that indirect and/or sandwich indirect ELISA tests, or immunoelectroblotting, could be more widely used to establish relationships between the different Ilarvirus serotypes. The success of these techniques in investigations involving the Bromo- and Cucumoviruses (see Sections III.A.1 and III.A.2) implies that similar insights into the breadth of serological interrelationships among Ilarviruses may be possible. One of the first uses of the techniques could be to attempt to demonstrate or to confirm serological relationships where these have been difficult to establish by DAS-ELISA or precipitin tests. As with the other "Tricornaviridae," it could be a good idea to test dissociated viruses with antisubunit sera using EIA techniques, rather than struggling to preserve intact particles for precipitin tests.

4. Alfalfa Mosaic Virus Group

It has been said elsewhere (van Regenmortel and Pinck, 1981) that the AMV group ". . . is a large conglomerate of strains with different biological properties but unified through a common morphology." The statement aptly describes a situation in which strains differing widely from "type" AMV in host range and symptomatology are classified as AMV, and not as separate viruses, on the basis of their shared (and unique) particle morphology. Serology has often been used to identify the viruses in the group, usually by means of the Ouchterlony double-diffusion test (van Vloten-Doting et al., 1968). Immunoelectrophoretic (Jaspers and Moed, 1966) and, more recently, EIA techniques (Marco and Cohen, 1979) have also been applied for AMV detection and characterization.

Serological evidence for strain differences between AMV isolates does exist, usually in the form of precipitin "spurs" in Ouchterlony tests (van Vloten-Doting et al., 1968; Roosien and van Vloten-Doting, 1983). However, there appear to be no AMV isolates that are readily distinguishable serologically (Jaspers and Bos, 1980). Strains from widely separated geographical areas, which produced very different symptoms, appeared closely related in precipitin ring interface tests (Bancroft et al., 1960). Two AMV strains from South Africa with differing biological properties were shown to be related but not antigenically identical in DAS-ELISA tests; their respective capsid subunit proteins cross-reacted strongly in IEB tests despite slight apparent molecular weight differences (Pietersen, 1983). No serological differences between morphologically different individual components of a single strain could be demonstrated by Ouchterlony or immunoelectrophoretic precipitin tests (Jaspers and Moed, 1966).

Although no comprehensive serological comparison of the best-known AMV isolates or strains has yet been published, the consensus appears to be that the group is essentially monotypic in that all isolates compared so far are closely related. In addition, AMV strains are not serologically related (in precipitin tests) to Ilarviruses such as TSV, and CVV and CLRV in Subgroups 1 and 2, respectively. This is surprising in view of the phenomenon of reciprocal coat protein activation of the purified RNA components in infectivity studies (van Vloten-Doting and Jaspars, 1977; see also Section III.A.3). However, perhaps the existence of monoclonal antibodies to AMV (Halk *et al.*, 1982; Hsu *et al.*, 1983)—and the proven power of indirect EIA techniques in demonstrating even very distant serological relationships—will prompt the reinvestigation of a possible antigenic relationship between Ilar- and alfalfa mosaic viruses.

To conclude, it is evident that very little comparative serology has been practiced on this virus group, apart from strain comparisons showing relatively minor serological differences. To our knowledge, no DAS-ELISA comparison of different strains has been performed; nor has there been any comprehensive listing of reciprocal SDI values for gel precipitin comparisons. It has previously been stated that all AMV strains are serologically related. This would mean the group is one big serogroup, which justifies classification of all "AMV-related" viruses as AMV if one ignores whether or not they are symptomatically homogeneous, or share similar host ranges. Much serological work needs to be done in this area, preferably by means of a combination of DAS-ELISA (for differentiation) and one of the indirect immune techniques (to determine strain relationships).

B. Relationships between Virus Strains

1. Bromoviruses

BMV isolates are usually serologically indistinguishable by precipitin-based serological techniques, even when their biological properties differ (Lane, 1981). However, some small differences between isolates have been demonstrated by means of precipitin tests (Proll *et al.*, 1972; Rybicki, 1979), and by DAS-ELISA (Rybicki, 1979; Rybicki and von Wechmar, 1981). The South African "type strain" (BMV-G9; von Wechmar, 1967), the USA type strain (ATCC No. e PV 47, 1981), and electrophoretic variants 2 and 4 (Lane and Kaesberg, 1971) were all shown to be antigenically distinguishable in DAS-ELISA tests. However, only two "serogroups"— consisting of variants 2 and 4 on the one hand, and electrophoretic variants 1 and 3, the USA type strain, and USA isolates "Nebraska" and "mild," and BMV-G9 on the other—could be distinguished by Ouchterlony precipitin tests with "early bleed" low titer virus-specific antisera (Rybicki, 1979). Recently, South African isolates from field infections collected from 1978 to 1981 were studied by means of DAS-ELISA tests, and were found to differ both from the original BMV-G9 field isolate collected in 1965, and in some cases from each other as well (von Wechmar, 1967; Rybicki, 1984, and unpublished). These data indicate that the biological variation of BMV is often reflected in the antigenic reactivity of different isolates. Further investigations on this subject are presently being pursued (E. P. Rybicki and M. B. von Wechmar, unpublished).

Serologically distinct strains of CCMV have been described (Fulton *et al.*, 1975). The three strains compared—CCMV cowpea strain, CCMV-A from beans, and bean yellow stipple virus (BYSV) from green beans—produced precipitin "spurs" against each other in Ouchterlony tests with antisera directed against each of the three viruses. Apparently, the strains produced some symptom differences when inoculated onto the same plant host range, but these were less convincing than the serological differences. Kuhn (1968) has isolated strains which were symptomatically distinguishable, but apparently serologically indistinguishable. On the basis of the published evidence available, CCMV strains appear to differ more widely than either BMV or BBMV strains or isolates.

BBMV strains have been described (Murant *et al.*, 1973) which are serologically slightly different on the basis of precipitin tests. Very little other information appears to be available on the existence of serological variants of this virus. However, the amino acid compositions of three different strains are known (Yamazaki and Kaesberg, 1963; Miki and Knight, 1965; Wittman and Paul, 1961), and these appear to be sufficiently different in a computer-generated dendogram for Bromovirus classification (Dale *et al.*, 1984) to perhaps allow serological differentiation with a method such as DAS-ELISA. Given the differentiation achieved for BMV using the same technique, such an approach would probably yield results for both BBMV and CCMV strains on the first attempt.

2. Cucumoviruses

As mentioned previously (Section III.A.2), there appears to be an almost bewildering variety of strains of variants of CMV. Many of these differ in biological properties, and many also differ quite widely in antigenic properties (Kaper and Waterworth, 1981). Devergne and Cardin (1973) and Devergne *et al.* (1981) have grouped a large number of isolates of CMV—on the basis of biological properties as well as precipitin and ELISA tests—into two main groups, namely, ToRS and DTL. A new CMV strain (CMV-Co; Devergne *et al.*, 1981) appears to be serologically distinct from both of these groups. Other CMV strains including Australian isolates can also be assigned in one or other of the above groups (Francki *et al.*, 1979). A number of Australian and Far Eastern strains of CMV have been serologically compared recently; namely, strains CMV-Q, -T, -U, -X, -K, and -M (Rao *et al.*, 1982; Tien-Po *et al.*, 1982). These strains may be regarded as a collection of "substrains" of serogroup ToRS as CMV strain Q has been previously assigned to this group (Francki *et al.*, 1979), and only minor serological differences were detectable between the strains. Musil and Richter (1983) have described an isolate of CMV from *Cucurbita pepo* that is closely serologically related to a ToRS strain, and less strongly to DTL strains. A serological map of the major serogroups of CMV is presented in Fig. 5.

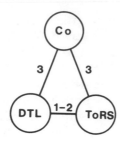

FIGURE 5. A serological map illustrating differences among serotypes of CMV. ToRS, DTL, and Co are all serotypes of CMV. Diagram drawn to scale, with lines indicating antigenic distance between serotypes in SDI units. Data from Devergne et al. (1981).

The serological difference between CMV-Co and the DTL and ToRS groups (SDI = 3) could be sufficient, in another taxonomic group, for the strain to be classified as a separate virus. For example, in the Tobamovirus group the average SDI between TMV-vulgare (type member) and ribgrass mosaic virus—with 56% coat protein amino acid sequence difference— is 2.1; the SDI between the biologically and physically different TMV-vulgare and soilborne wheat mosaic virus (SBWMV) is 4.0 (van Regenmortel, 1982). This argument notwithstanding, it is felt that the "Devergne" classification of CMV strains into distinct serogroups is of great value in the classification of new strains of the virus, especially as it correlates well with the biological properties of the viruses (Devergne et al., 1978, 1981; Kaper and Waterworth, 1981). Although all CMV strains appear to cross-react in precipitin tests, it is noteworthy that not all may be detected by DAS-ELISA using antiserum to a single strain. Thus, Devergne et al. (1981) could not detect strain Co using CMV-DTL antibodies, and Rao et al. (1982) obtained only a weak reaction with CMV strain M using antibodies to strains Q, X, and U.

As mentioned previously (Section III.A.2), sandwich-indirect ELISA (van Regenmortel and Burckard, 1980; Devergne et al., 1981) and DAS-ELISA with heterologous coating/conjugate combinations (Rao et al., 1982) have been used to good effect in the detection of distant strain relationships among CMV isolates and other Cucumoviruses. Immunoelectroblotting has also been used to detect serological similarities between a CMV isolate from Israel and a local tobacco isolate of CMV, and simultaneously show apparent coat protein size differences (P. Lupuwana and M. B. von Wechmar, unpublished results).

Distinct serotypic strains of PSV have also been recognized (Devergne and Cardin, 1975, 1976; Douine and Devergne, 1978; Diaz-Ruiz et al., 1979). Some virus isolates have been divided into two serotypes, designated as PSV-V and PSV-W, respectively (Devergne and Cardin, 1976) on the basis of reactions in precipitin tests. The "antigenic distance" between PSV serogroups is relatively large, as shown by the weakness of heterologous precipitin reactions and extent of precipitin "spurring" with the homologous reaction (Diaz-Ruiz et al., 1979). The latter authors also described an interesting phenomenon noticed in Ouchterlony tests with four strains of PSV (PSV-B, -V, -W, and -H) that were biologically different.

Their new isolate (PSV-B) was serologically identical to PSV-W, but did not cross-react reciprocally with CMV-DTL, whereas PSV-W did. Both PSV's did, however, cross-react with sera to CMV-To and TAV. This observation provoked the remark that ". . . some strains of each Cucumovirus are related to some strains of the other two Cucumoviruses" (Kaper and Waterworth,.1981). This is reinforced by the observation that there is a fairly close serological relationship between CMV-CpM and the clover blotch strain of PSV (Musil and Richter, 1983). It would be interesting to reinvestigate the differences (and similarities) between PSV strains using both DAS- and indirect ELISA techniques in order to bring PSV serotyping in line with that of the better-studied CMV cluster. It is worth noting that Devergne and Cardin (1975) showed, by means of intragel cross-absorption studies, that all the antibodies responsible for the TAV–CMV cross-reaction also reacted with PSV: this justifies the intermediate position of PSV between TAV and CMV in serological maps (van Regenmortel, 1982; see also Fig. 2, Section III.A.2).

Two viruses originally described as distinct entities—robinia mosaic and clover blotch—have been shown to be closely related both to each other, and to PSV (Richter et al., 1979). It is not clear which serotype they should be assigned to.

Many biological strains of TAV are known to occur, which differ widely in host range, symptomatology, aphid transmissibility, and in vitro stability (Hollings and Stone, 1971). However, only one serotype of the virus has so far been recognized, indicating a lack of correlation of difference in surface antigenicity with biological properties. An illustration of the serological closeness of TAV strains that are biologically different was given by Rao et al. (1982). They could not demonstrate any significant difference between the Australian V and N strains of TAV in DAS-ELISA tests with antibodies to either strain. This may be taken as indicating a very close relationship, as DAS-ELISA is far more strain-specific than most other serological techniques (see Section II.B.2). Thus, it may be possible to use single antisera for the detection of nearly all the biological strains of TAV, even when using the extremely strain-specific DAS-ELISA test.

3. Ilarviruses

Some comment was made in an earlier section (III.A.3) on strain relationships among specific Ilarviruses, and on apparent problems in the classification of strains. For example, Francki et al. (1984) grouped Danish plum line pattern virus with ApMV in Subgroup 3 (see Table III, Chapter 1), whereas Mink and Uyeda (1982) grouped it with PNRSV. As ApMV and PNRSV are serologically related, the apparent disagreement may be no more than semantics in the absence of reciprocal SDI values for all the viruses in the serogroup.

There are a large number of strains of certain of the Ilarviruses, and it appears that no one has conducted a comprehensive serological screening of them in order to better define the sometimes apparently nebulous strain differences. McMorran and Cameron (1983) (see also Section III.A.3) found that antisera to PNRSV-G and rose mosaic virus (Subgroup 3), and to PDV-B (Subgroup 4), reacted with 41 Ilarvirus isolates spanning the entire symptomatological and serological range of Subgroups 3 and 4. Thus, all PNRSV strains tested were detectable with one antiserum, as were all ApMV, and PDV strains, with their respective antisera. Barbara *et al.* (1978) found that some PNRSV strains reacted with homologous antibodies in DAS-ELISA tests but not with ApMV antibodies, and vice versa; also, that some strains of both viruses could be detected with both antibody preparations. They suggested that these strains should be referred to as anti-PNRSV-G-reactive (C), anti-ApMV-reactive (A), or intermediate (I). However, it would seem more likely that there is a continuous spectrum of antigenic variation between PNRSV and ApMV strains, from an "ApMV" to a "PNRSV" extreme. This would mean that any attempt to subdivide strains between the two "type" viruses will only lead to a hopeless muddle, given that different antisera to even "type" PNRSV and ApMV strains may well react differently in terms of recognition of strains of the other virus. It may be found, after further data become available, that the distinction between the two viruses is only arbitrary as far as serological differentiation is concerned. This may lead to problems in assigning of new strains; however, both viruses are in the same serological subgroup so classification should perhaps be on biological rather than on serological grounds.

Other DAS-ELISA studies on Ilarviruses that further illustrate this problem include those of Mink (1980) and Skotland and Kaniewski (1981). The first paper reported that there was a large serological difference between cherry rugose virus (a strain of PNRSV) and PNRSV, and that CRV and some PNRSV strains did not react with antisera to ApMV strains and a hop strain of PNRSV. The second paper reported the detection of three NRSV strains in hops, one of which was detectable only with an antiserum to a hop isolate, another that was only detectable using antisera to two ApMV strains, and another that could be detected using all three antisera. The three antisera did not recognize a cherry isolate of NRSV, or Danish plum line pattern virus. Such confusion could conceivably be eliminated by use of sandwich indirect ELISA (van Regenmortel and Burckard, 1980; Barbara and Clark, 1982). The far lower homologous specificity of the technique would probably enable detection and/or recognition of heterologous virus reactions far more reliably than has hitherto been the case.

A less confusing situation has emerged from serological studies of TSV strains (Subgroup 1). Uyeda and Mink (1983) tested combinations of various TSV strains, and antisera to different strains, in ring interface precipitin tests. They divided the strains into two serotypes, with RN-B and RN-S isolates in one group, and Ro-A and Ro-S isolates in another. SDI values within a serotype were below 1, while between members of the two serotypes the values were less than 4 (our calculations). The antisera used did not react with any of the Subgroup 2 viruses in similar precipitin tests.

IV. IMMUNOCHEMICAL STUDIES

A. Current Status

1. Bromoviruses

Immunochemical studies on the Bromoviruses have mainly been confined to BMV, except for mention of the effect of swelling and FA treatment on the antigenicity of CCMV and BBMV (Rybicki and von Wechmar, 1981), and of the apparent lack of effect of dimethyladipimidate treatment on CCMV and its protein (Bancroft and Smith, 1975). von Wechmar and van Regenmortel (1968) showed that intact BMV particles and dissociated coat protein had unique antigenic determinants (termed neotopes and cryptotopes, respectively), while also sharing some antigenic similarity (metatopes). The effect of swelling on the serology of the viruses has already been discussed in some detail (Section III.A.1). However, later experiments on the antigenic changes that accompany both the swelling and dissociation of BMV, and the chemical treatment of BMV (Rybicki and Coyne, 1983), deserve further discussion. DAS-ELISA tests were used to demonstrate antigenic differences between compact virus, swollen virus, and dissociated protein, and between empty capsids and compact (low-pH form) virus. Differences were also shown between native and FA-treated virus, both at pH 6 and 7.4. The intensity of the color reaction produced with the different virus antigens depended on how "free-protein-like" they were. Thus, free protein reacted more strongly than swollen virus, which reacted more strongly than compact virus at pH 6, and empty capsids reacted better than compact virus at pH 5 in 0.3 M NaCl. FA-treated BMV reacted less well than native BMV at both pH 6 and pH 7.4. A slight difference was shown between virus treated with FA at pH 6 and virus treated at pH 7.4, in tests at pH 7.4. This indicated that the virus may have been partially stabilized in a "semi-compact" state by FA.

The serological difference shown between BMV and "artificial top component," or empty capsids made from reassociated subunit protein, is in contradiction to a report by Bancroft et al. (1969). However, this probably reflects the use of a more discriminatory serological technique in the later work. The result with empty capsids as well as swollen and nonswollen virus particles is in agreement with the physical data of Chauvin et al. (1978) and Krüse et al. (1981). This indicates that the antigenic differences are due to differences in protein subunit tertiary structure, and capsid polymer quaternary structure, in the compact, swollen, RNA-free, and dissociated states.

Immunoelectroblotting studies on Bromoviruses have shown that both BBMV and BMV—and BMV in particular—are subjected to limited proteolysis in their plant hosts, and in purified preparations (Rybicki and von Wechmar, 1982; Rybicki, 1984). This produces characteristic one-dimensional peptide maps akin to those produced by deliberate limited proteolysis (Cleveland et al., 1977; Koenig et al., 1981). The less-than-monomer-sized polypeptide fragments are detectable only by gross ov-erloading of Coomassie blue-stained polyacrylamide gels, or by silver staining (E. P. Rybicki and D. H. du Plessis, unpublished results) or IEB peroxidase staining (Rybicki, 1984). Preliminary experiments have indi-cated that trypsin treatment of BMV results in a highly characteristic "immuno-peptide map"; as the IEB system is antigen-specific, it may be possible to identify specific "Tricornaviridae" (or other virus) strains by partial proteolysis of virus even in plant sap (E. P. Rybicki, unpublished results). A potentially serious implication of "in situ" proteolysis of viral capsids is that monoclonal antibodies specific for peptide sequences that are lost during proteolysis, may fail to recognize the virus. This potential problem needs to be investigated before monoclonals are routinely used for virus detection.

2. Cucumoviruses

Devergne and Cardin (1975) performed an exhaustive antigenic anal-ysis of viruses of the CMV serotypes -ToRS and -DTL, the TAV serogroup, and a strain of PSV (PSV-V), using Ouchterlony's test and immunoelec-trophoresis. Conclusions drawn from this study are: (1) that each sero-group has unique epitopes, recognized only by homologous antisera; (2) PSV and TAV have common epitopes that are not shared by CMV strains; (3) PSV and the CMV strains share epitopes not present in TAV isolates; (4) the TAV epitopes that are recognized by CMV-specific antisera are also present in PSV; (5) some TAV-specific antisera weakly recognize an epitope(s) in CMV-DTL not present in either CMV-ToRS or PSV. This analysis laid the basis of serological classification of the Cucumoviruses (see Section III.A.2). It also raises the possibility of producing monoclonal or affinity-purified polyclonal antibodies, reactive with similar epitopes present in all of the serogroups, that may be used to detect all of the known Cucumoviruses in routine serological screening of plants.

Earlier work by Devergne and Cardin (1970, 1973) dealt with the antigenic analysis of strains of CMV belonging to the DTL and ToRS serotypes. The virus particles of two ToRS strains (-R and -T) were found to possess unique capsid epitopes (neotopes), as well as shared epitopes present also on dissociated protein (metatopes). Free protein also ex-pressed epitopes not present on the capsids (cryptotopes); these were shared between the two strains. Some of the cryptotopes of one isolate appeared to be metatopes of the other (Devergne and Cardin, 1970), which may indicate differences in capsid structure. Devergne and Cardin (1973) laid the foundation for CMV serotypic classification. After examining the serology of 11 CMV isolates they defined four serotypes, each with unique epitopes: these were To, R, S, and DTL. Serotypes To, R, and S shared epitopes not present in DTL isolates, and vice versa. These analyses provide a basis for the preparation of antibodies which are either subtype (e.g., strain -To, or -R, or -DTL)-specific, or subgroup-specific (e.g., ToRS), or generally CMV-reactive.

3. Ilarviruses

To our knowledge, the Ilarviruses have never been the subject of as detailed antigenic analyses as have the Bromo- and Cucumoviruses; partly, perhaps, because of the greater complexity of the group. However, monoclonal antibodies specific for various Ilarvirus epitopes have been prepared at the American Type Culture Collection in Rockville, Maryland. Their reactivities with various Ilarviruses have been tested mainly by means of indirect ELISA (Halk et al., 1982, 1984; Halk and Franke, 1983; Hsu et al., 1983). It was found that some monoclonals did not cause immunoprecipitates with virus in Ouchterlony tests, but reacted strongly in indirect ELISA. One hybridoma clone produced antibody that reacted with all PNRSV and ApMV isolates tested (Halk et al., 1982), suggesting that at least one epitope is shared by the PNRSV–ApMV range of antigens. Other clones produced antibodies reactive with different combinations of PNRSV strains, reflecting antigenic differences between them. Halk et al. (1984) used seven monoclonal antibodies to screen a large number of PNRSV/ApMV isolates in indirect ELISA tests. They defined three PNRSV serotypes, and five ApMV types in Subgroup 3. In addition, TSV isolates (Subgroup 1) could be differentiated into four different serotypes by use of five monoclonal antibody preparations.

Although monoclonal antibodies will undoubtedly be used with increasing frequency in plant virus serology—and may ease the problem of specific virus detection enormously—they should not be used indiscriminantly, and some of the "strain differentiation" results obtained should be the subject of close scrutiny before extensive reclassification of viruses is embarked upon. As has been mentioned previously (Section II.A.4), certain anomalous results were obtained when the reactions of certain Tobamoviruses with different monoclonal preparations were analyzed. Al-Moudallal et al. (1982) found that single amino acid exchanges—which could be caused by point mutations in the coat protein gene—were sufficient to cause changes in antibody binding affinity. Moreover, the exchanges could affect the epitope even if they occurred outside of the sequence specifying the epitope. This phenomenon has also been noticed with poliovirus variants (Newmark, 1984). Thus, some of the "serotypes" within the Ilarvirus serogroups (e.g., Subgroup 3, containing PNRSV and ApMV) could be viruses which differ by a single amino acid residue in the coat protein. More important perhaps, is the possibility that mutations affecting a specific epitope may lead to nonbinding of a particular monoclonal antibody to virus of a strain that is closely related to the homologous strain. It is to be hoped that these possibilities will be taken into account before far-reaching revisions are made in virus classification systems.

4. Alfalfa Mosaic Virus Group

A remarkably original approach to investigating serological relationships among the "Tricornaviridae" was made by van Tol and van Vloten-Doting (1981). They compared the reactivity of the 3a nonstructural proteins encoded by the RNA 3 genome components of AMV, TSV, BMV, and CMV with antibodies specific for AMV 3a protein. ^{35}S-labeled 3a proteins of the four viruses were synthesized in cell-free translation mixtures, and treated with the antibodies to the AMV 3a protein. Immunoprecipitates were analyzed by SDS-PAGE. No serological relationship was detected between the different proteins, as only the AMV protein was precipitated. No apparent sequence homology was detectable between RNA 3 from AMV, BMV, CCMV, and TSV in other experiments (Bol et al., 1975); however, homology was assessed by competition hybridization and not by sequence comparisons, and short stretches of homologous sequences may have gone undetected. It is possible that a repeat of this exercise, using antibodies specific for each of the 3a proteins and using a technique such as IEB in addition to immunoprecipitation, could result in antigenic similarities being detected.

B. The Viruses as Model Tools for Immunochemical Research

Certain aspects of the immunochemical investigations that have been performed on the "Tricornaviridae" deserve extensive follow-up studies, in order to provide specific antibodies for detection of particular viruses as well as for the fundamental knowledge they may provide. For example, the study on the effects of the swelling phenomenon on the antigenicity of BMV (see Sections III.A.1 and IV.A.1) not only demonstrated that the morphological changes accompanying swelling and loss of RNA produce antigenic changes, but also provided information on which of the different physical forms (morphomers) of BMV was the most easily detectable antigen. The possibility of producing antibodies specific for serotype, serogroup, or taxonomic group epitopes in the Cucumoviruses was discussed previously (Section IV.A.2). Serogroup-specific antibodies are now a reality for the Ilarvirus group, thanks to the characterization of monoclonals reacting with PNRSV and ApMV strains (Section IV.A.3).

An exciting area for future immunochemical research on the "Tricornaviridae" is in the detection of possible antigenic similarities in the noncapsid viral proteins (NCVPs). This has been briefly discussed (Section IV.A.4); however, recent developments in genome sequencing of the viruses make the proposition appear quite feasible. The entire genomic sequence of BMV and AMV, and much of the sequence of CMV, are known (see Chapter 3). Knowledge of the sequences has enabled determination of, for example, deduced amino acid sequence homologies between the 3a proteins coded for by RNA 3 of BMV and CMV (Murthy, 1983). This study revealed over 35% homology between the sequences of the 3a proteins, but little homology between the coat proteins. The deduced amino acid sequences of the respective proteins coded for by the two largest RNAs of AMV and BMV also show striking similarities. In addition, proteins A1 (from AMV RNA 1) and B1 (from BMV RNA 1) show sequence homology with p126 of TMV and nsP1 and nsP2 of Sindbis virus (an animal Alphavirus), and A2 and B2 are similar to p187 of TMV and a read-through gene product of Sindbis virus (Haseloff et al., 1984). Apart from the obvious evolutionary considerations, these studies are fascinating from the point of view of possible serological studies: it is possible that the closer homologies could be reflected in antigenic similarities between the proteins mentioned, and, for example, between NCVP's of the sequenced viruses and the as yet unsequenced Ilarviruses.

Another potentially interesting avenue of research is to examine the serological relatedness of viral proteins that have dissimilar primary structures, but similar secondary and/or tertiary structures. For example, we have previously speculated that the antigenic similarities between BBMV and the other Bromoviruses may be due to similar structural features rather than to protein sequence homologies, given the very dissimilar tryptic peptide maps of BBMV and either BMV or CCMV (Rybicki and von Wechmar, 1981). Recent evidence from research on animal virus immunogenicity has shown that two serologically related Picornaviral proteins shared neither primary nor predicted secondary structure, but had strikingly similar predicted surface features (Newmark, 1984). Thus, given the use of sufficiently sensitive techniques, it may yet be possible to demonstrate the serological relatedness of viruses that only have surface features in common. This may include the virus groups under discussion in this Chapter.

With the determination of protein and/or genomic sequences of some of the "Tricornaviridae" has come the possibility of using the virus coat proteins as model antigens in research into the nature of the effect of sequence and conformational changes on antigenicity. Many mutants of the viruses have been made artificially, and many natural variants are known. The best studied of the groups in this respect is the alfalfa mosaic virus: van Vloten-Doting et al. (1980) have described 13 temperature-sensitive mutants; Roosien and van Vloten-Doting (1983) reported an interesting mutant with an altered coat protein (see Chapter 5). Detailed structural investigations have been performed on strain VRU (Driedonks et al., 1978; Castel et al., 1979), and AMV is known not to swell (Oostergetel et al., 1981). The capsid protein subunits of simple viruses have some unique advantages as experimental antigens: first, they are obtainable in relatively large amounts; and second, they are capable of polymerizing into defined quaternary structures—also studied in some detail (see Chapter 2)—which magnifies small or single-epitope changes by at least two orders of magnitude. A combination of physical and immunochemical studies on these viruses could yield a fund of data on the effect of protein sequence changes not only on the antigenicity of the protein, but also on the structural alterations accompanying the changes (Rybicki and Coyne, 1983).

V. TAXONOMIC IMPLICATIONS

There appears to be no conflict over the taxonomy of the established Bromoviruses BMV, CCMV, and BBMV on serological grounds. The viruses are sufficiently different to be classified as separate entities; for example, neither CCMV nor BBMV are detectable in DAS-ELISA tests using BMV antibodies, and only very weak cross-reactions have been shown between BMV and CCMV in precipitin tests. The only uncertainties in the comparative serology of the group concern the two more recently characterized viruses; namely, melandrium yellow fleck virus (MYFV) and cassia yellow blotch virus (CYBV). It is to be hoped that the two viruses will be tested with antisera to all of the other Bromoviruses in simple indirect ELISA tests, as these were the only means of detecting the relatedness of BMV, CCMV, and BBMV before the advent of the IEB technique.

Some mention has already been made of possible anomalies in the nomenclature of the Cucumovirus group, especially as concerns CMV strains (see Section III.B.2). However, such anomalies abound in the nomenclature and classification of plant viruses. The criteria for designating a given virus as a related strain of another, or as a separate entity, appear to depend on which taxonomic group it is assigned to. It is perhaps best to leave the final classification of newly discovered viruses in particular groups in the hands of researchers with the most experience of that group. For example, the "Devergne" classification of serotypes of Cucumoviruses, and of CMV in particular (Devergne and Cardin, 1975; Devergne et al., 1981) appears to correlate well with the symptom expression of the strains (see Section III.B.2).

The Ilarviruses appear to be in a chaotic state as regards nomenclature and classification. To paraphrase Dr. G. I. Mink (Mink and Uyeda, 1982), it appears that the widespread use of DAS-ELISA as a means of detecting and of identifying Ilarviruses can only worsen an already confused situation (see Section III.B.3). Perhaps a better idea of relationships between subgroups—if any exist—will be gained from indirect ELISA or electroblot studies, possibly using monoclonal antibodies. One problem needing examination is the question of whether the PNRSV and ApMV serotypes in Subgroup 3 represent a continuous spectrum of antigenic variation from "PNRSV-like" to "ApMV-like." Indirect EIA techniques could well be used to show relationships between supposedly unrelated serogroups such as the viruses in Subgroups 1 and 2, or 3 and 4, as well as allowing the classification of as yet ungrouped viruses such as American plum line pattern, spinach latent, and hydrangea mosaic (see Table III, Chapter 1).

The properties of alfalfa mosaic virus strains appear to qualify them for membership of the Ilarviruses, as another serologically and symptomatically unique serogroup on the basis of the criteria listed by Shepherd et al. (1976) for recognition of the Ilarviruses (see also Chapter 1). It would be pleasant indeed to have this talking point resolved by serological evidence gained from the use of one of the contemporary, broad-spectrum techniques.

VI. CONCLUDING REMARKS

It has been pointed out (Matthews, 1970) that 5% or less of the genome of a small RNA virus is involved in coding for the coat protein, which is usually the only part of the virus that is involved in determining the serological specificity of the whole virus. It could be expected that homology between viruses would be better reflected in comparisons of entire genomic or corresponding subgenomic nucleic acid sequences because more of the genetic information of the viruses could be used in the comparisons. However, only a few plant virus genomic sequences have been elucidated, and many will probably never be sequenced: these viruses will still have to be studied by serology and/or amino acid analysis in order to classify them.

Even in the case of viruses whose genome sequences are known, serological comparisons should not fall away entirely. For example, although BMV and CMV share over 35% homology in the (deduced) amino acid sequence of their 3a proteins (Murthy, 1983), the RNA 3 sequences that code for the proteins show far less homology due to codon redundancy. The same is true for the RNA sequences encoding the coat proteins of BMV and CCMV, which are known to share significant amino acid sequence homology and which are serologically related. Thus, direct comparison of RNA sequences could give a false impression of relatedness when compared to the actual gene product. Further evidence of this is given by Diaz-Ruiz and Kaper (1983), who compared the relatedness of nucleotide sequences of 30 PSV isolates by RNA competition hybridization. They found the isolates fell into two groups (PSV-V and -W; see Section III.B.2), with greater than 80% homology inside each group, and only about 10% homology between groups. Very little sequence homology was detected between PSV and CMV strains. These results are initially surprising in view of other close similarities between the viruses, especially antigenic similarities. However, coat protein amino acid (if not nucleotide) sequences could be expected to be strongly conserved through evolution, so perhaps the coat proteins of viruses are generally the best index of relatedness in an evolutionary sense. Gibbs and his co-workers (see Dale et al., 1984) have indeed made use of amino acid analyses of certain of the "Tricornaviridae" to construct a classification dendogram that is in good agreement with existing classifications based on other characteristics of the viruses. However, as such analyses have not been done for many of the viruses of the "family," and may be tedious to perform for every new isolate, the recourse should still be serology. As antigenic similarity between virus particles depends on coat protein sequence and structure, serological studies may still be the simplest reliable means of determining virus relationships in the absence of sophisticated facilities and molecular biological expertise.

ACKNOWLEDGMENTS.   We wish to thank R. G. Milne, G. I. Mink, H. R. Cameron, J. P. McMorran, L. van Vloten-Doting, and L. C. Lane for their cooperation in compiling the literature for this review, and R. I. B. Francki for his help and encouragement. We are also grateful to Dimitri Erasmus for his help in the compilation of this review.

REFERENCES

Al-Moudallal, Z., Briand, J. P., and van Regenmortel, M. H. V., 1982, Monoclonal antibodies as probes of the antigenic structure of tobacco mosaic virus, EMBO J. 1:1005.

Ball, E. M., 1973, Solid phase radioimmunoassay for plant viruses, Virology 55:516.

Bancroft, J. B., and Smith, D. B., 1975, The effect of dimethyladipimidate on the stability of cowpea chlorotic mottle and brome mosaic viruses, J. Gen. Virol. 27:257.

Bancroft, J. B., Moorhead, E. L., Tuite, J., and Liu, H. P., 1960, The antigenic characteristics and the relationships among strains of alfalfa mosaic virus, Phytopathology 50:34.

Bancroft, J. B., Wagner, G. W., and Bracker, C. E., 1969, The self-assembly of a nucleic-acid-free pseudo-top component for a small spherical virus, Virology 36:146.

Barbara, D. J., and Clark, M. F., 1982, A simple indirect ELISA using F(ab')₂ fragments of immunoglobulin, J. Gen. Virol. 58:315.

Barbara, D. J., Clark, M. F., Thresh, J. M., and Casper, R., 1978, Rapid detection and serotyping of Prunus necrotic ringspot virus in perennial crops by enzyme-linked immunosorbent assay, Ann. Appl. Biol. 90:395.

Bar-Joseph, M., and Malkinson M., 1980, Hen egg yolk as a source of antiviral antibodies in the enzyme-linked immunosorbent assay (ELISA): A comparison of two plant viruses, J. Virol. Methods 1:179.

Bar-Joseph, M., and Salomon, R., 1980, Heterologous reactivity of tobacco mosaic virus strains in enzyme-linked immunosorbent assays, J. Gen. Virol. 47:509.

Bol, J. F., Brederode, F. T., Janze, G. C., and Rauh, D. K., 1975, Studies on sequence homology between the RNA's of alfalfa mosaic virus, *Virology* **65**:1.

Castel, A., Kraal, B., de Graaf, J. M, and Bosch, L., 1979, The primary structure of the coat protein of alfalfa mosaic virus strain VRU, *Eur. J. Biochem.* **102**:125.

Chauvin, C., Pfeiffer, P., Witz, J., and Jacrot, B., 1978, Structural polymorphism of bromegrass mosaic virus: A neutron small angle scattering investigation, *Virology* **88**:138.

Clark, M. F., and Adams, A. N., 1977, Characteristics of the microplate method of enzyme-linked immunosorbent assay for the detection of plant viruses, *J. Gen. Virol.* **34**:475.

Cleveland, D. W., Fischer, S. G., Kirschner, M. W., and Laemmli, U. K., 1977, Peptide mapping by limited proteolysis in sodium dodecyl sulphate and analysis by gel electrophoresis, *J. Biol. Chem.* **252**:1102.

Cohen, J., Loebenstein, G., and Milne, R. G., 1982, Effect of pH and other conditions on immunosorbent electron microscopy of several plant viruses, *J. Virol. Methods.* **4**:323.

Dale, J. L., Gibbs, A. J., and Behncken, G. M., 1984, Cassia yellow blotch virus: A new Bromovirus from an Australian native legume, *Cassia pleurocarpa, J. Gen. Virol.* **65**:281.

Dasgupta, R., and Kaesberg, P., 1982, Complete nucleotide sequences of the coat protein messenger RNAse of brome mosaic virus and cowpea chlorotic mottle virus, *Nucleic Acids Res.* **10**:703.

Derrick, K. S., 1973, Quantitative assay for plant viruses using serologically specific electron microscopy, *Virology* **56**:652.

Devergne, J. C., 1975, A study of the serological behaviour of CMV: Relationship between CMV strains and other viruses, *Meded. Fac. Landbouwwet. Rijksuniv. Gent* **40**:19.

Devergne, J. C., and Cardin, L., 1970, Etude sérologique comparative de plusiers isolats du virus de la mosaique du concombre (CMV): Relations sérologiques au niveau du virus et de l'antigéne soluble, *Ann. Phytopathol.* **2**:639.

Devergne, J. C., and Cardin, L., 1973, Contribution a l'etude du virus de la mosaique du concombre (CMV). IV. Essai de classification de plusieurs isolats sur la base de leur structure antigénique, *Ann. Phytopathol.* **5**:409.

Devergne, J. C., and Cardin, L., 1975, Relations serologiques entre cucumovirus (CMV, TAV, PSV), *Ann. Phytopathol.* **7**:255.

Devergne, J. C., and Cardin, L., 1976, Characterisation de deux serotypes du virus du Rabougrissement de l'Arachide (PSV), *Ann. Phytopathol.* **8**:449.

Devergne, J. C., Cardin, L., and Quiot, J. B., 1978, Detection et identification serologiques des infections naturelles par le virus de la mosaique du concombre, *Ann. Phytopathol.* **10**:233.

Devergne, J. C., Cardin, L., Burckard, J., and van Regenmortel, M. H. V., 1981, Comparison of direct and indirect ELISA for detecting antigenically related Cucumoviruses, *J. Virol. Methods* **3**:193.

Diaz-Ruiz, J. R., and Kaper, J. M., 1983, Nucleotide sequence relationships among thirty peanut stunt virus isolates determined by competition hybridisation, *Arch. Virol.* **75**:277.

Diaz-Ruiz, J. R., Kaper, J. M., Waterworth, H. E., and Devergne, J. C., 1979, Isolation and characterisation of peanut stunt virus from alfalfa in Spain, *Phytopathology* **69**:504.

Douine, L., and Devergne, J. C., 1978, Isolement en France du virus du Rabougrissement de l'Arachide (peanut stunt virus, PSV), *Ann. Phytopathol.* **10**:79.

Driedonks, R. A., Krijgsman, P. C. J., and Mellema, J. E., 1978, Coat protein polymerisation of alfalfa mosaic virus strain VRU, *J. Mol. Biol.* **124**:713.

Erasmus, D. S., 1982, The association of brome mosaic virus with *Puccinia graminis tritici*, M.Sc. thesis, University of Cape Town.

Erasmus, D. S., Rybicki, E. P., and von Wechmar, M. B., 1983, The association of brome mosaic virus and wheat rusts. II. Detection of BMV in/on uredospores of wheat stem rust, *Phytopathol. Z.* **108**:34.

Ford, D. J., Radin, R., and Pesce, A. J., 1978, Characterisation of glutaraldehyde-coupled alkaline phosphatase–antibody and lactoperoxidase–antibody conjugates, *Immunochemistry* **15**:237.

Francki, R. I. B., and Habili, N., 1972, Stabilization of capsid structure and enhancement of immunogenicity of cucumber mosaic virus (Q strain) by formaldehyde, *Virology* **48**:309.

Francki, R. I. B., Mossop, D. W., and Hatta, T., 1979. Cucumber mosaic virus, *CMI/AAB Descriptions of Plant Viruses* No. 213.

Francki, R. I. B., Milne, R. G., and Hatta, T., 1984, *An Atlas of Plant Viruses*, Volume II, CRC Press, Boca Raton, Florida.

Fulton, R. W., 1968, Serology of viruses causing cherry necrotic ringspot, plum line pattern, rose mosaic, and apple mosaic, *Phytopathylogy* **58**:635.

Fulton, R. W., 1981, Ilarviruses, in: *Handbook of Plant Virus Infections and Comparative Diagnosis* (E. Kurstak, ed.), pp. 377–413, Elsevier/North-Holland, Amsterdam.

Fulton, R. W., 1982, Ilar-like characteristics of American plum line pattern virus and its serological detection in prunus, *Phytopathology* **72**:1345.

Fulton, J. P., Gamez, R., and Scott, H. A., 1975, Cowpea chlorotic mottle and yellow stipple viruses, *Phytopathology* **65**:741.

Ghabrial, S. A., and Shepherd, R. J., 1980, A sensitive radioimmunosorbent assay for the detection of plant viruses, *J. Gen. Virol.* **48**:311.

Gough, K. H., and Shukla, D. D., 1980, Further studies on the use of protein A in immune electron microscopy for detecting virus particles, *J. Gen. Virol.* **51**:45.

Habili, N., and Francki, R. I. B., 1974, Comparative studies on tomato aspermy and cucumber mosaic viruses. 1. Physical and chemical properties, *Virology* **57**:392.

Habili, N., and Francki, R. I. B., 1975. Comparative studies on tomato aspermy and cucumber mosaic viruses. IV. Immunogenic and serological properties, *Virology* **64**:421.

Halk, E. L., and Franke, J., 1983, Identification of serological types of apple mosaic, prunus necrotic ringspot and tobacco streak viruses with monoclonal antibodies, *Phytopathology* **73**:789 (abstract).

Halk, E. L., Hsu, H. T., and Aebig, J., 1982, Properties of virus specific monoclonal antibodies to prunus necrotic ringspot (NRSV), apple mosaic (ApMV), tobacco streak (TSV) and alfalfa mosaic (AMV) viruses, *Phytopathology* **72**:953 (abstract).

Halk, E. L., Hsu, H. T., Aebig, J., and Franke, F., 1984, Production of monoclonal antibodies against three Ilarviruses and alfalfa mosaic virus and their use in serotyping, *Phytopathology* **74**:367.

Haseloff, J., Goelet, P., Zimmern, D., Ahlquist, P., and Dasgupta, R., 1984, Striking similarities in amino acid sequence among nonstructural proteins encoded by RNA viruses that have dissimilar genomic organization, *Proc. Natl. Acad. Sci. USA* **81**:4358.

Hollings, M., and Horvath, J., 1978, *Rep. Glasshouse Crops Res. Inst. for 1977* p. 129.

Hollings, M., and Horvath, J., 1981, Melandrium yellow fleck virus, *CMI/AAB Descriptions of Plant Viruses* No. 236.

Hollings, M., and Stone, O. M., 1971, Tomato aspermy virus, *CMI/AAB Descriptions of Plant Viruses* No. 79.

Hsu, H. T., Halk, E. L., and Lawson, R. H., 1983, Monoclonal antibodies in plant virology [abstract], Proceedings, 4th International Congress of Plant Pathology, Melbourne, Australia, August 1983.

Hsu, H. T., Jordon, R. L., and Lawson, R. H., 1984, Monoclonal antibodies and viruses, *ASM News* **50**(3):99.

Incardona, N. L., and Kaesberg, P., 1964, A pH-induced structural change in bromegrass mosaic virus, *Biophys. J.* **4**:12.

Jaspars, E. M. J., and Bos, L., 1980, Alfalfa mosaic virus, *CMI/AAB Descriptions of Plant Viruses* No. 229 (No. 46 revised).

Jaspars, E. M. J., and Moed, J. R., 1966, The complexity of alfalfa mosaic virus, in: *Viruses of Plants* (A. B. R. Beemster and J. Dijkstra, eds.), pp. 188–195, North-Holland, Amsterdam.

Kaper, J. M., and Waterworth, H. E., 1981, Cucumoviruses, in: *Handbook of Plant Virus Infections and Comparative Diagnosis* (E. Kurstak, ed.), pp. 257–323, Elsevier/North-Holland, Amsterdam.

Koenig, R., 1978, ELISA in the study of homologous and heterologous reactions of plant viruses, *J. Gen. Virol.* **40:**309.

Koenig, R., 1981, Indirect ELISA for the broad specificity detection of plant viruses, *J. Gen. Virol.* **55:**53.

Koenig, R., Francksen, H., and Stegemann, H., 1981, Comparison of tymovirus capsid proteins in SDS polyacrylamide porosity gradient gels after partial cleavage with different proteases, *Phytopathol. Z.* **100:**347.

Krüse, J., Verduin, B. J. M., and Visser, A. J. W. G., 1981, Fluorescence of cowpea chlorotic mottle virus modified with pyridoxal-5'-phosphate, *Eur. J. Biochem.* **105:**395.

Kuhn, C. W., 1968, Identification of specific infectivity of a soybean strain of cowpea chlorotic mottle virus, *Phytopathology* **58:**1441.

Lane, L. C., 1981, Bromoviruses, in: *Handbook of Plant Virus Infections and Comparative Diagnosis* (E. Kurstak, ed.), pp. 333–376, Elsevier/North-Holland, Amsterdam.

Lane, L. C., and Kaesberg, P., 1971, Multiple genetic components in bromegrass mosaic virus, *Nature New Biol.* **232:**40.

Lommel, S. A., McCain, A. H., and Morris, T. J., 1982, Evaluation of indirect enzyme-linked immunosorbent assay for the detection of plant viruses, *Phytopathology* **72:**1018.

McMorran, J. P., and Cameron, H. R., 1983, Detection of 41 isolates of necrotic ringspot, apple mosaic, and prune dwarf viruses in *Prunus* and *Malus* by enzyme-linked immunosorbent assay, *Plant Dis.* **67:**536.

Marco, S., and Cohen, S., 1979, Rapid detection and titer evaluation of viruses in pepper by enzyme-linked immunosorbent assay, *Phytopathology* **69:**1259.

Matthews, R. E. F., 1970, *Plant Virology,* pp. 636–637, Academic Press, New York.

Matthews, R. E. F., 1979, Classification and nomenclature of viruses, *Intervirology* **12:**131.

Matthews, R. E. F., 1982, Classification and nomenclature of viruses, *Intervirology* **17:**1.

Miki, T., and Knight, C. A., 1965, Preparation of broad bean mottle virus protein, *Virology* **25:**478.

Milne, R. G., 1980, Some observations and experiments on immunosorbent electron microscopy of plant viruses, *Acta Hortic.* **110:**129.

Milne, R. G., and Luisoni, E., 1977, Rapid immune electron microscopy of virus preparations, in: *Methods in Virology* (K. Maramorosch and H. Koprowski, eds.), Volume VI, pp. 264–281, Academic Press, New York.

Mink, G. I., 1980, Identification of ringspot mosaic diseased cherry trees by enzyme-linked immunosorbent assay, *Plant Dis. Rep.* **64:**691.

Mink, G. I., and Uyeda, I., 1982, Ilarviruses: Suggested revision of subgroups [abstract], Proceedings, 4th International Conference on Comparative Virology, Banff, Alberta, August 1982.

Moosic, J. P., 1978, The primary structure of brome mosaic virus coat protein, Ph.D. thesis, University of Wisconsin, Madison.

Mossop, D. W., Francki, R. I. B., and Grivell, C. J., 1976, Comparative studies on tomato aspermy and cucumber mosaic viruses. V. Purification and properties of a cucumber mosaic virus inducing severe chlorosis, *Virology* **74:**544.

Murant, A. F., Abu-Salih, H. S., and Goold, R. A., 1973, *Rep. Scott. Hortic. Res. Inst.* p. 67.

Murthy, M. R. N., 1983, Comparison of the nucleotide sequences of cucumber mosaic virus and brome mosaic virus, *J. Mol. Biol.* **168:**469.

Musil, M., and Richter, J., 1983, Serological properties of cucumber mosaic virus isolated from *Cucurbita pepo* L. in Slovakia, *Biologia* (*Bratislava*) **38:**237.

Newmark, P., 1984, In search of novel immunogens, *Nature* (*London*) **311:**510.

O'Donnell, I. J., Shukla, D. D., and Gough, K. H., 1982, Electro-blot radio-immunoassay of virus-infected plant sap—A powerful new technique for detecting plant viruses, *J. Virol. Methods* **4:**19.

Olmsted, J. B., 1981, Affinity purification of antibodies from diazotised paper blots of heterogeneous protein samples, *J. Biol. Chem.* **256:**11955.

Oostergetel, G. T., Krijgsman, P. C. J., Mellema, J. E., Cusack, S., and Miller, A., 1981, Evidence for the absence of swelling of alfalfa mosaic virions, *Virology* **109**:206.

Ouchterlony, O., 1962, Diffusion-in-gel methods for immunological analysis. II, in: *Progress in Allergy* (P. Kallos and B. H. Waksman, eds.), Volume VI, pp. 30–154, Karger, Basel.

Pares, R. D., and Whitecross, M. I., 1982, Gold-labelled antibody decoration (GLAD) in the diagnosis of plant viruses by immuno-electron microscopy, *J. Immunol. Methods* **51**:23.

Pietersen, G., 1983, Properties and detection of alfalfa mosaic virus, M.Sc. thesis, University of Pretoria.

Polson, A., von Wechmar, M. B., and van Regenmortel, M. H. V., 1980a, Isolation of viral IgY antibodies from yolks of immunized hens, *Immunol. Commun.* **9**:475.

Polson, A., von Wechmar, M. B., and Fazakerley, G., 1980b, Antibodies to proteins from yolks of immunized hens, *Immunol. Commun.* **9**:495.

Proll, E., Richter, J., Hofferek, H., and Eisenbrandt, K., 1972, Untersuchungen zur Differenzierung von drei Staemmen des Trespenmosaik Virus. III. Charakterisierung einer pH-induzieten Aenderung der Konformation, *Zentralbl. Bakteriol.* (Abstr. II) **127**:573.

Purcifull, D. E., Christie, S. R., and Lima, J. A. A., 1981, Detection of four isometric plant viruses in sodium dodecyl sulfate immunodiffusion tests, *Phytopathology* **71**:1221.

Rao, A. L. N., Hatta, T., and Francki, R. I. B., 1982, Comparative studies on tomato aspermy and cucumber mosaic viruses. VII. Serological relationships reinvestigated, *Virology* **116**:318.

Richter, J., Eisenbrandt, K., Hofferek, H., and Proll, E., 1972, Untersuchungen über den Einfluss von Aldehyden auf das Trespenmosaik Virus. I. Versuche mit Formaldehyd, *Arch. Pflanzenschutz* **8**:253.

Richter, J., Eisenbrandt, K., Proll, E., and Hofferek, H., 1973, Untersuchungen über den Einfluss von Aldehyden auf das Trespenmosaik Virus. II. Versuche mit Glutaraldehyd, *Arch. Pflanzenschutz* **9**:211.

Richter, J., Proll, E., and Musil, M., 1979, Serological relationships between robinia mosaic, clover blotch and peanut stunt viruses, *Acta Virol.* **23**:489.

Roberts, I. M., Milne, R. G., and van Regenmortel, M. H. V., 1982, Suggested terminology for virus/antigen interactions observed by electron microscopy, *Intervirology* **18**:147.

Rochow, W. F., and Carmichael, L. E., 1979, Specificity among barley yellow dwarf viruses in enzyme immunosorbent assays, *Virology* **95**:415.

Roosien, J., and van Vloten-Doting, L., 1983, A mutant of alfalfa mosaic virus with an unusual structure, *Virology* **126**:155.

Rybicki, E. P., 1979, The serology of the Bromoviruses, M.Sc. thesis, University of Cape Town.

Rybicki, E. P., 1984, Investigations of viruses affecting South African small grains, Ph.D. thesis, University of Cape Town.

Rybicki, E. P., and Coyne, V. E., 1983, Serological differentiation of brome mosaic virus morphomers, *FEMS Microbiol. Lett.* **20**:103.

Rybicki, E. P., and von Wechmar, M. B., 1981, The serology of the Bromoviruses, 1. Serological interrelationships of the Bromoviruses, *Virology* **109**:309.

Rybicki, E. P., and von Wechmar, M. B., 1982, Enzyme-assisted immune detection of plant virus proteins electroblotted onto nitrocellulose paper, *J. Virol. Methods* **5**:267.

Scott, H. A., and Slack, S. A., 1971, Serological relationship of brome mosaic and cowpea chlorotic mottle viruses, *Virology* **46**:490.

Shepard, J. F., Secor, G. A., and Purcifull, D. E., 1974a, Immunological cross-reactivity between the dissociated capsid proteins of PVY group plant viruses, *Virology* **58**:464.

Shepard, J. F., Gaard, G., and Purcifull, D. E., 1974b, A study of tobacco etch virus-induced inclusions using indirect immunoferritin procedures, *Phytopathology* **64**:418.

Shepherd, R. J., Francki, R. I. B., Hirth, L., Hollings, M., Inouye, T., Macleod, R., Purcifull, D. E., Sinha, R. C., Tremaine, J. H., and Valenta, V., 1976, New groups of plant viruses approved by the International Committee on Taxonomy of Viruses, *Intervirology* **6**:181.

Shukla, D. D., and Gough, K. H., 1979, The use of protein A, from *Staphylococcus aureus*, in immune electron microscopy for detecting plant virus particles, *J. Gen. Virol.* **45**:533.

Shukla, D. D., O'Donnell, I. J., and Gough, K. H., 1983, Further studies on the electro-blot radioimmunoassay (EBRIA) for detecting plant viruses, *Acta Phytopathol. Acad. Sci. Hung.* **18**:79.

Singer, S. J., and Schick, A. F., 1961, The properties of specific stains for electron microscopy prepared by the conjugation of antibody molecules with ferritin, *J. Biophys. Biochem. Cytol.* **9**:519.

Skotland, C. B., and Kaniewski, W., 1981, Viruses in hop (*Humulus lupulus*), *Phytopathology* **71**:255 (Abstract).

Thomas, B. J., Barton, R. J., and Tuszynski, A., 1983, Hydrangea mosaic virus, a new Ilarvirus from *Hydrangea macrophylla* (Saxifragaceae), *Ann. Appl. Biol.* **103**:261.

Tien-Po, Rao, A. L. N., and Hatta, T., 1982, Cucumber mosaic virus from corn flower in China, *Plant Dis.* **66**:337.

Torrance, L., and Jones, R. A. C., 1981, Recent developments in serological methods suited for use in routine testing for plant viruses, *Plant Pathol.* **30**:1.

Towbin, H., Staehelin, T., and Gordon, J., 1979, Electrophoretic transfer of proteins from polyacrylamide gels to nitrocellulose sheets: Procedure and some applications, *Proc. Natl. Acad. Sci. USA* **76**:4350.

Uyeda, I., and Mink, G. I., 1983, Relationships among some Ilarviruses: Proposed revision of subgroup A, *Phytopathology* **73**:47.

van Balen, E., 1982, The effect of pretreatments of carbon-coated formvar films on the trapping of potato leafroll virus particles using immunosorbent electron microscopy, *Neth. J. Plant Pathol.* **88**:33.

van Regenmortel, M. H. V., 1967, Serological studies on naturally occurring strains and chemically induced mutants of tobacco mosaic virus, *Virology* **31**:467.

van Regenmortel, M. H. V., 1981, Tobamoviruses, in: *Handbook of Plant Virus Infections and Comparative Diagnosis* (E. Kurstak, ed.), pp. 541–564, Elsevier/North-Holland, Amsterdam.

van Regenmortel, M. H. V., 1982, *Serology and Immunochemistry of Plant Viruses*, Academic Press, New York.

van Regenmortel, M. H. V., and Burckard, J., 1980, Detection of a wide spectrum of tobacco mosaic virus strains by indirect enzyme immunosorbent assays (ELISA), *Virology* **106**:327.

van Regenmortel, M. H. V., and Pinck, L., 1981, Alfalfa mosaic virus group, in: *Handbook of Plant Virus Infections and Comparative Diagnosis* (E. Kurstak, ed.), pp. 415–421, Elsevier/North-Holland, Amsterdam.

van Regenmortel, M. H. V., and von Wechmar, M. B., 1970, A reexamination of the serological relationship between tobacco mosaic virus and cucumber virus 4, *Virology* **41**:330.

van Tol, R. G. L., and van Vloten-Doting, L., 1981, Lack of serological relationship between the 35K nonstructural protein of alfalfa mosaic virus and the corresponding proteins of three other plant viruses with a tripartite genome, *Virology* **109**:444.

van Vloten-Doting, L., and Jaspars, E. M. J., 1977, Plant covirus systems: Three component systems, in: *Comprehensive Virology*, Volume 11 (H. Fraenkel-Conrat and R. R. Wagner, eds.), pp. 1–53, Plenum Press, New York.

van Vloten-Doting, L., Kruseman, J., and Jaspars, E. M. J., 1968, The biological function and mutual dependence of bottom component and top component *a* of alfalfa mosaic virus, *Virology* **34**:728.

van Vloten-Doting, L., Hasrat, J. A., Oosterwijk, E., van't Sant, P., Schoen, M. A., and Roosien, J., 1980, Description and complementation analysis of 13 temperature-sensitive mutants of alfalfa mosaic virus, *J. Gen. Virol.* **46**:415.

van Vloten-Doting, L., Francki, R. I. B., Fulton, R. W., Kaper, J. M., and Lane, L. C., 1981, Tricornaviridae—a proposed family of plant viruses with tripartite, single-stranded RNA genomes, *Intervirology* **15**:431.

von Wechmar, M. B., 1967, A study of viruses affecting Gramineae in South Africa, Ph.D. thesis, University of Stellenbosch.

von Wechmar, M. B., and van Regenmortel, M. H. V., 1968, Serological studies on bromegrass mosaic virus and its protein fragments, *Virology* **34**:36.

von Wechmar, M. B., Kaufmann, A., Desmarais, F., and Rybicki, E. P., 1984, Detection of seed-transmitted brome mosaic virus by ELISA, radial immunodiffusion and immunoelectroblotting tests, *Phytopathol. Z.* **109**:341.

Wittman, H. G., and Paul, H. L., 1961, Vergleich der Aminosaurenzusammensetzung der Proteine des Echten Ackerbohnenmosaik-Virus, des broad bean mottle-Virus, und des Tobakmosaikvirus, *Phytopathol. Z.* **41**:74.

Yamazaki, H., and Kaesberg, P., 1963, Isolation and characterisation of a protein subunit of broad bean mottle virus, *J. Mol. Biol.* **6**:465.

CHAPTER 8

Virus Transmission

R. I. HAMILTON

I. INTRODUCTION

I wish at the outset to make a distinction between natural and experimental transmission of plant viruses. By natural transmission, I refer to the process of virus spread in the natural state, which in most instances would be in the out-of-doors and usually without the direct mediation by man. In this environment, transmission (i.e., introduction to and infection of the suscept) is governed by the interaction between the suscept, the virus, and those factors, biotic and abiotic, which impinge on this interaction. The biotic factors include biological vectors such as insects, nematodes, and fungi and may include man to the extent that he facilitates transmission by mechanical means in the normal course of crop husbandry, e.g., mechanical transmission of certain viruses to perennial plants in pastures by mowing machines and virus spread by horticultural processes involving grafting. The abiotic factors include temperature, soil fertility, and plant density insofar as they affect the physiology of the suscept and the activity of vectors.

Experimental transmission, on the other hand, would represent the introduction of viruses to suscepts in a controlled state, i.e., under circumstances where transmission is a desired result. Thus, the usual types of experiments whereby the host range of a virus is determined by mechanical inoculation or by the use of specific vectors would be examples of experimental transmission. It is important to recognize that the results of experimental transmission are sometimes at odds with what is observed in the natural state.

The purpose of this chapter is to compare and contrast the modes of transmission which occur in the four virus groups considered in this

R. I. HAMILTON • Agriculture Canada Research Station, Vancouver, British Columbia V6T 1X2, Canada.

245

volume and to assess their importance to the survival of the respective viruses.

II. MODES OF TRANSMISSION

A. Bromoviruses

The Bromoviruses consist of three definitive members, bromegrass mosaic virus (BMV) (type member), broad bean mottle virus (BBMV), and cowpea chlorotic mottle virus (CCMV); a fourth virus, melandrium yellow fleck (MYFV), is considered a possible member. These viruses typically have narrow host ranges: BMV infects primarily grasses; BBMV and CCMV are primarily viruses of legumes; and MYFV occurs naturally only in *Melandrium album*, a member of the Caryophyllaceae. The Bromoviruses are readily transmitted mechanically and by beetles; recent reports indicate that BMV may also be vectored by aphids. BBMV has been reported to be seed-transmitted to a low extent but there are conflicting reports about the seed transmission of BMV.

1. Mechanical Transmission

All members of the Bromovirus group are efficiently transmitted by mechanical inoculation under experimental conditions. At least one of these viruses, BMV, can be transmitted mechanically under field conditions. In an experiment in which people walked on alternate plots of BMV-infected and virus-free bromegrass and orchardgrass plants to simulate pasture conditions, BMV spread to 50% of the bromegrass plants and to 25% of the orchardgrass plants within a year (McKinney, 1953). Spread of BMV in a grass lawn may occur during mowing (Valverde, 1983). Although several cereals are experimental hosts of BMV, naturally infected cereals are rarely found in North America (Moline, 1973), presumably because the opportunities are rare for mechanical transmission of BMV from perennial grasses to cereals. Whether other Bromoviruses are transmitted mechanically under field conditions is not known.

BMV was experimentally transmitted by electroendosmosis (Polson and von Wechmar, 1980), a method which might have application in certain circumstances.

2. Vector Transmission

a. Nematodes

Three species of *Xiphinema* (i.e., *X. diversicaudatum*, *X. paraelongatum*, and *X.* n. sp.) have been demonstrated to transmit BMV under experimental (laboratory) conditions (Schmidt *et al.*, 1963; Fritzsche,

TABLE I. Beetle Vectors of Bromoviruses

Virus	Vector	Reference
Bromegrass mosaic	*Chaetocnema aridula*	Panarin (1978)
	Lema sexpunctata	Valverde (1983)
	Oulema melanopus	Panarin and Zabavina (1977a,b), Panarin (1978), Proeseler (1978)
	Phyllotreta vittula	Panarin (1977, 1978)
Broad bean mottle	*Acalymima trivittata*	Walters and Surin (1973)
	Colaspis flavida	Walters and Surin (1973)
	Diabrotica undecimpunctata	Walters and Surin (1973)
Cowpea chlorotic mottle	*Ceratoma ruficornis*	Gamez (1976), Blanco and Schumann (1979)
	C. trifurcata	Walters and Dodd (1969), Hobbs and Fulton (1979)
	Diabrotica balteata	Gamez (1976), Blanco and Schumann (1979)
	D. undecimpunctata	Walters and Dodd (1969), Hobbs and Fulton (1979)

1975) but there is no evidence that nematodes are natural vectors of BMV or of the other Bromoviruses.

b. Insects

Several chrysomelid beetles can act as vectors of representative Bromoviruses. Numerous vector experiments using aphids have, with one exception, failed; those involving chrysomelid beetles have demonstrated a consistent but low rate of transmission of representative Bromoviruses (Table I). The design of most of these experiments was to allow potential vector species, usually those known to feed on the plant in question, an acquisition feeding of about 24 hr on virus-infected plants and then to transfer them directly to test plants for an additional 24 hr. Beetles transmitted at low rates, usually less than 10%, but they rarely transmitted for longer than 1 to 2 days after acquisition, indicating that the mechanism of transmission was a noncirculative one. The regurgitant from beetle vectors contains a factor which prevents infection by non-beetle-transmitted viruses (e.g., tobacco mosaic and tobacco ringspot viruses) but it has no effect on two beetle-transmitted viruses (Gergerich *et al.*, 1983).

There is little information on the role that beetles play in the epidemiology of the Bromoviruses. No evidence of spread of CCMV in soybean or cowpea was observed in a field experiment lasting 5 years (Demski and Chalkley, 1979) even though infected source plants, two vector species, *Diabrotica undecimpunctata howardii* (Mann.) and *Ceretoma trifurcata* (Forst.), and susceptible plants were present.

As indicated above, aphids have not been found to act as vectors of Bromoviruses under laboratory conditions. However, von Wechmar and Rybicki (1981) have reported that at least four aphid species (*Diuraphis noxia, Rhopalosiphum padi, R. maidis,* and *Schizaphis graminum*) were able to acquire and transmit BMV from filter paper soaked with concentrated virus, through Parafilm membranes and from seedlings systemically infected following mechanical inoculation. Moreover, circumstantial evidence indicated that *D. noxia,* an aphid newly arrived in South Africa, was associated with outbreaks of Free State streak disease of wheat which has been shown to be due to a complex of BMV, barley yellow dwarf virus (BYDV), and an uncharacterized aphid-infecting virus. One of the characteristics of the Luteoviruses, of which BYDV is the type member, is the capacity of their coat proteins to encapsidate heterologous single-stranded viral RNAs (Rochow, 1977), including RNAs of unrelated viruses (Falk *et al.,* 1978; Waterhouse and Murant, 1981). Thus, the natural spread of BMV associated with Free State streak disease may be enhanced by mixed infection of wheat by BMV and BYDV.

c. Fungal Spores

An association between BMV and the uredospores of *Puccinia graminis tritici,* the stem rust fungus of wheat, was reported by von Wechmar (1980) who presented evidence that uredospores, from wheat doubly infected with stem rust and BMV, could transmit the virus to wheat seedlings when such spores were "dusted" onto wheat seedlings in the usual procedure for initiating stem rust infections. The type of association between BMV and the uredospore has not been unequivocally determined but serological studies (i.e., ELISA, fluorescent antibody binding, and immunosorbent electron microscopy) indicate that BMV is carried externally (Erasmus *et al.,* 1983). Further work with uredospores of another rust fungus, *P. recondita* f. sp. *tritici* (leaf rust), demonstrated a similar mode of transmission of BMV to wheat (von Wechmar *et al.,* 1982).

3. Seed and Pollen Transmission

There is only one documented instance of seed transmission of a Bromovirus. A strain of BBMV, one of two viruses associated with broad bean mosaic in Sudan, was found to be seed-transmitted (< 1%) in a Sudanese broad bean cultivar (Murant *et al.,* 1974). More recently, von Wechmar and Rybicki (1981) have claimed that BMV is seed-transmitted in the South African wheat cultivar Scheepers and they suggest that seed-transmitted BMV is the principal source of the virus for acquisition and transmission by aphids to other wheat cultivars.

There is little published information about the virus–seed relationship in plants infected with Bromoviruses. Gay (1969) demonstrated, by infectivity assays, that CCMV was present in immature seed coats but not in embryos of infected cowpea and that the virus in seed coats, as is common with many other viruses, was inactivated during seed maturation. No seed transmission was obtained following planting of 3000 seeds stored for a month after harvest and the same result was obtained with seeds of bean infected with bean yellow stipple virus, a strain of CCMV (Gamez, 1976). Similar results were obtained in a study of the virus–seed relationship in BMV-infected Atlas barley (Ednie, 1970). Although the report by Murant et al. (1974) does not provide information on the distribution of BBMV in seeds of broad bean, there is no reason to conclude that the distribution of BBMV in seeds is different from that of other Bromoviruses. The report by von Wechmar and Rybicki (1981) is at odds with the general observation for BMV (Lane, 1981) but the possibility of cultivar difference may explain their results. M. B. von Wechmar (personal communication) found that BMV was more readily detected by ELISA in embryos from some cultivars than from others. Whether these results indicate the presence of BMV antigen in situ or as a contaminant from infected seed coats, as proposed by McDonald and Hamilton (1972) for southern bean mosaic virus in French (common) bean, remains to be determined as does the relationship between the presence of BMV antigen and the seed transmissibility of the virus. Research at the cellular level is obviously required to determine the virus–seed relationship of the Bromoviruses with respect to seed transmission, especially in view of the fact that most instances of virus transmission by seed result from infection of the embryo (Baker and Smith, 1966).

B. Alfalfa Mosaic Virus

Alfalfa mosaic virus (AMV) occurs worldwide, and its various strains have been found in natural infections of about 150 plant species representing 22 families (Schmelzer et al., 1973) but the combined experimental and natural host range include over 600 species in 70 families (Horvath, 1981). Most of its hosts are herbaceous, but several woody hosts have recently been included in its natural host range (Schwenk et al., 1971; Bercks et al., 1973). The virus is transmitted mechanically, by grafting, by aphids, and by seed and pollen.

1. Mechanical Transmission

The virus is readily transmitted manually to a wide host range; interestingly, considerable difficulty has been reported in attempting to transmit isolates to alfalfa, one of its natural hosts, under greenhouse conditions (Hull, 1969). However, field-grown plants, or greenhouse plants which have been kept in darkness at 32–36°C for 24 hr before inoculation, occasionally become infected (Gibbs and Tinsley, 1961). Transmission may be affected by natural inhibitors or inactivators in extracts of the source plant (Hull, 1969) which can often be neutralized by supplementing inocula with additives, e.g., sodium diethyldithiocarbamate, thioglycolic acid, or nicotine (Gibbs and Harrison, 1976; Matthews, 1981). Alternatively, one can inoculate a host that is not affected by these inhibitors and use this host as a source of virus for transmission to other test plants. It is well to recognize, however, that there are many isolates of AMV whose host range is conditioned by viral genotype (Frosheiser, 1969).

2. Graft and Dodder Transmission

AMV occurs in a number of woody hosts, e.g., grape (Bercks *et al.*, 1973; Beczner and Lehoczky, 1981), and in *Ilex* sp., *Vibernum* spp., and *Hebe* sp. (Schwenk *et al.*, 1971) and thus can be propagated by bud or chip grafting. Because of inhibitors in AMV-infected peppermint (*Mentha piperita*), the virus could only be experimentally transmitted to other plant species by grafting (Lovisolo and Luisoni, 1963). The virus can also be transmitted between plants by the use of parasitic dodder, *Cuscuta* spp. (Schmelzer, 1956), which, when established as a bridge between an AMV-infected plant and a healthy one, allows passage of the virus from one plant to the other.

3. Vector Transmission

AMV has been reported to be transmitted by at least 15 aphid species (Hull, 1969) and this is the principal means of natural transmission. Transmission is nonpersistent (Swenson, 1952), i.e., characterized by rapid acquisition (10–30 sec) of the virus during aphid feeding on a source plant or from purified preparations, followed by immediate transmission, without a latent period, to test plants during a subsequent brief feeding period. Aphids lose the ability to continue transmission of the virus in less than an hour unless there is renewed acquisition access.

Although the evidence suggests that aphid transmission is associated with virus contamination of the maxillary stylets of the insect (Bradley, 1966), that evidence and evidence from more recent experiments have been interpreted to implicate an "ingestion and egestion" process involving the foregut in the transmission of viruses such as AMV (see review by Harris, 1977) Individual aphid species vary greatly in their capacity to transmit different AMV strains (Swenson, 1952). These differences may be associated with changes in the coat protein of the virion which, for some viruses, e.g., cucumber mosaic virus (Mossop and Francki, 1977), appear to govern the interaction between the virion and the stylet or foregut as well as the host preferences of the particular aphid species.

4. Seed and Pollen Transmission

Transmission of AMV by seeds has been reported for *Capsicum annuum* (Sutic, 1959), *Chenopodium quinoa, Datura stramonium,* and *Solanum nigrum* (Quantz, 1968, cited by Tosic and Pesic, 1975), *Medicago lupulina* (Paliwal, 1982), *M. sativa* (Belli, 1962; Zschau, 1964; Frosheiser, 1964; Tosic and Pesic, 1975; Ekbote and Mali, 1978; Paliwal, 1982), *Nicandra physaloides* (Gallo and Ciampor, 1977), and *Trifolium alexandrinum* (Mishra *et al.*, 1980). Rates of transmission vary from 0.1 to 50%, depending, probably, on the species or variety of the host and the virus strain. Cross-pollination experiments using infected and virus-free parents have established that the rate of transmission to developing seeds is about three times more frequent in pollen than via the ovule (Frosheiser, 1974; Hemmati and McLean, 1977). Aggregates of AMV particles were detected in thin sections of the ovary wall, tapetum, pollen cytoplasm, and cells of the embryonic cotyledon (Wilcoxson *et al.*, 1975). No virions were seen in 50 ovaries taken from 10 unfertilized ovules, a result which is surprising. However, without data on the rate of seed transmission to seedlings in this particular experiment, it is difficult to interpret the results. The aforementioned lower frequency of AMV transmission in ovules than in pollen might decrease the likelihood of detecting virions in the female gametophyte. Transmissibility of AMV can apparently remain stable in alfalfa seed stored for up to 5 years at temperatures between 18 and 27°C (Frosheiser, 1974).

Although transmission of AMV to the developing seed in alfalfa is mediated by infected pollen, there is no evidence that the seed-producing parent becomes infected as is the case with certain Ilarviruses (George and Davidson, 1963).

C. Cucumoviruses

Three viruses, cucumber mosaic (CMV), peanut stunt (PSV), and tomato aspermy (TAV), constitute the definitive Cucumoviruses. Two others, cowpea ringspot (Phatak *et al.*, 1976) and robinia mosaic (Schmelzer, 1971), have been considered probable members. The Cucumoviruses are transmitted by all of the known modes of transmission; their natural host ranges are narrow except that of CMV which is very wide. Kaper and Waterworth (1981) list 100 of the more than 470 species in 67 families that are natural hosts of CMV; a number of these are woody hosts, e.g., grape and various species of *Prunus*.

1. Mechanical Transmission

The Cucumoviruses are readily transmissible mechanically to a wide range of hosts. Depending on the source plant, receptor plant, and virus isolate, the inoculum may have to be amended with various additives to facilitate transmission. A general buffer for transmission from herbaceous hosts is 0.025 M phosphate buffer, pH 7.2, containing 0.02 M DIECA; for transmission from woody hosts to herbaceous hosts, the same buffer with higher ionic strength (0.05–0.1 M) is recommended by Kaper and Waterworth (1981).

2. Graft and Dodder Transmission

There are few authenticated reports of graft transmission of Cucumoviruses. CMV, which was part of a complex of viruses in *Prunus* spp., was graft-transmitted to peach trees which were then used as source trees for a number of experiments (Willison and Weintraub, 1957; Tremaine, 1966). Other reports of CMV in *Prunus* (Nyland, 1960; Kishi *et al.*, 1973) indicate that the virus, which may be latent in *Prunus*, is occasionally graft-transmitted.

Several species of dodder can transmit Cucumoviruses between herbaceous hosts (Schmelzer, 1956; Miller and Troutman, 1966); some members appear to be transmitted more readily than are others. CMV replicates in dodder, whereas tobacco mosaic virus (TMV), although acquired and transmitted by dodder, apparently does not. By placing dodder, previously cultured on plants infected with both viruses, on immune plants, CMV was separated from TMV and transmitted to susceptible hosts (Bennett, 1940).

3. Vector Transmission

Cucumoviruses are efficiently transmitted in the nonpersistent manner by over 60 aphid species (Kennedy *et al.*, 1962) and new vectors have been reported (Quiot *et al.*, 1982). There is more known, perhaps, about the virus–vector relationship of CMV than of the other Cucumoviruses. The rate of transmission is affected by the host on which the aphid species has been raised, the virus source plant, the test plant, the aphid species, and the virus strain. Simons (1955) demonstrated a difference in ability of three aphid species, the cotton aphid (*Aphis gossypii*), the bean aphid (*A. rumicis*), and the green peach aphid (*Myzus persicae*), to transmit the same CMV strain to test plants. About 35% of the test plants inoculated via the cotton and green peach aphids were infected while only 10% of those inoculated via the bean aphid became diseased. In the same experiment, the effect of source plant was demonstrated: three times as many test plants were infected when the green peach aphid acquired the virus from pepper compared to infections following acquisition from chard. Similar results were obtained in a later study (Simons, 1957) involving three CMV strains and two aphid species. The green peach aphid was about twice as efficient as the cotton aphid regardless of the source plant (pepper or chard), but either aphid transmitted the virus more efficiently when pepper was also the test plant. Normand and Pirone (1968) demonstrated that apparent differences in the transmissibility of CMV strains from source plants were abolished when aphids acquired the virus strains from purified preparation via membrane feeding techniques. These experiments (Simons, 1955, 1957; Normand and Pirone, 1968) and others (see reviews by Watson and Plumb, 1972; Pirone and Harris, 1977) demonstrate the complexity which attends what would appear to be a relatively simple biological phenomenon.

One aspect of the specificity and perhaps the efficiency of the transmission process by aphids is the interaction between the virion and those surfaces in the stylet (Taylor and Robertson, 1974; Lim et al., 1977) or foregut of the aphid that are thought to be involved in transmission. Pseudorecombinants constructed from the genome segments of the Q strain of CMV, which is transmissible by the green peach aphid (Habili and Francki, 1974), and the M strain, which is not (Mossop et al., 1976), have demonstrated that RNA 3 codes for a factor, most likely the coat protein, which is associated with aphid transmissibility (Mossop and Francki, 1977). Moreover, in in vitro reassembly experiments, transencapsidation of the RNA of strain CMV-6, which is inefficiently transmitted by A. gossypii, by the protein of a strain (CMV-T) which is readily transmitted, resulted in high transmission of strain 6, while encapsidation of the RNA of CMV-T by CMV-6 protein depressed transmission (Gera et al., 1979). The observed loss in aphid transmissibility of strains of CMV (Badami, 1958; Mossop and Francki, 1977) of PSV (Tolin, 1972), and of TAV (Hollings and Stone, 1971) may have resulted from changes in the amino acid sequences of the coat proteins as a consequence of mutation and the effect that these amino acid substitutions would have on the virus–vector relationship. Maintenance of an aphid-transmitted culture of CMV by continual mechanical inoculation (Badami, 1958) may allow for selection of non-aphid-transmitted strains from the bulk population, leading to the isolation of strains that are no longer aphid-transmitted.

4. Seed and Pollen Transmission

Each of the definitive Cucumoviruses is transmitted through seeds of at least one of its hosts, and most references to seed transmission of the Cucumoviruses concern transmission of CMV in both cultivated crops and weeds (Table II). Considerable attention has been paid recently to seed transmission of CMV in legumes, primarily Phaseolus vulgaris (common bean). Percent seed transmission varies from less than 1% to over 50%, depending on the virus–host combination. PSV and TAV ap-

TABLE II. Seed Transmission of Cucumoviruses

Virus	Host	Reference
Cowpea ringspot	Vigna unguiculata	Phatak et al. (1976)
Cucumber mosaic	Benincasa hispida	Sharma and Chohan (1973)
	Cerastium holosteoides	Tomlinson and Carter (1970b)
	Cucumis melo	Kendrick (1934)
	C. pepo	Reddy and Nariani (1963), Sharma and Chohan (1973)
	C. sativus	Doolittle and Gilbert (1919)
	Cucurbita moschata	Sharma and Chohan (1973)
	Echinocystis lobata	Doolittle and Gilbert (1919), Doolittle and Walker (1925), Lindberg et al. (1956)
	Lamium purpureum	Tomlinson and Carter (1970b)
	Lupinus luteus	Troll (1957), Poremskaya (1964)
	Phaseolus vulgaris	Bos and Maat (1974), Provvidenti (1976), Marchoux et al. (1977), Meiners et al. (1977), Davis et al. (1981)
	Spergula arvensis	Tomlinson and Carter (1970b)
	Stellaria media	Hani et al. (1970), Tomlinson and Carter (1970a,b), Hani (1971), Tomlinson and Walker (1973)
	Vigna cylindrica	Brantley et al. (1965)
	V. radiata	Phatak (1974), Iwaki (1978), Purivirojkul et al. (1978)
	V. sesquipedalis	Anderson (1957)
	V. unguiculata	Anderson (1957), Meiners et al. (1977), Iwaki (1978)
Peanut stunt	Arachis hypogaea	Troutman et al. (1967), Kuhn (1969)
	Glycine max	Iizuka and Yunoki (1974)
Tomato aspermy	Stellaria media	Noordam et al. (1965)

pear to be seed-transmitted much less efficiently than does CMV. CMV is readily seed-transmitted in *Stellaria media*, a common weed in agricultural lands (Tomlinson and Carter, 1970a). Seeds from infected plants can retain seed-transmissible CMV for many months in soil (Tomlinson and Walker, 1973) and seedlings arising from such seeds are a likely source of virus for transmission to crops by aphid vectors (Tomlinson and Carter, 1970b; Hani, 1971; Quiot et al., 1979).

There is little information on the relationship between Cucumoviruses and the gametes of their hosts which is surprising considering the prevalence of CMV in agricultural crops. Crowley (1957) detected CMV in seedcoats (27%) and embryos (0.7%) of wild cucumber in which seed transmission was 0.25 percent. Cucumoviruses are not readily distinguished from ribosomes *in situ*, thus making it difficult to study the gamete—virus relationship by electron microscopy. Differential hydrolysis of the RNA in ribosomes without affecting virion RNA has been utilized for detecting CMV in leaf tissues (Hatta and Francki, 1981), and its application to seed and pollen is warranted. It is likely, however, that Cucumoviruses infect the embryos of their hosts although the statements by Phatak (1980) and Mandahar (1981), based on the report by Tomlinson and Carter (1970b) that CMV occurs in pollen of *S. media*, should be interpreted in the light of studies pointing to contamination of pollen surfaces by some plant viruses (Hamilton *et al.*, 1977).

D. Ilarviruses

The Ilarviruses are an expanding group of quasi-isometric viruses comprised of 14 members (see Table III, Chapter 1), several of which are characterized by a marked instability of infectivity in crude sap. They infect a very wide range of hosts; tobacco streak virus (TSV), the type member and probably the most common of the group, infects many species in over 30 families (Fulton, 1971). As a group, the Ilarviruses are transmissible by all of the known modes of transmission. All are transmissible mechanically, albeit with difficulty in some cases. Many are seed-borne and some of these are also transmitted to the seed-bearing plant by pollen. Only two Ilarviruses, TSV and prunus necrotic ringspot virus (PNRSV), have been transmitted by invertebrate vectors.

1. Mechanical Transmission

The infectivity of several Ilarviruses isolated from woody plants is very labile and usually requires stabilization either by adjusting the pH to about 8.0–8.7 and/or by the addition of chemical compounds to inocula (Fulton, 1957, 1966). For example, prune dwarf virus (PDV) in sap from infected cucumber can be stabilized for several hours by the addition of a mixture of 0.01 M Na-DIECA and 0.005 M cysteine hydrochloride (Hampton and Fulton, 1961) whereas about 50% of the infectivity is lost in less than a minute in the absence of these compounds. Other additives, e.g., sodium thioglycolate and 2-mercaptoethanol (Fulton, 1965), caffeine, polyvinylpyrrolidone, and polyethylene glycol (Ramaswamy and Posnette, 1971), have been used and it is now standard practice to incorporate a chemical stabilizer in the inoculum during mechanical transmission of most Ilarviruses.

Mechanical transmission of Ilarviruses from herbaceous plants back to their woody hosts is generally difficult (Fulton, 1966) but inoculation of very young seedlings usually results in infection (Fulton, 1958). Experimental transmission of citrus leaf rugose virus to citrus by contaminated cutting tools has been reported (Garnsey and Gonsalves, 1976).

2. Graft and Dodder Transmission

The Ilarviruses infecting woody hosts are graft-transmitted in these hosts, usually by bud or bark grafting, and this mode of transmission is eminently suited to the wide distribution of the viruses in planting stock. Three Ilarviruses (plum line pattern virus, PDV, and PNRSV) were able to pass through graft unions in *Prunus* spp. and achieve 100% transmission in 72 hr as opposed to other viruses which required between 100 and 150 hr of bud contact for maximum transmission (Fridlund, 1967). Root-grafts have also been implicated in the transmission of some of these viruses between neighboring trees.

Two species of dodder (*Cuscuta campestris* and *C. subinclusa*) transmitted tulare apple mosaic virus from tobacco to apple (Yarwood, 1955) and TSV was transmitted by *C. campestris* from tobacco to pea and red clover (Fulton, 1948). Experiments with these and several other dodder species failed to demonstrate transmission of apple mosaic virus (Gilmer, 1958), elm mottle virus (Schmelzer, 1969), or PNRSV (Fulton, 1970).

3. Vector Transmission

Reports of vector transmission of Ilarviruses are very few and are limited to nematode and thrips vectors.

a. Nematodes

Transmission of PNRSV by *Longidorus macrosoma* from cucumber to cucumber in laboratory experiments has been reported by Fritzsche and Kegler (1968) but the role of this nematode in the field spread of PNRSV is not known. The reported slow spread of PNRSV from infected hop to adjacent plants (Anonymous, 1968, 1969; Thresh, 1980, 1981) suggests the possibility of a nematode vector.

b. Insects

TSV has been transmitted experimentally by thrips (*Frankliniella* sp.) from *Ambrosia polystachya*, a wild reservoir host, to soybean and tobacco in Brazil (Costa and Lima Neto, 1976) and more recently in North America, from infected white sweet clover to *Chenopodium quinoa* and white sweet clover by *Thrips tabaci* and/or *F. occidentalis* (Kaiser *et al.*, 1982). In the latter study, about 15% of cowpea seedlings growing as a trap crop near TSV-infected sweet clover became infected with the virus as did sweet clover plants growing near infected, thrips-infested sweet clover in a lath house (shade house).

No other insect has been reported to transmit TSV, despite numerous attempts (Converse and Lister, 1969; Kaiser *et al.*, 1982; Salazar *et al.*, 1982). Similarly, no vector of PNRSV was established in experiments involving many species of aphids, leaf hoppers, other insects and mites (Phillips, 1951; Swenson and Milbrath, 1964) although transmission by a mite (*Vasates fockeui*) and by an aphid (*Amphororphora rubitoxica* has been reported by Proeseler (1968) and by Swenson and Milbrath (1964), respectively.

4. Seed and Pollen Transmission

A high proportion of the Ilarviruses are seed-transmitted, often in their important natural hosts (Table III). Seed transmission is a particularly serious problem in *Prunus* spp., where seedlings have traditionally been the source of rootstocks used in the propagation of commerical fruit trees (Gilmer and Kamalsky, 1962; Fleisher *et al.*, 1964). Moreover, pollen transmission to seeds on healthy trees has been reported for PDV (Gilmer and Way, 1960; George, 1962) and PNRSV (Way and Gilmer, 1958; Gilmer and Way, 1960). There is little published information on virus–seed relationships among the seed-transmitted Ilarviruses, and what little information there is was obtained by infectivity assays of whole seeds or seed parts. PNRSV was detected in cotyledons and/or seed coats of Montmorency cherry (George, 1962; Megahed and Moore, 1967), but not in the plumules, hypocotyls, or radicles of seeds of the same variety (Megahed and Moore, 1967). Infectious TSV was recovered from 90–100% of the seed coats, embryos, and cotyledons of mature Bragg soybean seed in which seed transmission was about 90% (Kaiser *et al.*, 1982) but not from seed coats and mature embryos of two other varieties in which seed transmission was about 30% (Ghanekar and Schwenk, 1974). Spinach latent virus was not detected in mature seed coats of spinach but about 45% of the embryos contained infective virus (Bos *et al.*, 1980). In this latter instance, seed transmission was only 20%, suggesting that not all infected embryos produced infected seedlings.

A characteristic of some Ilarviruses, especially those of woody hosts, is their transmission from plant to plant by pollen (Das and Milbrath, 1961; Gilmer and Way, 1963; George and Davidson, 1963, 1964; Gilmer, 1965; Converse and Lister, 1969; Cameron *et al.*, 1973; Smith and Stubbs, 1976; Davidson, 1976). Circumstantial evidence suggests that the route of infection is through the flower because deblooming of healthy plants effectively prevents or markedly decreases the rate of spread. For viruses such as PDV and PNRSV the major method of plant-to-plant spread in woody hosts is probably via pollen from infected donors (Davidson, 1976). Spread of TSV in *Rubus* spp. may also occur in a similar manner (Converse, 1977) although the recent reports of thrips transmission of TSV to other crops (Costa and Lima Neto, 1976; Kaiser *et al.*, 1982) and the evidence for aerial spread of the virus in *Rubus* (Converse, 1980) suggest that a route other than that by pollen may be operative in field spread of this virus in this genus.

Infective PNRSV (Ehlers and Moore, 1957; Williams *et al.*, 1962) and PDV (Ehlers and Moore, 1957) have been detected in homogenates of pollen from infected *Prunus* as has infective TSV in pollen homogenates from *Rubus* (Converse and Lister, 1969) and the assumption has been

TABLE III. Seed Transmission of Ilarviruses

Virus	Host	Reference
Asparagus virus II	*Asparagus officinalis*	Uyeda and Mink (1981)
Black raspberry latent	*Rubus occidentalis*	Converse and Lister (1969), Lister and Converse (1972)
Elm mottle	*Ulmus glabra*	Jones and Mayo (1973)
Lilac ring mottle	*Celosia argentea*	Van der Meer *et al.* (1976)
	Chenopodium amaranticolor	Van der Meer *et al.* (1976)
	C. quinoa	Van der Meer *et al.* (1976)
Prune dwarf	*Prunus cerasus*	Gilmer and Way (1960)
	P. mahaleb	Cation (1949, 1952)
	P. persica	Cochran (1950)
Prunus necrotic ringspot	*Cucurbita maxima*	Das and Milbrath (1961), Das *et al.* (1961)
	Prunus americana	Hobart (1956)
	P. avium	Cochran (1946), Megahed and Moore (1967)
	P. cerasus	Cation (1949, 1952), Megahed and Moore (1967), Davidson (1976)
	P. mahaleb	Cation (1949), Megahed and Moore (1967)
	P. pennsylvanica	Megahed and Moore (1957)
	P. persica	Cochran (1950), Millikan (1959), Wagnon *et al.* (1960)
Spinach latent	*Celosia cristata*	Bos *et al.* (1980)
	Chenopodium quinoa	Bos *et al.* (1980)
	Nicotiana rustica	Bos *et al.* (1980)
	N. tabacum	Bos *et al.* (1980)
Tobacco streak	*Chenopodium amaranticolor*	Shukkla and Gough (1983)
	C. quinoa	Brunt (1969), Kaiser *et al.* (1982), Shukkla and Gough (1983)
	Datura stramonium	Brunt (1969)
	Fragaria × Ananassa	R. H. Converse (personal communication)
	Glycine max	Ghanekar and Schwenk (1974), Kaiser *et al.* (1982)
	Gomphrena globosa	Ghanekar and Schwenk (1974)
	Melilotus alba	Kaiser *et al.* (1982)
	Nicandra physaloides	Salazar *et al.* (1982)
	Nicotiana clevelandii	Ghanekar and Schwenk (1974)
	Phaseolus vulgaris	Thomas and Graham (1951)
	Vigna unguiculata	Shukkla and Gough (1983)

that such pollen and the pollen associated with field spread are infected. A consequence of this assumption is that fertilization of the egg by infected pollen would lead to infection of the seed-bearing plant. This is not likely to occur because of the isolation of the zygote from direct plasmodesmatal connections with the surrounding nucellar tissues as observed in barley infected with barley stripe mosaic virus (Carroll and Mayhew, 1976) and in tobacco ringspot virus-infected soybean (Yang and Hamilton, 1974). More recently, evidence obtained by infectivity tests and ELISA suggests that PNRSV is located primarily, if not exclusively, on the surface of the pollen in sweet cherry (Cole et al., 1982). On the other hand, Kelley and Cameron (1983) obtained evidence by ELISA and electron microscopy of both infection and surface contamination of pollen from sweet cherry infected with either PNRSV or PDV.

The role of surface-borne virus in the etiology of diseases caused by these pollen-transmitted viruses remains to be determined. The circumstantial evidence implicating the flower as the site of natural infection of *Prunus* spp. by PNRSV and PDV (Gilmer and Way, 1960; Cameron *et al.*, 1973; Davidson, 1976) suggests that surface-contaminated pollen is involved in plant-to-plant transmission of these viruses, probably via a mechanical inoculation process (Cole *et al.*, 1982), while pollination with infected pollen would lead to transmission of these viruses to the egg and consequent seed infection. The report by Williams *et al.* (1963) that infection of three *Prunus* spp. with PNRSV and PDV was obtained by inserting pollen from infected trees under the bark of test trees suggests a mechanical inoculation process involving virus-contaminated pollen.

III. CONCLUDING REMARKS

The modes of transmission which appear to be important to the survival of plant viruses with tripartite ssRNA genomes are largely dependent on the member groups and in some instances on the particular virus. Although all of the viruses are mechanically transmissible, this mode is not particularly suitable for their survival. On the other hand, transmission by aphid vectors coupled with seed transmission affords the Cucumoviruses and alfalfa mosaic virus with excellent survival opportunities as well as the circumstances for genetic interchange between some related viruses. The Ilarviruses, many of which infect woody perennials, are readily transmissible to suscepts by grafting, by seed, and, in some instances, directly by pollen. By contrast, the Bromoviruses appear to be more dependent on mechanical transmission than do the other members. However, the process of transmission by beetles, although there is specificity in the virus–vector relationship, is probably analogous to mechanical inoculation. Specificity in the transmission process may markedly enhance survival of these viruses when non-beetle-transmitted viruses and Bromoviruses are acquired by beetles feeding on plants in-

fected by a mixture of such viruses. It is also noteworthy that the Bromoviruses reach high concentrations in infected cells of their hosts, and
thus the transmission of BMV by aphids and contaminated seeds (von
Wechmar and Rybicki, 1981; von Wechmar *et al.*, 1982) and by uredospores (von Wechmar, 1980) may be related to exceptionally high inoculum
concentrations.

Many aspects of the tranmission of the viruses discussed here require
further study. Among these are the identification of viral genes which
affect seed transmission (alfalfa mosaic virus and Cucumoviruses); the
site(s) of infection associated with plant-to-plant transmission by pollen
from infected plants (Ilarviruses); the details of the virus–vector relationship in beetle-transmitted viruses (Bromoviruses); and the specific
interaction between BMV and the various "vectors" associated with its
spread in cereals in South Africa.

REFERENCES

Anderson, C. W., 1957, Seed transmission of three viruses in cowpea, *Phytopathology*
    47:515.
Anonymous, 1968, Annual Report for 1967, p. 43, East Malling Research Station, Maidstone,
    U.K.
Anonymous, 1969, Annual Report for 1968, p. 40, East Malling Research Station, Maidstone,
    U.K.
Badami, R. S., 1958, Changes in the transmissibility by aphids of a strain of cucumber mosaic
    virus, *Ann. Appl. Biol.* **46**:554.
Baker, K. F., and Smith, S. H., 1966, Dynamics of seed transmission of plant pathogens,
    Annu. Rev. Phytopathol. **4**:311.
Beczner, L., and Lehoczky, J., 1981, Grapevine disease in Hungary caused by alfalfa mosaic
    virus infection, *Acta Phytopathol. Acad. Sci. Hung.* **16**:119.
Belli, G., 1962, Rilievi ed esperienze sulla transmissione per seme del virus de mosaico
    dellerba medica e dimostrazione della sulla exclusione in cloni vite virosati, *Ann. Fac.
    Agric. Milano* **10**.
Bennett, C. W., 1940, Acquisition and transmission of viruses by dodder (*Cuscuta subin-
    clusa*), *Phytopathology* **30**:2 (abstract).
Bercks, R., Lesemann, O., and Querfurth, G., 1973, Uber den Nachweis des alfalfa mosaic
    virus in einer Weinrebe, *Phytopathol. Z.* **76**:166.
Blanco, N., and Schumann, K., 1979, Vectors of bean yellow stipple virus in Cuba, *Cola-
    boracion Cientifico-Tecnica Cuba-R.D.A.* **1979**:28.
Bos, L., and Maat, D. Z., 1974, A strain of cucumber mosaic virus seed-transmitted in beans,
    Neth. J. Plant Pathol. **80**:113.
Bos, L., Huttinga, H., and Maat, D. Z., 1980, Spinach latent virus, a new Ilarvirus seed-borne
    in *Spinacia oleracea, Neth. J. Plant Pathol.* **86**:79.
Bradley, R. H. E., 1966, Which of an aphid's stylets carry transmissible virus?, *Virology*
    29:396.
Brantley, B. B., Kuhn, C. W., and Sowell, G., Jr., 1965, Effect of cucumber mosaic virus on
    southern pea (*Vigna sinensis*), *Proc. Am. Soc. Hortic. Sci.* **87**:355.
Brunt, A. A., 1969, Annual Report for 1968, p. 104, Glasshouse Crops Research Institute,
    Littlehampton, U.K.
Cameron, H. R., Milbrath, J. A., and Tate, L. A., 1973, Pollen transmission of prunus ringspot
    in prune and sour cherry orchards, *Plant Dis. Rep.* **57**:241.

Carroll, T. W., and Mayhew, D. E., 1976, Occurrence of virions in developing ovules and embryo sacs of barley in relation to the seed transmissibility of barley stripe mosaic virus, *Can. J. Bot.* **54**:2497.

Cation, D., 1949, Transmission of cherry yellows virus complex through seeds, *Phytopathology* **39**:37.

Cation, D., 1952, Further studies on transmission of ringspot and cherry yellows viruses through seeds, *Phytopathology* **42**:4 (abstract).

Cochran, L. C., 1946, Passage of the ringspot virus through mazzard cherry seeds, *Science* **104**:269.

Cochran, L. C., 1950, Passage of the ringspot virus through peach seeds, *Phytopathology* **40**:964 (abstract).

Cole, A., Mink, G. I., and Regev, S., 1982, Location of prunus necrotic ringspot virus on pollen grains from infected almond and cherry trees, *Phytopathology* **72**:1542.

Converse, R. H., 1977, *Rubus* viruses in the United States, *Hort. Science* **12**:471.

Converse, R. H., 1980, Transmission of tobacco streak virus in *Rubus, Acta Hortic.* **95**:53.

Converse, R. H., and Lister, R. M., 1969, The occurrence and some properties of black raspberry latent virus, *Phytopathology* **59**:325.

Costa, A. S., and Lima Neto, V. da C., 1976, Transmissao do virus da necrose branca do fumo por *Frankliniella* sp., *IX Congr. Soc. Bras. Fitopatol.*

Crowley, N. C., 1957, Studies on the seed transmission of plant virus diseases, *Aust. J. Biol. Sci.* **10**:449.

Das, C. R., and Milbrath, J. A., 1961, Plant-to-plant transfer of stone fruit ringspot virus in squash by pollination, *Phytopathology* **51**:489.

Das, C. R., Milbrath, J. A., and Swenson, K. G., 1961, Seed and pollen transmission of prunus ringspot virus in buttercup squash, *Phytopathology* **51**:64.

Davidson, T. R., 1976, Field spread of prunus necrotic ringspot in sour cherries in Ontario, *Plant Dis. Rep.* **60**:1080.

Davis, R. F., Weber, Z., Pospieszny, H., Silbernagel, M., and Hampton, R. O., 1981, Seedborne cucumber mosaic virus in selected *Phaseolus vulgaris* germ plasm and breeding lines in Idaho, Washington and Oregon, *Plant Dis.* **65**:492.

Demski, J., and Chalkley, J., 1979, Non-movement of cowpea chlorotic mottle virus from cowpea and soybean, *Plant Dis. Rep.* **63**:761.

Doolittle, S. P., and Gilbert, W. W., 1919, Seed transmission of cucurbit mosaic by the wild cucumber, *Phytopathology* **9**:326.

Doolittle, S. P., and Walker, M. N., 1925, Further studies on the overwintering and dissemination of cucurbit mosaic, *J. Agric. Res.* **31**:1.

Ednie, A. B., 1970, Investigations into the non-seed transmissibility of bromegrass mosaic virus in barley, M.Sc. thesis, McGill University.

Ehlers, C. G., and Moore, J. D., 1957, Mechanical transmission of certain stone fruit viruses from *Prunus* pollen, *Phytopathology* **47**:519 (abstract).

Ekbote, A. U., and Mali, V. R., 1978, Occurrence of alfalfa mosaic virus in alfalfa in India, *Indian Phytopathol.* **31**:171.

Erasmus, D. S., Rybicki, E. P., and von Wechmar, M. B., 1983, The association of brome mosaic virus and wheat rusts. II. Detection of BMV in/on uredospores of wheat stem rust, *Phytopathol. Z.* **108**:34.

Falk, B. W., Duffus, J. E., and Morris, T. J., 1978, Transmission, host range, and serological properties of the viruses that cause lettuce speckles disease, *Phytopathology* **69**:612.

Fleisher, Z., Blodgett, E. C., and Aichele, M. D., 1964, Presence of virus (necrotic ringspot group) in mazzard and mahaleb cherry seedlings grown in the Pacific Northwest from various seed sources, *Plant Dis. Rep.* **48**:280.

Fridlund, P. R., 1967, The relationship of inoculum-receptor contact period to the rate of graft transmission of twelve *Prunus* viruses *Phytopathology* **57**:1296.

Fritzsche, R., 1975, Übertragung des Trespenmosaik - und Arabis-Mosaik-Virus durch Nematoden in Abhängigkeit von der Infektiosität der Wurzeln der Wirtspflanzen, *Arch. Phytopathol. Pflanzenschutz* **11**:197.

Fritzsche, R., and Kegler, H., 1968, Nematodes as vectors of virus diseases of fruit plants, *Dtsch. Akad. Landwirtsch. Wiss. Berlin Tagungsber* **97**:289.

Frosheiser, F. I., 1964, Alfalfa mosaic virus transmitted through alfalfa seed, *Phytopathology* **54**:893 (abstract).

Frosheiser, F. I., 1969, Variable influence of alfalfa mosaic virus on growth and survival of alfalfa and on mechanical and aphid transmission, *Phytopathology* **59**:857.

Frosheiser, F. I., 1974, Alfalfa mosaic virus transmission to seed through alfalfa gametes and longevity in alfalfa seed, *Phytopathology* **64**:102.

Fulton, R. W., 1948, Hosts of the tobacco streak virus, *Phytopathology* **38**:421.

Fulton, R. W., 1957, Properties of certain mechanically transmitted viruses of *Prunus*, *Phytopathology* **47**:683.

Fulton, R. W., 1958, Identity of and relationships among certain sour cherry viruses mechanically transmitted to *Prunus* species, *Virology* **6**:499.

Fulton, R. W., 1965, A comparison of two viruses associated with plum line pattern and apple mosaic, *Zast. Bilja* **16**:427.

Fulton, R. W., 1966, Mechanical transmission of viruses of woody hosts, *Annu. Rev. Phytopathol.* **4**:79.

Fulton, R. W., 1970, Prunus necrotic ringspot virus, *CMI/AAB Descriptions of Plant Viruses* No. 5.

Fulton, R. W., 1971, Tobacco streak virus, *CMI/AAB Descriptions of Plant Viruses* No. 44.

Gallo, J., and Ciampor, F., 1977, Transmission of alfalfa mosaic virus through *Nicandra physaloides* seeds and its location in embryo cotyledons, *Acta Virol.* **21**:344.

Gamez, R., 1976, Bean viruses in Central America. IV. Some properties of bean yellow stipple virus and its transmission by chrysomelids, *Turrialba* **26**:160.

Garnsey, S. M., and Gonsalves, D., 1976, Citrus leaf rugose virus, *CMI/AAB Descriptions of Plant Viruses* No. 164.

Gay, J. D., 1969, Effect of seed maturation on the infectivity of cowpea chlorotic mottle virus, *Phytopathology* **59**:802.

George, J. A., 1962, A technique for detecting virus-infected Montmorency cherry seeds, *Can. J. Plant Sci.* **42**:193.

George, J. A., and Davidson, T. R., 1963, Pollen transmission of necrotic ringspot and sour cherry yellows viruses from tree to tree, *Can. J. Plant Sci.* **43**:276.

George, J. A., and Davidson, T. R., 1964, Further evidence of pollen transmission of necrotic ringspot and sour cherry yellows viruses in sour cherry, *Can. J. Plant Sci.* **44**:383.

Gera, A., Loebenstein, G., and Raccah, B., 1979, Protein coats of two strains of cucumber mosaic virus affect aphid transmission by *Aphis gossypii*, *Phytopathology* **69**:396.

Gergerich, R. C., Scott, H. A., and Fulton, J. P., 1983, Regurgitant as a determinant of specificity in the transmission of plant viruses by beetles, *Phytopathology* **73**:936.

Ghanekar, A. M., and Schwenk, F. W., 1974, Seed transmission and distribution of tobacco streak virus in six cultivars of soybeans, *Phytopathology* **64**:112.

Gibbs, A. J., and Harrison, B. D., 1976, *Plant Virology: The Principles*, Arnold, London.

Gibbs, A. J., and Tinsley, T. W., 1961, Lucerne mosaic virus in Great Britain, *Plant Pathol.* **10**:61.

Gilmer, R. M., 1958, Two viruses that induce mosaic of apple, *Phytopathology* **48**:432.

Gilmer, R. M., 1965, Additional evidence of tree-to-tree transmission of sour cherry yellows virus by pollen, *Phytopathology* **55**:482.

Gilmer, R. M., and Kamalsky, L. R., 1962, The incidence of necrotic ringspot and sour cherry yellows viruses in commercial mazzard and mahaleb cherry rootstocks, *Plant Dis. Rep.* **46**:583.

Gilmer, R. M., and Way, R. D., 1960, Pollen transmission of necrotic ringspot and prune dwarf viruses in sour cherry, *Phytopathology* **50**:624.

Gilmer, R. M., and Way, R. D., 1963, Evidence for tree-to-tree transmission of sour cherry yellows virus by pollen, *Plant Dis. Rep.* **47**:1051.

Habili, N., and Francki, R. I. B., 1974, Comparative studies on tomato aspermy and cucumber mosaic viruses. III. Further studies on relationship and construction of a virus from parts of the two viral genomes, *Virology* **61**:443.

Hamilton, R. I., Leung, E., and Nichols, C., 1977, Surface contamination of pollen by plant viruses, *Phytopathology* **67**:395.

Hampton, R. E., and Fulton, R. W., 1961, The relation of polyphenol oxidase to instability *in vitro* of prune dwarf and sour cherry necrotic ringspot viruses, *Virology* **13**:44.

Hani, A., 1971, Zur Epidemiologie des Gurkenmosaikvirus in Tessin, *Phytopathol. Z.* **72**:115.

Hani, A., Pelet, F., and Kern, H., 1970, Zur Bedeutung von *Stellaria media* (L.) Vill. in der Epidemiologie des Gurkenmosaikvirus, *Phytopathol. Z.* **68**:81.

Harris, K. F., 1977, An ingestion–egestion hypothesis of non-circulative virus transmission, in: *Aphids as Vectors* (K. F. Harris and K. Maramorosch, eds.), pp. 165–220, Academic Press, New York.

Hatta, T., and Francki, R. I. B., 1981, Identification of small polyhedral virus particles in thin sections of plant cells by an enzyme cytochemical technique, *J. Ultrastruct. Res.* **74**:116.

Hemmati, K., and McLean, D. L., 1977, Gamete-seed transmission of alfalfa mosaic virus and its effect on seed germination and yield in alfalfa plants, *Phytopathology* **67**:576.

Hobart, O. F., 1956, Introduction and spread of necrotic ringspot virus in sour cherry nursery trees, *Iowa State Coll. J. Sci.* **30**:381.

Hobbs, H. A., and Fulton, J. P., 1979, Beetle transmission of cowpea chlorotic mottle virus, *Phytopathology* **69**:255.

Hollings, M., and Stone, O. M., 1971, Tomato aspermy virus, *CMI/AAB Descriptions of Plant Viruses* No. 79.

Horvath, J., 1981, New artificial hosts and non-hosts of plant viruses and their role in the identification and separation of viruses. XV. Monotypic (almovirus) group: Alfalfa mosaic virus, *Acta Phytopathol. Acad. Sci. Hung.* **16**:315.

Hull, R., 1969, Alfalfa mosaic virus, *Adv. Virus Res.* **15**:365.

Iizuka, N., and Yunoki, T., 1974, Peanut stunt virus isolated from *Glycine max* Merr., *Bull. Tohoku Natl. Agric. Exp. Stn.* **47**:1.

Iwaki, M., 1978, Seed transmission of cucumber mosaic virus in mungbean (*Vigna radiata*), *Ann. Phytopathol. Soc. Jpn.* **44**:337.

Jones, A. T., and Mayo, M. A., 1973, Purification and properties of elm mottle virus, *Ann. Appl. Biol.* **75**:347.

Kaiser, W. J., Wyatt, S. D., and Pesho, G. R., 1982, Natural hosts and vectors of tobacco streak virus in eastern Washington, *Phytopathology* **72**:1508.

Kaper, J. M., and Waterworth, H. E., 1981, Cucumoviruses, in: *Handbook of Plant Virus Infections and Comparative Diagnosis* (E. Kurstak, ed.), pp. 257–332, Elsevier/North-Holland, Amsterdam.

Kelley, R. D., and Cameron, H. R., 1983, Location of prune dwarf and prunus necrotic ringspot viruses in sweet cherry pollen and fruit, *Phytopathology* **73**:791 (abstract).

Kendrick, J. B., 1934, Cucurbit mosaic transmitted by muskmelon seed, *Phytopathology* **24**:820.

Kennedy, J. S., Day, M. F., and Eastop, V. F., 1962, *A Conspectus of Aphids as Vectors of Plant Viruses*, Commonw. Inst. Entomol. London.

Kishi, K., Abiko, K., and Takanashi, K., 1973, Studies on the virus diseases of stone fruit. VII. Cucumber mosaic virus isolated from *Prunus* trees, *Ann. Phytopathol. Soc. Jpn.* **39**:297.

Kuhn, C. W., 1969, Effects of peanut stunty virus alone and in combination with peanut mottle virus on peanut, *Phytopathology* **59**:1513.

Lane, L. C., 1981, Bromoviruses, in: *Handbook of Plant Virus Infections and Comparative Diagnosis* (E. Kurstak, ed.), pp. 333–376, Elsevier/North-Holland, Amsterdam.

Lim, W. L., de Zoeten, G. A., and Hagedorn, D. J., 1977, Scanning electron microscope evidence for attachment of a non-persistently transmitted virus to its vector's stylets, *Virology* **79**:121.

Lindberg, G. D., Hall, D. H., and Walker, J. C., 1956, A study of melon and squash mosaic viruses, *Phytopathology* **46**:489.

Lister, R. M., and Converse, R. H., 1972, Black raspberry latent virus, *CMI/AAB Descriptions of Plant Viruses* No. 106.

Lovisolo, O., and Luisoni, E., 1963, A new virosis of peppermint and the presence in this plant of a virus inhibitor, *Atti Accad. Sci. Torino Cl. Sci. Fis. Mat. Nat.* **98**:213.

McDonald, J. G., and Hamilton, R. I., 1972, Distribution of southern bean mosaic virus in the seed of *Phaseolus vulgaris*, *Phytopathology* **62**:387.

McKinney, H. H., 1953, Virus diseases of cereal crops, in: *Plant Diseases, the Yearbook of Agriculture* (A. Stefferud, ed.), pp. 350–360, U.S. Department of Agriculture, Washington.

Mandahar, C. L., 1981, Virus transmission though seed and pollen, in: *Plant Diseases and Vectors: Ecology and Epidemiology* (K. Maramorosch and K. R. Harris, eds.), pp. 241–292, Academic Press, New York.

Marchoux, G., Quiot, J. B., and Devergne, J. C., 1977, Characterisation d'un isolat du virus de la mosaique du concombre transmis par les graines du haricot (*Phaseolus vulgaris* L.), *Ann. Phytopathol.* **9**:421.

Matthews, R. E. F., 1981, *Plant Virology*, 2nd ed., Academic Press, New York.

Megahed, E.-S., and Moore, J. D., 1967, Differential mechanical transmission of *Prunus* viruses from seed of various *Prunus* spp. and from different parts of the same seed, *Phytopathology* **57**:821 (abstract).

Meiners, J. P., Waterworth, H. E., Smith, F. F., Alconero, R., and Lawson, R. H., 1977, A seed-transmitted strain of cucumber mosaic virus isolated from bean, *J. Agric. Univ. P.R.* **61**:137.

Miller, L. I., and Troutman, J. L., 1966, Stunt disease of peanuts in Virginia, *Plant Dis. Rep.* **50**:139.

Millikan, D. F., 1959, The incidence of the ringspot virus in peach nursery and orchard trees, *Plant Dis. Rep.* **43**:82.

Mishra, M. D., Raychaudhuri, S. P., Ghosh, A., and Wilcoxson, R. D., 1980, Berseem mosaic, a seed-transmitted virus disease, *Plant Dis.* **64**:490.

Moline, H. E., 1973, Mechanically tranmissible viruses from corn and sorghum in South Dakota, *Plant Dis. Rep.* **57**:373.

Mossop, D. W., and Francki, R. I. B., 1977, Association of RNA-3 with aphid transmission of cucumber mosaic virus, *Virology* **81**:177.

Mossop, D. W., Francki, R. I. B., and Grivell, C. J., 1976, Comparative studies on tomato aspermy and cucumber mosaic viruses. V. Purification and properties of a cucumber mosaic virus inducing severe chlorosis, *Virology* **74**:544.

Murant, A. F., Abu-Salih, H. S., and Goold, R. A., 1974, 20th Annual Report for the year 1973, p. 67, Scottish Horticultural Research Institute, Dundee, U.K.

Noordam, D., Bijl, M., Overbeek, S. C., and Quiniones, S. S., 1965, Virussen uit *Campanula rapunculoides* en *Stellaria media* en hun relatie tot komkommozaiekvirus en tomaat-"aspermy"-virus, *Neth. J. Plant Pathol.* **71**:61.

Normand, R. A., and Pirone, T. P., 1968, Differential transmission of strains of cucumber mosaic virus by aphids, *Virology* **36**:538.

Nyland, G., 1960, Juice transmission of cucumber mosaic virus to mazzard and mahaleb cherry, *Phytopathology* **50**:85 (abstract).

Paliwal, Y. C., 1982, Virus diseases of alfalfa and biology of alfalfa mosaic virus in Ontario and western Quebec, *Can. J. Plant Pathol.* **4**:175.

Panarin, I. V., 1977, Cereal flee beetles as vectors of Hungarian brome grass mosaic virus, *Sb. Nauchn. Tr. Krashodar NII Skh.* **13**:158.

Panarin, I. V., 1978, Vectors and the transmission mechanism of smooth bromegrass mosaic virus, *Skh. Biol.* **13**:230.

Panarin, I. V., and Zabavina, E. S., 1977a, Interrelation between Hungarian bromegrass mosaic virus and the vector *Oulema melanopa* L., *Sb. Nauchn. Tr. Krashodar NII Skh.* **13**:156.

Panarin, I. V., and Zabavina, E. S., 1977b, Criteria for assessing the resistance of wheat to awnless bromegrass mosaic virus in the glasshouse, *Skh. Biol.* **13**:160.

Phatak, H. C., 1974, Seed-borne plant viruses—Identification and diagnosis in seed health testing, *Seed Sci. Technol.* **2**:3.

Phatak, H. C., 1980, The role of seed and pollen in the spread of plant pathogens, particularly viruses, *Trop. Pest Manage.* **26**:278.

Phatak, H. C., Diaz-Ruiz, J. R., and Hull, R., 1976, Cowpea ringspot virus: A seed-transmitted Cucumovirus, *Phytopathol. Z.* **87**:132.

Phillips, J. H. H., 1951, An annotated list of *Hemiptera* inhabiting sour cherry orchards in the Niagara Peninsula, Ontario, *Can. Entomol.* **83**:194.

Pirone, T. P., and Harris, K. F., 1977, Non-persistent transmission of plant viruses by aphids, *Annu. Rev. Phytopathol.* **15**:55.

Polson, A., and von Wechmar, M. B., 1980, A novel way to transmit plant viruses, *J. Gen. Virol.* **51**:179.

Poremskaya, N. B., 1964, Seed transmission of virus diseases of lupins, *Tr. Vses. Inst. Zashch. Rast.* **20**:54.

Proeseler, G., 1968, Ubertragungsversuche mit dem latenten Prunus-Virus und der Gall-milbe *Vasates fockeui* Nal., *Phytopathol Z.* **63**:1.

Proeseler, G., 1978, Transmission of brome mosaic virus by cereal beetles (Coleoptera, Chrysomelidae), *Arch. Phytopathol. Pflanzenschutz* **14**:267.

Provvidenti, R., 1976, Reaction of *Phaseolus* and *Macroptilium* species to a strain of cucumber mosaic virus, *Plant Dis. Rep.* **60**:289.

Purivirojkul, W., Sittiyos, P., Hsu, C. H., Poehlman, J. M., and Sehgal, O. P., 1978, Natural infection of mungbean (*Vigna radiata*) with cucumber mosaic virus, *Plant Dis. Rep.* **62**:530.

Quiot, J. B., Marchoux, G., Douine, L., and Vigouroux, A., 1979, Ecologie et epidemiologie du virus de la mosaique du concombre dans le sud-est de la France. V. Role des especes spontanees dans la conservation du virus, *Ann. Phytopathol.* **11**:325.

Quiot, J. B., Labonne, G., and Marrou, J., 1982, Controlling seed and insect-borne viruses, in: *Pathogens, Vectors and Plant Diseases: Approaches to Control* (K. F. Harris and K. Maramorosch, eds.), pp. 95–122, Academic Press, New York.

Ramaswamy, S., and Posnette, A. F., 1971, Properties of cherry ring mottle, a distinctive strain of prune dwarf virus, *Ann. Appl. Biol.* **68**:55.

Reddy, K. R. C., and Nariani, T. K., 1963, Studies on mosaic diseases of vegetable marrow (*Cucurbita pepo* L.), *Indian Phytopathol.* **16**:260.

Rochow, W. F., 1977, Dependent virus transmission from mixed infections, in: *Aphids as Virus Vectors* (K. F. Harris and K. Maramorosch, eds.), pp. 253–273, Academic Press, New York.

Salazar, L. F., Abad, J. A., and Hooker, W. J., 1982, Host range and properties of a strain of tobacco streak virus from potatoes, *Phytopathology* **72**:1550.

Schmelzer, K., 1956, Contribution to the knowledge of the transmissibility of viruses by *Cuscuta* species, *Phytopathol. Z.* **28**:1.

Schmelzer, K., 1969, The elm mottle virus, *Phytopathol. Z.* **64**:39.

Schmelzer, K., 1971, Robinia mosaic virus, *CMI/AAB Descriptions of Plant Viruses* No. 65.

Schmelzer, K., Schmidt, H. E., and Beczner, L., 1973, Spontaneous host plants of alfalfa mosaic virus, *Biol. Zentralbl.* **92**:211.

Schmidt, H. B., Fritzsche, R., and Lehmann, W., 1963, Transmission of ryegrass mosaic virus by nematodes, *Naturwissenschaften* **50**:386.

Schwenk, F. W., Smith, S. H., and Williams, H. E., 1971, Component ratio differences in strains of alfalfa mosaic virus, *Phytopathology* **61**:1159.

Sharma, Y. R., and Chohan, J. S., 1973, Transmission of cucumis viruses 1 and 3 through seeds of cucurbits, *Indian Phytopathol.* **26**:596.

Shukkla, D. D., and Gough, K. H., 1983, Tobacco streak, broad bean wilt, cucumber mosaic and alfalfa mosaic viruses associated with ringspot of *Ajuga reptans* in Australia, *Plant Dis.* **67**:221.

Simons, J. N., 1955, Some plant–vector–virus relationships of southern cucumber mosaic virus, *Phytopathology* **45**:217.

Simons, J. N., 1957, Three strains of cucumber mosaic virus affecting bell pepper in the Everglades area of south Florida, *Phytopathology* **47**:145.

Smith, P. R., and Stubbs, L. L., 1976, Transmission of prune dwarf virus by peach pollen and latent infection in peach trees, *Aust. J. Agric. Res.* **27**:839.

Sutic, D., 1959, Die Rolle des Paprikasamens bei der Virusubertragung, *Phytopathol. Z.* **36**:84.

Swenson, K. G., 1952, Aphid tranmission of a strain of alfalfa mosaic virus, *Phytopathology* **42**:261.

Swenson, K. G., and Milbrath, J. A., 1964, Insect and mite transmission tests with prunus ringspot virus, *Phytopathology* **54**:399.

Taylor, C. E., and Robertson, W. M., 1974, Electron microscopy evidence for the association of tobacco etch virus with the maxillae in *Myzus persicae* (Sulz.), *Phytopathol. Z.* **80**:257.

Thomas, W. D., Jr., and Graham, R. W., 1951, Seed transmission of red node in pinto beans, *Phytopathology* **41**:959.

Thresh, J. M., 1980, Annual Report for 1979, p. 103, East Malling Research Station, Maidstone, U.K.

Thresh, J. M., 1981, Annual Report for 1980, p. 86, East Malling Research Station, Maidstone, U.K.

Tolin, S. A., 1972, Aphid transmissibility of two isolates of peanut stunt virus, *Int. Virol.* **2**:253.

Tomlinson, J. A., and Carter, A. L., 1970a, Seed transmission of cucumber mosaic virus in chickweed, *Plant Dis. Rep.* **54**:150.

Tomlinson, J. A., and Carter, A. L., 1970b, Studies on the seed transmission of cucumber mosaic virus in chickweed (*Stellaria media*) in relation to the ecology of the virus, *Ann. Appl. Biol.* **66**:381.

Tomlinson, J. A., and Walker, V. M., 1973, Further studies on seed transmission in the ecology of some aphid-transmitted viruses, *Ann. Appl. Biol.* **73**:292.

Tosic, M., and Pesic, Z., 1975, Investigation of alfalfa mosaic virus transmission through alfalfa seed, *Phytopathol. Z.* **83**:320.

Tremaine, J. H., 1966, Serological identification of a strain of cucumber mosaic virus from *Prunus, Phytopathology* **56**:152 (abstract).

Troll, H. J., 1957, Zur Frage der Braunevirus-Ubertragung durch das Saatgut bei *Lupinus luteus, Nachrichtenbl. Dtsch. Pflanzenschutzdienst (Berlin)* **11**:218.

Troutman, J. L., Bailey, W. K., and Thomas, C. A., 1967, Seed transmission of peanut stunt virus, *Phytopathology* **57**:1280.

Uyeda, I., and Mink, G. I., 1981, Properties of asparagus virus II, a new member of the Ilarvirus group, *Phytopathology* **71**:1264.

Valverde, R. A., 1983, Brome mosaic virus isolates naturally infecting *Commelina diffusa* and *C. communis, Plant Dis.* **67**:1194.

Van der Meer, F. A., Huttinga, H., and Maat, D. Z., 1976, Lilac ring mottle virus: Isolation from lilac, some properties, and relation to lilac ringspot disease, *Neth. J. Plant Pathol.* **82**:67.

von Wechmar, M. B., 1980, Transmission of brome mosaic virus by *Puccinia graminis tritici, Phytopathol. Z.* **99**:289.

von Wechmar, M. B., and Rybicki, E. P., 1981, Aphid transmission of three viruses causes Freestate streak disease, *S. Afr. J. Sci.* **77**:488.

von Wechmar, M. B., Erasmus, D. S., and Rybicki, E. P., 1982, Aphid-fungal- and seed-transmission of brome mosaic virus and barley stripe mosaic virus cause atypical symptoms in small grains, IV Int. Conf. Comp. Virol., Banff, Canada, p. 157 (abstract).

Wagnon, H. K., Traylor, J. A., Williams, H. E., and Weinberger, J. H., 1960, Observations on the passage of peach necrotic leaf spot and peach ringspot viruses through peach and nectarine seeds and their effects on the resulting peach seedlings, *Plant Dis. Rep.* **44**:117.

Walters, H. J., and Dodd, N. L., 1969, Identification and beetle transmission of an isolate of cowpea chlorotic mottle virus from *Desmodium, Phytopathology* **59**:1055 (abstract).

Walters, H. J., and Surin, P., 1973, Transmission and host range of broad bean mottle virus, *Plant Dis. Rep.* **57**:833.

Waterhouse, P. E., and Murant, A. F., 1981, Purification of carrot red leaf virus and evidence from four serological tests for its relationship to Luteoviruses, *Ann. Appl. Biol.* **97**:191.

Watson, M. A., and Plumb, R. T., 1972, Transmission of plant-pathogenic viruses by aphids, *Annu. Rev. entomol.* **17**:425.

Way, R. D., and Gilmer, R. M., 1958, Pollen transmission of necrotic ringspot virus in sweet cherry, *Plant Dis. Rep.* **42**:1222.

Wilcoxson, R. D., Johnson, L. E. B., and Frosheiser, F. I., 1975, Variation in the aggregation forms of alfalfa mosaic virus strains in different alfalfa organs, *Phytopathology* **65**:1249.

Williams, H. E., Traylor, J. A., and Wagnon, H. K., 1962, Recovery of virus from refrigerated fruit tree and grapevine pollen collections, *Phytopathology* **52**:367 (abstract).

Williams, H. E., Traylor, J. A., and Wagnon, H. K., 1963, The infectious nature of pollen from certain virus-infected stone fruit trees, *Phytopathology* **53**:1144.

Willison, R. S., and Weintraub, M., 1957, Properties of a strain of cucumber mosaic virus isolated from *Prunus* hosts, *Can. J. Bot.* **35**:763.

Yang, A. F., and Hamilton, R. I., 1974, The mechanism of seed transmission of tobacco ringspot virus in soybean, *Virology* **62**:26.

Yarwood, C. E., 1955, Mechanical transmission of an apple virus, *Hilgardia* **23**:613.

Zschau, K., 1964, Ein Beitrag zum Auftreten des Luzernemosaikvirus in Deutschland, *Nachrichtenbl. Dtsch. Pflanzenschutzdienstes (Berlin)* **18**:44.

CHAPTER 9

Virus Epidemiology and Control

R. G. GARRETT, J. A. COOPER, AND P. R. SMITH

I. INTRODUCTION

For a virus to survive in nature, it must either acquire the ability to spread from host to host, or it must convey an ecological advantage upon the infected host. Otherwise, a virus could at best only maintain its original (presumably small) proportion of infected plants. Vegetative reproduction, or the acquisition of seed transmission, would not change the situation; in the absence of spread from plant to plant, the survival of a virus would depend entirely upon the competitive ability of the infected host.

Among the Tymoviruses, there is an example of a virus conveying an ecological advantage upon its host; kennedya yellow mosaic virus, for which no vector has yet been described, reduces the palatability of *Kennedya rubicunda* to rabbits, and so healthy plants are preferentially grazed (Gibbs, 1980). There is no known example of this among the viruses considered in this volume, all of which have evolved means of spread from plant to plant. The Cucumoviruses and alfalfa mosaic virus (AMV) have adapted mainly to nonpersistent transmission by aphids, the Ilarviruses and AMV to spread by pollen and pollinating insects, whereas the Bromoviruses have perhaps only recently adapted to inefficient transmission by vectors.

The mechanisms of transmission are discussed in Chapter 8. Here, we will concentrate on the consequences of the transmission mechanisms on the ecology and control of viruses in the four groups. However, some aspects of vector behavior, especially of aphids and pollinating in-

R. G. GARRETT, J. A. COOPER, AND P. R. SMITH • Plant Research Institute, Burnley Gardens, Burnley, Victoria 3121, Australia.

sects, will also be discussed, although some of the data were obtained in studies on viruses of other groups.

II. POLLEN-BORNE VIRUSES

Pollen transmission is an important characteristic of some Ilarviruses. It has been reported for prunus necrotic ringspot virus (PNRSV) (George and Davidson, 1964; Davidson, 1976), prune dwarf virus (PDV) (George and Davidson, 1964; Smith and Stubbs, 1976), and black raspberry latent virus (BRLV) (Converse and Lister, 1969). The presence of elm mottle virus (Schmelzer, 1969) and tobacco streak virus (TSV) (Converse, 1976) in pollen have also been reported although evidence of pollen transmission of these viruses has not been established. The major significance of pollen transmission in the epidemiology of the Ilarviruses lies in the ability of infected pollen to transmit virus to healthy plants, thus creating new sources of infection in the field. The effectiveness of pollen transmission as a means of virus dissemination depends largely on the activity of pollinating insects.

A. Behavior of Insects and Patterns of Virus Spread

Pollen of apple and cherry is distributed by wind to some exent (Langridge, 1969), but the pollen of fruit trees is generally moist and sticky and relies mainly on insects for transfer (Free, 1960). Therefore, for those viruses which are pollen-transmitted, the behavior of pollen-bearing insects plays an important role in virus transmission, and a study of their behavior will lead to greater understanding of the patterns of virus spread observed in the field.

Many insects visit blossoms and may play a small part in pollen transfer. However, the honeybee is the most common insect visiting the flowers of fruit trees (Free, 1960) and its adaptation to pollination gives it a major role in the epidemiology of pollen-borne viruses. Bees tend to remain constant to one flower species during pollen collection, and will generally only move to another species when the pollen of the first species becomes unavailable. Studies on purity of pollen loads have shown that less than 11% of honeybee pollen loads contain mixed pollen (Free, 1970).

In fruit tree orchards, bees forage locally and either keep to a single tree during a foraging trip or move only between adjacent trees (Free, 1970). Thus, virus transmission in an orchard is more likely to occur between adjacent trees than between distant trees. Smith *et al.* (1977a) studied the epidemiology of peach rosette and decline (PRD), caused by dual infection by two pollen-borne viruses, PDV and PNRSV, and found that healthy peach trees adjacent to PRD-affected trees were more likely to become infected than those further away. Similar results were reported by Demski and Boyle (1968) who found that the majority of new PNRSV infections in a sour cherry orchard were located next to previously infected trees. Data on sour cherry yellows virus (= PDV) (Willison et al., 1948) also showed that infection spreads more readily to adjacent trees and that the natural spread of PNRSV in sour cherry is generally limited to trees 6–12 m from the source of inoculum (Davidson and George, 1964).

Climatic conditions have a major influence on bee activity, and during unfavorable weather bees travel only short distances from their hives. For example, Hootman and Cale (1930) noticed few bees further than 60 m from their hives in cool windy weather, and Nevkryta (1957) found that bees did not visit sweet cherry trees more than 125 m from their hives when temperatures dropped to 12–15°C. This is reflected in the spread of pollen-borne virus in orchards. Thus, in Michigan, Cation (1961, 1967) found that PDV and PNRSV did not spread naturally, and Lazar and Fridlund (1967) recorded slow spread of PDV and PNRSV in Washington peach orchards. In contrast, Smith et al. (1977a) found that PRD infection of peaches in Victoria rose from 0.9% to 91.5% over 10 years. The warm spring experienced in southern Australia was thought to encourage honeybee activity, resulting in more virus spread there than in the northern United States.

B. The Epidemiology of Specific Diseases

Ilarviruses are not widespread in nature, but have been found infecting species in at least seven different plant families. Most research on the epidemiology of Ilarviruses has been concentrated on those of greatest economic importance such as PNRSV, PDV, TSV, and apple mosaic virus (ApMV). Little information is available on the epidemiology of diseases caused by citrus leaf rugose, elm mottle, lilac ring mottle, and spinach latent viruses. Thus, this section is necessarily confined to exploring the epidemiology of the better studied Ilarviruses and drawing comparisons, where possible, with other members of the group.

The major factors involved in the epidemiology of the Ilarviruses include:

1. The common horticultural practice of vegetative propagation
2. The characteristics of seed and pollen transmissibility
3. Strain variation of the viruses

1. Vegetative Propagation

The most important factor in the dissemination of Ilarviruses is vegetative propagation of plants for commercial purposes. The majority of these viruses occur in crops of horticultural significance such as fruit trees, roses, and berries, with budding, grafting, runners, or rooted cuttings as the main methods of propagation. These techniques enable the production of thousands of virus-infected plants from only a few infected plants, which may then be widely distributed by the nursery industry.

The survey conducted by Fischer and Baumann (1980) on the incidence of PNRSV infection in almond plantings in Morocco provides an illustration of the role of vegetative propagation in the epidemiology of virus diseases. Almonds in Morocco are grown either intensively in orchards or extensively in mountainous regions where trees grow spontaneously from seed. In the areas of intensive cultivation, where trees of imported cultivars were distributed by the nursery industry, high percentages of PNRSV infection occurred. This was attributed to virus infection of the imported scionwood. In areas where almonds grew extensively from seed, very little infection with PNRSV was detected, and only occurred following the introduction of new propagating material.

Vegetative propagation may also contribute to the relative abundance of one virus compared with another. In roses, while both ApMV and PNRSV have been detected in the U.S.A. (Fulton, 1967; Casper, 1973) and in the U.K. (Ikin and Frost, 1974), PNRSV is the more widespread. This may be attributed to the fact that ApMV induces severe mosaic symptoms in roses, which are less likely to be propagated by nurserymen than material infected with PNRSV, which often shows very indistinct symptoms (Sweet, 1980). TSV has also been reported as a causal agent of rose mosaic in America (Fulton, 1970b; Converse and Bartlett, 1979) although its incidence appears low. The disease is more severe than other forms of rose mosaic, causing stunting and leaf twisting (Fulton, 1970b), and hence it seems unlikely that infected plants would be used for propagation.

2. Pollen and Seed Transmission

Seed transmission has been reported for many of the Ilarviruses (see Table III of Chapter 8). The level of seed transmission is frequently a property of particular virus strains or host varieties (Shepherd, 1972). For example, the transmission of virus from the mother tree to 30% of the progeny has been recorded for PNRSV in *Prunus cerasus* (Cation, 1949), 6% in *Prunus avium* (Cochran, 1946), and up to 35.7% in *Prunus persica* (Stubbs and Smith, 1971). Seed from *Chenopodium amaranticolor* infected with lilac ring mottle virus produced 91% infected seedlings, although no seed transmission of this virus has been detected in lilac (Van der Meer and Huttinga, 1979). However, lack of transmission in experimental trials does not preclude the possibility of a rare instance of transmission in nature.

Seed transmission provides important sources of virus spread in successive seasons. Some viruses may retain infectivity for many years in dormant seeds (Fulton, 1964). This is of particular importance in annual crops, but has also provided an effective means of virus dispersal in many perennial hosts. Horticultural methods of propagation have sometimes helped make seed transmission an important source of virus infection. Gilmer (1955) reported that up to 16% of *Prunus mahaleb* seeds imported from France for use as rootstock material in the U.S.A. were infected with PNRSV. The practice of propagating cultivars onto seedling rootstocks, some of which may be infected, has resulted in widespread distribution of viruses through local and international trade in nursery material.

Pollen and seed transmission are closely related factors in virus epidemiology. Most studies on pollen transmission of viruses indicate that a greater percentage of seed contains virus when the mother plant is infected than when pollen is the sole source of infection (Bennett, 1969). Vértesy (1976) demonstrated that in Montmorency sour cherry, 28% seed transmission was obtained when only the pollen was infected with PNRSV, 53% when only the mother plant was infected, and 88% when both pollen and the mother plant were infected.

Seed transmission of PNRSV appears more efficient than that of PDV. Stubbs and Smith (1971) tested the progeny of seeds produced by trees infected with PRD and found that seed transmission of PNRSV ranged from 8.2% to 35.7%, compared with 2.0% to 6.8% for PDV. PDV has only been recorded in *Prunus* species to date, whereas PNRSV has been found to infect *Prunus*, *Rosa*, and *Humulus* species. PDV occurs less frequently and appears to spread more slowly than PNRSV in cherry orchards (Willison et al., 1948; Klos and Parker, 1960; Gerginova, 1980). Klos and Parker (1960) measured the increase in incidence of PNRSV over 5 years as 2.2% to 78.4% in a Montmorency sour cherry orchard while the incidence of PDV increased from 3.6% to 49.6% over the same period (Table I). The annual rate of spread of the two viruses varied from year to year and differed most in the period 1949/1950. Willison et al. (1948) compared the spread of PNRSV and PDV in six Montmorency orchards and found that, on average, 8.9% of healthy trees became infected with PDV each year, while the apparent incidence of PNRSV was 13.2%.

The effect of the virus on pollen may play a role in these differences observed between PNRSV and PDV. Marénaud and Saunier (1974) found that infection of peach trees with PDV and PNRSV decreased the number of morphologically normal pollen grains, reduced germination *in vitro*, and retarded pollen tube growth. Nyéki and Vértesy (1974) also found that sour cherry pollen infected with PNRSV or PDV was smaller than healthy pollen, and that viability and pollen-tube formation were reduced. They also reported that PNRSV had a more serious effect on pollen size, morphology, viability, and pollen-tube formation than did PDV. However, the effect of virus on pollen is not sufficient to affect the rate of spread, for PNRSV spreads faster than PDV in all species studied. Also, viability of pollen may not be necessary for infection to occur, for PNRSV and PDV have been detected on both the surface and the interior of cherry pollen grains (Kelley and Cameron, 1983), and PNRSV has been detected on the surface of cherry and almond pollen (Cole et al., 1982). More im-

TABLE I. Cumulative
Percentage of Trees in
Montmorency Cherry Orchard
Showing PNRSV and PDV
Symptoms, over a Period of 8
Years[a]

Year[c]	PNRSV[b] % infection	PDV[b] % infection
1945		3.6
1949	2.2	3.6
1950	13.7	7.9
1951	44.6	19.4
1952	73.4	35.3
1953	78.4	49.6

[a] Data from Klos and Parker (1960).
[b] Calculated on the basis of 139 trees in the study.
[c] No additional symptomatic trees were found in 1946 and 1947. No survey was made in 1948.

portant, perhaps, is the ability of the virus to systemically invade the mother tree. Infection of a tree does not always occur as a result of pollination with infective pollen (Gilmer and Way, 1960). The faster spread of PNRSV observed in the field could be explained if this virus were more efficient either in passing from the embryo to the maternal tree, or in systemically invading the tree, and comparative studies of the efficiencies of invasion by these two viruses would provide insight into this possibility.

3. Strain Variation

Strain variation, based on herbaceous and woody host reactions, has been reported for many of the Ilarviruses. For example, Nyland *et al.* (1976) described five isolates of PNRSV distinct enough to be called strains, where "strains" were defined as variants which are sufficiently definable to be recovered and identified (Cochran, 1944). Most cultivated and wild species of *Prunus* are susceptible to one or more strains of PNRSV (Gilmer, 1955; Nyland *et al.*, 1976). However, little information is available on whether particular strains of PNRSV are found more commonly in one species than another.

More information is available on the occurrence of PNRSV compared with that of ApMV, which can be considered a serotype of PNRSV (De Sequeira, 1967; Fulton, 1968). In hop, two distinct virus isolates were reported (Bock, 1967), one very closely related to ApMV and the other intermediate in serological behavior between ApMV and PNRSV. Bock suggested that the intermediate type represents a more basic necrotic ringspot virus type, from which the two distinct serotypes of PNRSV and ApMV have diverged. PNRSV has been commonly found in *Prunus* species, roses, and hops (Fulton, 1970a). ApMV appears to infect a more diverse range of hosts, being found in apples (Fulton, 1972), roses (Fulton, 1968) hops (Bock, 1967), *Rubus* species (Baumann *et al.*, 1982), horse chestnut and hazel (Sweet and Barbara, 1979), plum (Gilmer, 1956; Barbara *et al.*, 1978), and cherry (Barbara, 1980). How ApMV has occurred in such a wide host range remains an unanswered question, for no evidence of pollen transmission (Sweet, 1980) or insect transmission (Fulton, 1972; Wood *et al.*, 1975) has yet been found.

C. Perennial Hosts and Virus Survival

The narrow host range of most Ilarviruses, and the restriction of spread to a short flowering season, result in potentially slow rates of spread. In most cases this is compensated for by the longevity of the woody perennial hosts. For viruses such as ApMV, which are neither pollen- nor seed-borne, and are not insect-transmitted, perennial hosts are essential for the survival of the virus. Without commercial vegetative propagation from infected material and the occasional transmission through root grafting, ApMV would soon be eliminated, were it not able to infect perennial hosts. Tulare apple mosaic virus (TApMV) is an interesting case of a virus found in a single apple tree in California (Yarwood, 1955), which has since died. No further natural occurrence of TApMV was known until it was detected in a hazelnut tree in France (Cardin and Marénaud, 1975). The epidemiology of the disease in hazelnut has not been studied, but a rare occurrence of transmission from an unrelated host may have caused the infection in the apple tree. No spread of the disease is known, but had the infected tree been used as a source of material for commercial propagation, TApMV may have become established in commercial apple orchards.

Even for those Ilarviruses which are pollen- and seed-borne, perennial hosts are important, since the rate of pollen and seed transmission is usually low. Without the perennial host remaining in the field as a continued source of infection, the natural rate of spread of the viruses would decline. The exception is TSV, which naturally infects a wide host range including annuals and a number of weed species (Fulton, 1971). The survival of this virus is in part due to its high rate of seed transmission. TSV is seed-transmitted in soybeans (Ghanekar and Schwenk, 1974), pinto beans (Thomas and Graham, 1951), *Chenopodium quinoa*, and *Datura stramonium* (Brunt, 1969). Transmission rates of 100% were recorded for *C. quinoa* (Brunt, 1970). However, the recent reports that thrips can also transmit TSV (Costa and Lima Neto, 1976; Kaiser *et al.*, 1982) provide a far more satisfactory explanation of spread and survival, and probably explain the wider host range. Such a virus, with seed, pollen, and vector transmission, and capable of infecting both annual and perennial crops, is well adapted for survival.

D. Spread of Viruses within the Tissues of Woody Hosts

Little work has been done on tracing the distribution of viruses within the tissues of their woody hosts. However, even the limited information available provides some insight into the problems of sampling techniques for detecting a virus during indexing procedures.

Hampton (1963) compared the distribution of PDV in three different cherry cultivars in the first season after inoculation with infected buds. All rapidly growing terminals in the most sensitive cultivar were infected in the first season, whereas intervening tissues largely became infected during subsequent seasons. Terminals showing little growth became infected only after two or three seasons. This was explained in terms of the movement of PDV from the points of inoculation in the direction of predominant translocation streams. Using scinctured peach seedlings, Smith et al. (1977b) demonstrated that PDV was not translocated in the xylem, but was presumably translocated up and down the plant in the phloem. The results of Hampton's work suggested that the maximum chance for detection of PDV in cherry trees in the first season would be achieved when samples were taken from branches with actively growing terminals. The mean rate of movement of PDV measured following inoculation was 1–2 in./day.

Uneven distribution of PDV in cherries has also been reported by Gilmer and Brase (1963), in Muir peach by Milbrath (1961), and in Golden Queen peach by Smith et al. (1977b). Tests by Smith et al. (1977b) showed that 2 months after initial symptoms appeared in 10 naturally infected trees, 65% of limbs adjacent to the initially infected limb were infected, while only 30% of the limbs furthest away were shown to contain virus (Fig. 1, Table II). The low probability of detecting PDV in peaches in the first year of infection was similar to that in cherries (Hampton, 1966). When screening important propagating material for recent infections, the detection of virus is complicated by its uneven distribution in the plant. Smith et al. (1977b) recommended sampling a minimum of eight buds per tree from different limbs for testing peach trees suspected of recent PDV infection. The efficiency of detection of virus in a single tree can be examined by the strategy used to detect virus in potato crops (Moran et al., 1983). The number of samples that should be taken depends on the chance that any sample will contain no virus, and on the failure rate that can be tolerated. Thus, if 3 months after infection, 50% of the samples are likely to carry virus, then the chance of detecting virus with four samples per tree is given by $p = (1 - 0.5^4)$, i.e., $p = 0.938$. Using this approach it would be possible to predict the effectiveness of different sampling strategies, for infections of different antiquities.

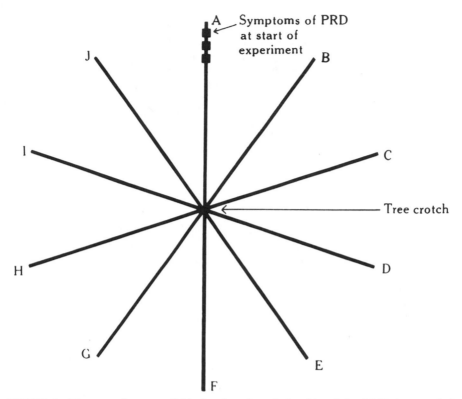

FIGURE 1. Diagram of tree scaffold showing the relationship of the 10 limbs sampled monthly to determine the distribution of PDV within peach trees in the first season following natural infection. Redrawn after Smith *et al.* (1977b).

Fleisher *et al.* (1971) found that uneven distribution of PNRSV in roses limited the reliability of using Shirofugen cherry as a method for detecting the virus, since budwood from the middle and terminal parts of infected branches gave a lower degree of positive reactions than budwood from the basal parts. Similarly, testing peach trees inoculated with PNRSV near the branch terminals, Holland and Cooper (unpublished results) found the highest percentage detection of virus at the bases of the branches, indicating movement of the virus down the branch from the point of inoculation. However, care must be taken in extrapolating data of this kind to other virus–host combinations; for example, Hampton (1963) found that PDV inoculated at either the trunk or the branch terminals of cherry trees moved toward the growing terminals. An extreme example of the irregular distribution of ApMV in apple trees was reported by Luckwill (1954). Two apple varieties, Crawley Beauty and Cap of Liberty, displayed symptoms of mosaic for 8 years on one or two branches only, with little apparent spread to other branches. Buds taken from symp-

TABLE II. Number of Peach
Tree Limbs Newly Infected with
PDV at Different Times in the
First Season of Peach Rosette
and Decline (PRD) Symptom
Expression[a]

Limb[b]	Number of limbs infected				
	Oct.	Nov.	Dec.	Jan.	Feb.
A[c]	7	10	10	10	10
B	3	6	7	8	8
C	5	5	9	9	10
D	3	3	6	7	8
E	2	3	6	9	9
F	2	3	4	5	5
G	4	6	8	9	9
H	3	5	7	7	8
I	4	7	8	9	9
J	4	7	9	9	9

[a] Data from Smith *et al.* (1977b).
[b] Ten limbs sampled monthly from 10 affected trees. Each limb tested for the presence of PDV.
[c] Initial affected limb.

tomless branches of these trees produced no symptoms of ApMV when tested on the indicator Lord Lambourne.

III. APHID-BORNE VIRUSES

The adaptation to nonpersistent spread by aphids appears to have resulted in very large host ranges. For example, cucumber mosaic virus (CMV) has been found naturally infecting 476 species in 67 families, and is capable of transmission to a further 536 species in 53 families (Horváth, 1979, 1980). Although the host range of alfalfa mosaic virus (AMV) is also diverse, including 70 families, it is not as large as that of CMV, with only 600 species recorded as hosts (Horváth, 1981). Several properties have helped to establish a flux of feeding activity which has exposed the nonpersistent viruses to very wide ranges of potential hosts. These include the very wide range of plants which can support aphids in pest proportions, the very high reproductive rates of aphids, the almost random behavior of aphids as they disperse to find new hosts, and the probing behavior of aphids as they explore surfaces on which they land.

In the laboratory and the field, winged and wingless aphid forms may act as vectors. Although only the winged forms can spread virus long distances, they do not necessarily do so, and their role in virus spread to adjacent plants should not be underestimated. An understanding of the ecology of aphid-borne viruses depends on an understanding of the behavior of the aphid vectors themselves and the mechanisms of virus transmission. A full discussion is inappropriate here, and the reader is referred to Johnson (1969) and to reviews such as those by Van Emden (1972), Harris and Maramorosch (1977), and Gibbs (1977). However, in the next section, some aspects of the production of alatae and their behavior are discussed.

Most aphids are the offspring of parthenogenetic viviparous females, which may or may not bear wings. However, sexual and oviparous forms of most species occur in the autumn in suitable climates. Forms which develop from eggs (the fundatriginae) may be morphologically distinct, albeit only slightly, from the viviparously produced forms. The life cycles of aphids are therefore complex, involving several distinct physiological and morphological forms.

All the aphid-borne viruses discussed in this review are nonpersistent in the vector (Watson and Roberts, 1939) and all forms of most aphid species can to some degree transmit the viruses. That is to say, vector *specificity* is low, although vector *efficiency* may vary considerably.

A. Aphid Behavior

1. Production of Alatae

Both apterae and alatae produce nymphs which through four nymphal stages may develop into winged or wingless adults. The proportions of aphids producing wings has been called the "migratory tendency" (MacKay and Lamb, 1979) and is affected by day length (Lees, 1964), food plant quality (Johnson, 1966), crowding (Dixon *et al.*, 1968), birth sequence (Mittler and Dadd, 1966; MacKay and Wellington, 1977), and the previous flight behavior of the parent (Burns, 1972). These factors lead to seasonal changes in the aerial aphid numbers by dispersal from crowded colonies, from deteriorating host plants during autumn senescence, or during summer desiccation in hot dry climates. When spring migrants colonize plants, they produce only apterae for two generations. Thus, spring migrants form a brief peak in aerial numbers of aphids, followed by a lull in aphid activity before the summer population expansion results in crowding and an increase in alatae production.

Although alatae that have flown produce few nymphs which develop wings, alatae that have not flown produce alatiform nymphs as frequently as apterae according to the stimuli described above. However, effective stimuli are often small. For examples, flights of only a few seconds were sufficient to inhibit production of alatae; and 10 aphids in a 5×25-mm tube was sufficient crowding to stimulate alatae production by *Megoura viciae* (Burns, 1972).

2. Migration

Winged aphids do not necessarily fly. Flight can be discouraged by low temperature (Taylor, 1963), wind speeds above 1 m/sec (Müller and Unger, 1951), and low light (Johnson and Taylor, 1957), whereas flight can be encouraged by crowding (Dixon *et al.*, 1968). Many of these processes have been reviewed and they are often complex. For example, in an experiment by Kidd (1977), none of 200 aphids flew from previously uninfested leaves whereas 29 of 200 aphids flew from leaves which had previously borne large aphid populations. Thus, the effect of crowding on flight is in fact through effects on the plant.

3. Flight

The effect of flight speed can be simulated by moving a ground pattern beneath aphids suspended in a wind tunnel. During flight, aphids minimize the movement of such a ground pattern beneath them by flying in the same direction (Kennedy and Thomas, 1974). When the ground speed is less than 0.8 m/sec, the result is a net movement of 0.07 m/sec in the same (i.e., "upwind") direction. This behavioral pattern is called station-keeping, and may be important in determining the pattern of virus spread. Aphids fly mostly under conditions of low wind (Müller and Unger, 1951) which, with their station-keeping behavior, ensures that they do not move far from their current environment, including suitable host plants. Indeed, aphids may actually cease flight when apparent station-keeping is no longer possible. However, aphids carried to greater heights may need to respond to larger ground patterns and hence much higher wind speeds before angular changes become sufficient to show that station-keeping was not actually possible. Such aphids may inadvertently continue flying for long times, and over considerable distances.

These processes could reconcile the behavioral studies of Kennedy and colleagues (Kennedy and Ludlow, 1974; Kennedy and Thomas, 1974) with the observations that under some circumstances aphids can be trapped far from land (Hardy and Milne, 1937). Also, it is known that aphid behavior is not uniform, and that under some circumstances migration (or dispersal) does occur. Thus, the aphids studied by Kennedy and Thomas (1974) were reared under conditions unlikely to induce mass emigration, and so represent aphids exploiting their environment. Under other conditions, aphids may have been produced which responded to a different set of values. In the U.K., potatoes may support large aphid populations in the summer, and the crop populations can be monitored by the production of alatae trapable on yellow cylindrical traps. However, the autumn dispersal can result in a sudden, almost complete disappearance of alatae from the crop with few aphids in the traps. Clearly, the flight behavior of the emigrating aphids must have been quite different to that of the summer populations.

4. Landing

Aphids respond differently to the presentation of a moving target while flying in a wind tunnel, according to their previous flight history (Kennedy and Ludlow, 1974). After only a flight of 90 sec, most aphids flew toward the target when it was presented. Under these conditions, with young virginoparae *Aphis fabae*, flight duration was very short and, if artificially extended in a wind tunnel, flight rapidly ceased with target presentation. Thus, for such aphids to indulge in long flights, they would need to be carried to a sufficient height so that the ground, and perhaps movement relative to the ground, was not recognized.

Aphids on landing do not merely settle from the air. When presented with a moving target after a suitable flight, aphids will orient themselves to the target and land on it (Kennedy and Thomas, 1974). Such attraction to targets has been shown in the effect of trap color (Moericke, 1950) on the numbers of different species caught. Also, Moericke (1957) and A'-Brook (1968) showed that some aphid species landed mostly on traps or plants with surrounding bare earth, whereas Halbert and Irwin (1981) caught *Myzus persicae* and *Aphis citricola* largely in traps over closed canopies. Thus, different species respond differently to color of trap and trap background. This has been studied recently by M. E. Irwin (personal communication) who found a strong interaction between trap color and background color for some aphid species but not others. It is thus not surprising that some aphid-borne viruses occur more frequently on spaced plants, than in continuous stands (Catherall and Hoen, in A'Brook, 1973).

The results of experiments such as these raise many questions about the interpretation of the rates of spread of virus in the field. In particular, it seems important to know which aphid species are indeed the likely field vectors, and how they respond to different crop environments. Only then can realistic attempts be made to correlate virus spread with simple aphid trap catches, obtained by attempts to mimic the landing of aphids on the crop. It is for these reasons that aphid flight behavior may be more important than mere aphid numbers in determining rate of spread of nonpersistent viruses by even efficient vectors (Lockhart and Fischer, 1976; Mali and Rajegore, 1980).

B. Epidemiology of Aphid-Borne Viruses

1. Reservoirs of Aphid-Borne Viruses

Viruses may be introduced into crops with planting materials, or may spread from sources outside the crop. Clearly, vegetative reproduction of virus-infected sources is of critical importance, such as tomato aspermy virus (TAV) in chrysanthemums. Seed transmission of AMV in lucerne is important and can be as high as 10% (Hemmati and McLean, 1977).

The perennial growth form of lucerne, with both pollen and ovule transmission to seed (14 and 6%, respectively) together with aphid transmission, ensures that the virus persists in infected crops which act as reservoirs for spread of AMV into new plantings and other susceptible crops.
However, because the leaf symptoms are often transitory, the disease in
lucerne is usually ignored.

CMV was known for a long time before its seed-borne properties were
recognized. Now, CMV is known to be seed-borne in *Stellaria media*
(Tomlinson and Walker, 1973) and wild cucumber (Lindberg *et al.*, 1956)
while legume-infecting strains of the virus are known to be seed-borne
in mung bean to 0.6% (Purivirojkul *et al.*, 1978) and in *Phaseolus vulgare*
cv. The Prince to about 1/300 (Provvidenti, 1976). Two reasons why the
seed-borne nature of CMV was not recognized earlier have been the failure
to recognize the importance of low incidence of infected seed in the presence of much more effective means of spread; and the need to examine
large numbers of seed. For example, to reliably detect a seed-borne virus
at a rate of 1 in 1000, at least 3000 seeds must be tested. Studies of seed-
borne viruses rarely involve numbers of this size.

The wide host ranges of CMV and AMV have already been mentioned, and wherever these viruses are important to crop production, the
possibility of local reservoirs should be considered. Examples of weed
hosts as reservoirs of CMV are many. In banana plantations, weed hosts
of CMV include species of *Celosia, Datura, Passiflora*, and *Canna*, while
the natural host range could also include species of *Amaranthus, Viscosa,
Chenopodium, Commelina, Thodex, Xanthium, Physalis*, and *Solanum*
(Mali and Rajegore, 1980). *Celosia argentea* is an important ornamental
in commercial greenhouses in New York State and CMV caused a severe
disease in 70% of plants in a greenhouse crop (Provvidenti, 1975). The
source of virus was identified as the weed *Stellaria media*, in which the
virus is seed-borne. Seed-borne CMV in *S. media* is also the source of
virus spread in the spring into glasshouse cucumber in Czechoslovakia
(Kazda and Hervert, 1977).

Infection of perennial weeds by CMV, and seed transmission in them,
ensures that reservoirs are maintained (Havranek and Laska, 1972). Local
reservoirs are especially important (Marrou *et al.*, 1979b) because CMV
does not usually spread much beyond 100 m from its source. Although
spread may occur over longer distances, the local reservoirs are usually
sufficient for the maintenance of virus in that environment. Indeed,
changes in agricultural practice may result in the establishment of large
new reservoirs. For example, *Lespedeza* and *Coronilla* species are increasingly used as forage and cover crops and may extend the reservoirs
of peanut stunt virus (PSV) for spread into other crops (Tolin and Miller,
1975; Milbrath and Tolin, 1977).

2. Aphid Numbers

It is to be expected that some general relationship will exist between
numbers of aphids flying about a crop, and the amount of spread of aphid-

borne virus. Indeed, such relationships have been shown (e.g., Broadbent, 1950; de Wijs, 1974). However, to be biologically meaningful, the relationships should be adjusted for the proportions of infected and healthy plants, and the efficiencies of different vectors. In one study of the spread of CMV by aphids, this was partly taken into account by relating rate of spread with aphid flights for crops with only 75% of the plants infected (Marrou et al., 1979a).

By far the most detailed study of CMV spread in melon and tomato in recent years is that by Quiot and his colleagues in the south of France (Quiot et al., 1976, 1979a–f; Marrou et al., 1979a,b). Bait plants were used to show that the rate of spread of CMV depended on both the numbers of aphids caught in water traps, and the amount of infection in nearby melon crops (Quiot et al., 1976; Marrou et al., 1979a). Although the bait plant system proved useful to measure the rate of CMV spread during short discrete periods, the method was not sensitive enough to measure early spread. Also, using melon as the bait plant, the C-serotypes were measured but the legume strains were ignored (Marrou et al., 1979a).

Rates of spread of CMV into tomato were found to be slower than into melon crops (Quiot et al., 1979f) and high incidence followed the occurrence of high aphid numbers in water traps. Two reactions were observed in tomato: a filiform reaction and a necrotic (X) reaction. Both spread into tomato at the same time, but the X reaction spread more slowly. Using Van der Plank's doublet analysis, clusters of infected plants were detected, but mostly with those showing the necrotic reaction.

Bait plants have been used to study the spread of CMV in the vicinity of windbreaks (Marrou et al., 1979b) which were found to reduce the rate of spread of CMV provided aphid numbers were not large. In general, the rate of spread of CMV was reduced close to and to the lee of the windbreaks, and the effect decreased with distance downwind. Similarly, catches of aphids in yellow pan traps were smaller close to and downwind from the windbreaks, than far from and upwind of the windbreaks (Quiot et al., 1979b). High wind speeds and eddies that occur close to windbreaks may have reduced aphid activity there. Windbreaks had more effect on the spread of CMV in tomato in which spread was slower than in melon.

3. Strain Variation in the Field

There has not been a great deal of work done on virus strain determination in epidemiological studies. This is in contrast to the extensive studies of strain variation in common animal viruses (such as influenza) which have evolved in the presence of immunological pressures. However, with increasing diversity of cropping, and the use of restricted genotypes in agriculture, selection pressures on plant viruses may increase to the extent that the ability to differentiate between virus strains may become important for the detection of active reservoirs, or for strategies in breeding for resistance.

For example, Wyatt and Kuhn (1980) recovered a new strain (R) of cowpea chlorotic mottle virus from the type strain, and found it could infect cowpeas resistant to the type strain. Also, a strain of CMV (NY78-55) has been found which infects *Lactuca saligna* (PI261653), an important source of resistance against common CMV strains (Provvidenti *et al.*, 1980). Using PI261653, a survey revealed that two populations of CMV could be recognized: CMVLsS, which infected PI261653, and CMVLsR, which did not infect PI261653. The recognition of "thermosensitive" and "thermostable" strains of CMV (Quiot *et al.*, 1976; Devergne *et al.*, 1978) in southern France has shown that different strains may spread in different seasons. Thus, CMV-B, the thermosusceptible strain, was most frequently encountered in the spring, but CMV-C, the thermostable type, most frequently in the summer (Quiot *et al.*, 1979d,e). The different occurrences of CMV may well be due to different multiplication rates of the two strains under different field temperature regimes, causing different rates of spread by aphids.

The difficulty of recognizing strains with sufficient ease to apply to field populations, is probably the main reason why strain differences have not been used to detect active reservoirs of viruses. However, virus strain analysis may have wide application in plant virus epidemiology. For example, the CMV-B and CMV-C strains were found to occur in different weed reservoirs (Quiot *et al.*, 1979d). In surveys, 404 plants were found with CMV-B alone, 507 with CMV-C alone, and only 12 with both CMV-B and CMV-C.

IV. EPIDEMIOLOGY OF BROMOVIRUSES

There is very little to be said as yet about the epidemiology of the Bromoviruses. Clearly, they spread in the field, at least to the extent necessary for their survival. There are, however, insects of two orders associated with field spread, and with subsequent laboratory transmission. *Sitonia lineatus* var. *viridifrons* was observed (Borges and Louro, 1974) on broad beans infected with broad bean mottle virus, and transfer of beetles from field plants resulted in transmission to only 5–6% of healthy test plants. M. B. von Wechmar (personal communication) has associated field spread of brome mosaic virus (BMV) with the presence of *Diuraphis noxia* and has obtained transmissions in the laboratory using the aphid; BMV was also transmitted with stem rust and leaf rust of wheat, and through seed. The importance of these insects in the field has yet to be established, as also the reports of transmission of BMV by chrysomelids (Panarin, 1977; Proeseler, 1978).

V. CONTROL

The basic principles involved in controlling virus diseases in plants are generally similar, regardless of the type of virus involved. These control measures include:

1. Propagation from plants free of known virus disease
2. Reduction of virus spread into and within crops by:
    a. Isolation of healthy plantings from infected sources
    b. Suppression of vector activity and other agents of virus transmission
    c. Use of cultivars resistant or tolerant to infection

A. Production of Virus-Free Plants

The simplest method used to obtain disease-free material is the indexing of numerous plants of the same variety to find at least one which is not virus infected. The healthy plants found are used as sources of propagating material to produce healthy clones. Where no healthy plant is detected, virus elimination may generally be achieved by propagating new planting material from new growth of plants maintained at elevated temperatures (heat therapy), by meristem culture, or a combination of the two.

The mechanism of heat therapy is not well understood. One hypothesis is that heat therapy results in a shift of balance from virus synthesis to degradation (Kassanis and Posnette, 1961), so that the virus is eventually eliminated. An alternative hypothesis (Nyland and Goheen, 1969) is that high temperature causes the destruction of biochemical activities essential for both virus and host, but that the host is better able to recover from the damage. Information available on the heat treatment of Ilarviruses suggests that they are relatively easy to eliminate. By the application of heat therapy to infected plants, apple plants free of ApMV were obtained in 20 days at 37°C (Thomsen, 1961), cherry buds free of PNRSV and PDV in 17 days at 38°C, and peach free of PNRSV and PDV in 24 days at 38°C (Nyland, 1960). In some cases the entire plant is freed of virus (Posnette and Cropley, 1956; Nyland, 1960); in other cases only the shoot tips are freed (Campbell, 1962). Cucumoviruses can also be readily eliminated from plants by heat treatment; for example, growth of chrysanthemum infected with TAV for 30 days at 37°C is sufficient to produce planting material free of the virus (Ignash, 1977).

Meristem tip culture is often useful when heat therapy alone fails to eliminate virus infection. Viruses may invade the meristem tip to a varying extent, depending on the type of virus and the host species (Mori, 1977). In some instances, a meristem tip may be excised which is free of virus. In other cases the meristems are infected at the time of excision, but the virus is eradicated during tissue culture (Hollings and Stone, 1964). It has been suggested that virus eradication is caused by the metabolic disruption resulting from cell injury during excision (Mellor and Stace-Smith, 1977). The use of heat treatment and meristem culture is a particularly effective combination, and has been used to eliminate TAV from chrysanthemum (Asatani, 1972) and for a wide range of other virus–host combinations. Another application of tissue culture was demonstrated by Vértesy (1976), who reported that cotyledons were the source of PNRSV infection in young cherry seedlings, and successfully eliminated PNRSV by removal of cotyledons from the embryos and subsequent growth of embryos in tissue culture. Vértesy did not regard this experiment as complete because the embryos were not propagated to produce whole plants and then tested for virus. Nevertheless, the results indicate an interesting avenue to be explored in virus elimination through tissue culture.

The potential of virus-inhibiting chemicals such as ribavirin has also been investigated. Cheplick and Agrios (1983) attempted to eliminate ApMV and other apple diseases by injecting branches of apple trees with various synthetic antiviral and antibiotic compounds. Results showed that ribavirin can suppress the expression of ApMV symptoms, but does not completely inactivate the virus. Ribavirin was not successful in eliminating PNRSV or PDV from infected *Chenopodium quinoa* (Hansen and Green, 1983).

The production of virus-free rootstocks is an important aspect of fruit-tree culture because infected rootstocks are a common source of spread of PDV and PNRSV in nurseries and orchards. Virus-tested seed-source blocks and stoolbeds are therefore established and maintained by thorough indexing and inspection to ensure that virus-free rootstocks are available for propagation of healthy trees (Aichele, 1983; Smith, 1983).

B. Reduction of Virus Spread

1. Isolation of Healthy Plantings from Sources of Disease

Isolation of nuclear propagating material is of primary importance where pollen transmission of viruses is a factor to consider. Stone fruit cultivated for production of virus-tested seed and scionwood should be planted in areas isolated from possible sources of infection (Fridlund, 1976) to reduce the possibility of PNRSV and PDV transmission. ApMV is only known to spread naturally through root grafting (Hunter et al., 1958), so does not present the same problem. Healthy material should nevertheless be planted some distance from infected material.

In commercial nurseries and orchards, isolation is usually impractical, but interplanting of healthy new material among older infected trees should be avoided to minimize the risk of reinfection (Welsh, 1976). Rogueing of diseased trees is effective against viruses such as ApMV, which induces distinct leaf symptoms and spreads only very slowly in the field. Smith et al. (1977a) showed that rogueing of obviously diseased peach trees decreased the rate of spread of PDV by 29.4% in 2 years, but did not eradicate the disease because symptomless trees remained as sources of inoculum. Attempts to control spread of the cherry rugose mosaic strain of PNRSV in cherries were unsuccessful due to the similarity of disease symptoms to the effects of gibberellin which is often applied to cherries to retard fruit ripening, making growers reluctant to remove trees until symptoms were well advanced (Mink, 1980). Removal of diseased limbs of peach trees with PRD disease has also been used in attempts to eliminate the viruses (Smith *et al.*, 1977a), but was unsuccessful in preventing movement of virus into the rest of the tree.

With aphid-borne viruses, isolation becomes difficult because of the possibility of virus spread from afar. However, the main difficulty in the control of CMV and AMV is not so much the possible long-distance movement of virus, but the very wide range of hosts which can capture the rare viruliferous aphid, and allow an epiphytotic to develop in weed or crop hosts before expansion into crops planted later. Thus, in southern France, good weed control in the spring reduced the spread of CMV into melon crops, but the epiphytotics developed in nearby weeds and subsequent crops in the summer (Marrou *et al.*, 1979b).

Provided the virus does not infect common plants near the crop to be protected, disease-free stock schemes can generally be expected to succeed, as shown by work on aphid-borne viruses of other groups infecting potato and lettuce. Where good weed control is economically practicable, such as near elite stocks or in greenhouses (Provvidenti, 1975), control of even CMV is possible.

2. Suppression of vector activity and other agents of virus transmission

Although thrips have been implicated as vectors of TSV (Costa and Lima Neto, 1976; Kaiser *et al.*, 1982), insects are generally not considered important in the transmission of the Ilarviruses. The risk of pollen transmission of PNRSV and PDV in fruit tree repositories via honeybees can be minimized by manual deblossoming and chemical sprays. Similarly, removal of blossoms from certified *Rubus* plantings reduced the possibility of TSV infection either through pollen or through flower-visiting thrips (Converse, 1980). In commercial orchards, planting healthy trees as far as practicable from diseased trees is the most effective means of minimizing pollen transmission.

Generally, nonpersistent plant viruses are difficult to control using aphicides against the vectors. This is in part because viruses such as CMV and AMV can be transmitted by aphids during brief probes, which may not be sufficient for the aphid to acquire a lethal dose of the insecticide. However, an additional reason for the poor control of such viruses is their low vector specificity, and their transmission by aphids which do not colonize the host. Consequently, control measures must be applied beyond the crop boundaries, if they are to be effective. Problems of these kinds have led to searches for unorthodox methods of interfering with vector activity, some of which have been applied to the control of CMV and AMV.

Following studies on the transmission of virus from or to leaf surfaces covered by stretched Parafilm (Bradley, 1956), it was found that mineral oils on plant surfaces reduced transmission frequency (Bradley et al., 1962). Since then, a number of studies have shown that oil sprays can be used to control CMV (Loebenstein et al., 1964, 1970), whereas others (Nawrocka et al., 1975; Devaux, 1977) have shown poor control of CMV in lettuce or cucumber crops. Indeed, loss of yield and marketable quality has been attributed to damage by oil sprays (Nawrocka et al., 1975). The reasons for such variable results are not known and may in part be due to the use of different oils, or different conditions during application. For virus control the oils should be rich in unsulfonated residues, of suitable viscosity, and should be applied under "precise" conditions (Zitter, 1980). However, much of the work on which these ideas are based has not been published.

Where appropriate for production, plastic mulches have proved to be effective in reducing virus incidence at harvest, or reducing the amount of virus spread early in crop growth, and thereby serious losses. Both black and reflective mulches have been found to reduce aphid numbers and incidence of CMV in lettuce from 25% to 40%. However, in the absence of aphicides to control direct damage by aphids, mulches may give an actual loss in yield of marketable produce (Nawrocka et al., 1975). Aluminum foil has been used to reduce CMV infections in gladiolus (Bing and Johnson, 1971).

The plastic mulches used to control nonpersistent viruses function by repelling aphids. However, yellow sticky plastic sheets have been used to attract aphids and remove them from the crop. This has proved effective for the control of CMV in pepper (Cohen and Marco, 1973) and AMV in seed potatoes (Zimmerman-Gries, 1979).

3. Use of Cultivars Resistant or Tolerant to Infection

It is rare, in fruit trees at least, to find a cultivar which shows total resistance to virus infection (Welsh, 1976). Tolerance to infection is more common, and can be used in some instances to control the effects of disease on crop yield. Extensive testing of 122 Prunus species, representing all botanical sections of the genus, showed that all were susceptible to infection by PNRSV (Gilmer, 1955). Varying degrees of tolerance in Prunus species have been recorded (Gilmer, 1955), but the usefulness of this tolerance in breeding programs is of dubious benefit. Tolerant cultivars could still act as reservoirs of infective pollen to more sensitive cultivars in an orchard. The possibility of incorporating tolerance into all stone-fruit cultivars is one solution. However, complete elimination of the virus from the cultivars is a simpler and more sensible alternative.

The feasibility of protecting fruit trees against the effects of severe virus strains by inoculation with mild strains has also been investigated. Wood et al. (1975) found that yields of Jonathan apple trees infected with a severe strain of ApMV were almost doubled when these trees were topworked with scionwood containing a mild strain of ApMV, and symptoms were alleviated. However, the yields obtained were still substantially lower than yields from mosaic-free trees, suggesting that replanting with healthy trees would be more economical in the long term. Mild-strain protection of cherries against the cherry rugose mosaic strain of PNRSV may be a future possibility. Mink (1983) reported a form of PNRSV which produces no detectable effect on either tree vigor or fruit quality. Circumstantial evidence suggests that this symptomless form may provide protection against later infection of trees by severe strains of the virus.

The limited control of aphid-borne viruses by insecticides and by the several methods described above, emphasizes the need to breed for resistance against those viruses that cannot be controlled using certified planting material grown in isolation from infected reservoirs. Sources of immunity as opposed to tolerance are often hard to find but are available for important crops such as lettuce and the cucurbits. However, tolerance should not be belittled, especially if virus titers in tolerant plants are much lower than in "susceptible" plants. Low titers may be expected to provide plants which are less effective sources of virus resulting in slower growth of epiphytotics, with fewer severely affected plants. Such resistance is known in melon (Karchi et al., 1975) although epidemiological studies have not been done. Also, studies on the transmission of AMV by Acyrthosiphon pisum have shown that virus titer in the plant correlated with vector efficiency in the laboratory and in the field (Matisova, 1971a,b).

An important additional form of resistance has been reported for CMV spread by A. gossypii in melon (Lecoq et al., 1980; Pitrat and Lecoq, 1980, 1982). A gene has been found (rat) which renders the plant unacceptable to A. gossypii in such a way that virus transmission does not occur. The gene has no effect on the transmission of CMV by A. citricola, A. fabae, or A. craccivora. In one study (Pitrat and Lecoq, 1980), 22 cultivars or collections of Cucumis melo were investigated for susceptibility to CMV transmission by A. gossypii and for preference by the aphid species. Sixteen of the lines showed absolute correlation between resistance to inoculation by aphids and infestation. Of the other 6 lines involving 120 plants, only 14 plants were resistant, but preferred by aphids. A similar phenomenon has been found in plum, for transmission of the Potyvirus, plum pox virus, by M. persicae but not by Brevicaudus helicrysi (Maison and Massonié, 1982), and it may be more widespread than at present known. Resistance of this kind could be especially useful if a particular aphid species is found to be responsible for most spread, for it overcomes the common problem of virus-strain-dependent resistance. However, Annis et al. (1982) suggested that M. persicae may move between plants more often on nonpreferred than on preferred host, a trait not shown by A. gossypii.

VI. DISCUSSION

The epidemiology of plant viruses has not received the attention it warrants, and there have in recent years been few new ideas of value to disease control which can be applied on a field scale. Oils have not proved to be the panacea they first promised to be, although this may in part be due to the problems of defining an effective oil (Zitter, 1980). By far the most valuable control method has been the development of virus-tested stock schemes.

Pathogen-tested stock schemes have proved especially valuable for the control of the Ilarviruses, largely because of their narrow host ranges outside commercial crops, their slow rates of spread, and the short distances over which natural spread occurs. Aphid-borne viruses with few hosts common near commercial crops, such as LMV and many of the potato viruses, can be likewise controlled. For example, the use of certified lettuce seed has resulted in excellent control of the seed-borne, nonpersistently aphid-borne Potyvirus LMV, whereas CMV and AMV are both similarly aphid-borne but have wide host ranges and cannot be easily controlled. However, even CMV was effectively controlled in greenhouse crops where isolation and weed control were effective (Provvidenti, 1975). Clearly in some environments where weed hosts abound, control of CMV is much more difficult (Mali and Rajegore, 1980).

The spectre of long-distance spread is perhaps the biggest inhibitor of new attempts to control nonpersistent aphid-borne viruses by eradication, or by disease-free stock schemes; that is, spread from sources more than several kilometers from the crop. Certainly aphids can move such distances. However, as a rule, this is not important for nonpersistent viruses unless both vector and virus have hosts which are common near or within the crop; to provide either a nearby reservoir of virus and increase to large numbers the encounters between virus and vector; or to increase the chance that the rare aphid from afar carrying virus can inoculate a susceptible host.

Good epidemiological studies, designed to either allow modeling of the system, or the identification of the most important vectors or reservoirs, form the best basis of disease control programs. Thus, the certainty that local reservoirs do not exist, allows eradication programs or virus-free stock schemes to be initiated with confidence. The clear understanding of the mechanism of spread of CMV in banana (Allen, 1983) was gained by modeling the method of virus spread which demonstrated that eradication was possible. The information has been used to eradicate the virus from some areas in New South Wales. Studies on the vector species actually spreading virus in the field may be coupled with breeding programs for resistance to virus transmission by a particular aphid species. The identification of specific strains of virus spreading in the field may help the evaluation of breeding programs, while the location of the active reservoirs may provide sufficient information to initiate sanitation programs.

The use of certified stocks, whether internationally in the form of plant quarantine, or nationally in the form of elite stock schemes, still provides the best basis for disease control. Indeed, despite the movement of aphids throughout Europe (sic), it was not until plant quarantine was relaxed, that the plum pox virus (Potyvirus) became a serious problem in Europe and, eventually, Great Britain.

REFERENCES

A'Brook, J., 1968, The effects of plant spacing on the numbers of aphids trapped over the groundnut crop, *Ann. Appl. Biol.* **61**:289.

A'Brook, J., 1973, The effect of plant spacing on the number of aphids trapped over cocksfoot and kale crops, *Ann. Appl. Biol.* **74**:279.

Aichele, M. D., 1983, Testing for virus in registered seed and scion orchard blocks in Washington, *Acta Hortic.* **130**:267.

Allen, R. N., 1983, Spread of banana bunchy top and other plant virus diseases in time and space, in: *Plant Virus Epidemiology, the Spread and Control of Insect-Borne Viruses* (R. T. Plumb and J. M. Thresh, eds.), pp. 51–59, Blackwell, Oxford.

Annis, B., Berry, R. E., and Tamaki, G., 1982, Host preferences of the green peach aphid, *Myzus persicae* (Hemiptera:Aphididae), *Environ. Entomol.* **11**:824.

Asatani, M., 1972, Freeing chrysanthemums from the rod-shaped leaf mottling viruses and tomato aspermy virus by a combination of meristem tip culture with heat-treatment, *Ber. Ohara Inst. Landwirtsch. Biol. Okayama Univ.* **15**:169.

Barbara, D. J., 1980, Detecting prunus necrotic ringspot virus in Rosaceous hosts by enzyme-linked immunosorbent assay, *Acta Phytopathol. Acad. Sci. Hung.* **15**:329.

Barbara, D. J., Clark, M. F., Thresh, J. M., and Casper, R., 1978, Rapid detection and serotyping of prunus necrotic ringspot virus in perennial crops by enzyme-linked immunosorbent assay, *Ann. Appl. Biol.* **90**:395.

Baumann, G., Converse, R. H., and Casper, R., 1982, The occurrence of apple mosaic virus in red and black raspberry and in blackberry cultivars, *Acta Hortic.* **129**:13.

Bennett, C. W., 1969, Seed transmission of plant viruses, *Adv. Virus Res.* **14**:221.

Bing, A., and Johnson, G. V., 1971, Aluminium foil mulch reduces aphid transmission of cucumber mosaic virus in gladiolus, *Acta Hortic.* **2**:286.

Bock, K. R., 1967, Strains of prunus necrotic ringspot virus in hop (*Humulus lupulus L.*), *Ann. Appl. Biol.* **59**:437.

Borges, M. de L. V., and Louro, P., 1974, A biting insect as vector of broad bean mottle virus?, *Agron. Lusit.* **36**:215.

Bradley, R. H. E., 1956, Effect of depth of stylet penetration on aphid transmission of potato virus Y, *J. Microbiol.* **2**:539.

Bradley, R. H. E., Wade, C. V., and Wood, F. A., 1962, Aphid transmission of potato virus Y inhibited by oils, *Virology* **18**:327.

Broadbent, L., 1950, The correlation of aphid numbers with the spread of leaf roll and rugose mosaic in potato crops, *Ann. Appl. Biol.* **37**:58.

Brunt, A. A., 1969, Dahlia viruses, *Glasshouse Crops Res. Inst. Annu. Rep.* 1968, p. 104.

Brunt, A. A., 1969, Dahlia, *Glasshouse Crops Res. Inst. Annu. Rep.* p. 129.

Burns, M., 1972, Effect of flight on the production of alatae by the vetch aphid (*Megoura viciae*), *Entomol. Exp. Appl.* **15**:319.

Campbell, A. I., 1962, Apple virus inactivation by heat therapy and tip propagation, *Nature* (*London*) **195**:520.

Cardin, L., and Marénaud, C., 1975, Relations entre le virus de la mosaïque du noisetier et le tulare apple mosaic virus, *Ann. Phytopathol.* **7**:159.

Casper, R., 1973, Serological properties of prunus necrotic ringspot and apple mosaic virus isolates from rose, *Phytopathology* **62**:238.

Cation, D., 1949, Transmission of cherry yellows virus complex through seed, *Phytopathology* **39**:37.

Cation, D., 1961, A determination of necrotic ringspot virus in Michigan peach orchards and nursery stock, *Plant Dis. Rep.* **45**:109.

Cation, D., 1967, Maintaining virus-free cultivars of peach in Michigan, *Plant Dis. Rep.* **51**:261.

Cheplick, S. M., and Agrios, G. N., 1983, Effect of injected antiviral compounds on apple mosaic, scar skin and dapple apple diseases of apple trees, *Plant Dis.* **67**:1130.

Cochran, L. C., 1944, The "complex concept" of the peach mosaic and certain other stone fruit viruses, *Phytopathology* **34**:934.

Cochran, L. C., 1946, Passage of ringspot virus through Mazzard cherry seed, *Science* **104**:269.

Cohen, S., and Marco, S., 1973, Reducing the spread of aphid-transmitted viruses in peppers by trapping the aphids on sticky yellow polythene sheets, *Phytopathology* **63**:1207.

Cole, A., Mink, G. I., and Regev, S., 1982, Location of prunus necrotic ringspot virus on pollen grains from infected almond and cherry trees, *Phytopathology* **72**:1542.

Converse, R. H., 1976, Serological detection of viruses in *Rubus* sap, *Acta Hortic.* **66**:53.

Converse, R. H., 1980, Transmission of tobacco streak virus in *Rubus*, *Acta Hortic.* **95**:53.

Converse, R. H., and Bartlett, A. B., 1979, Occurrence of viruses in some wild *Rubus* and *Rosa* species in Oregon, *Plant Dis. Rep.* **63**:441.

Converse, R. H., and Lister, R. H., 1969, The occurrence and some properties of black raspberry latent virus, *Phytopathology* **59**:325.

Costa, A. S., and Lima Neto, V. da C., 1976, Transmissao do virus da necrose branca do fumo por *Frankliniella* sp., *IX Congr. Soc. Bras. Fitopatol.*

Davidson, T. R., 1976, Field spread of prunus necrotic ringspot in sour cherries in Ontario, *Plant Dis. Rep.* **60**:1080.

Davidson, T. R., and George, J. A., 1964, Spread of necrotic ringspot and sour cherry yellows viruses in Niagara Peninsula orchards, *Can. J. Plant Sci.* **44**:471.

Demski, J. W., and Boyle, J. S., 1968, Spread of necrotic ringspot virus in a sour cherry orchard, *Plant Dis. Rep.* **52**:972.

De Sequeira, O. A., 1967, Purification and serology of an apple mosaic virus, *Virology* **31**:314.

Devaux, A., 1977, Essais de moyens de lutte contre la mosaïque du concombre an Québec, *Phytoprotection* **58**:18.

Devergne, J. C., Cardin, L., and Quiot, J. B., 1978, Contribution à l'étude du virus de la mosaïque du concombre. IV. - Essai de classification de plusieurs isolats sur la base de leur structure antigénique, *Ann. Phytopathol.* **5**:409.

de Wijs, J. J., 1974, The correlation between the transmission of passionfruit ringspot virus and populations of flying aphids, *Neth. J. Plant Pathol.* **80**:133.

Dixon, A. F. G., Burns, M. D., and Wangboon Kong, S., 1968, Migration in aphids: Response to current adversity, *Nature* (*London*) **220**:1337.

Fischer, H., and Baumann, G., 1980, Selection of virus-free almond clones and investigations on prunus viruses in Morocco, *Acta Phytopathol. Acad. Sci. Hung.* **15**:191.

Fleisher, Z., Drori, T., and Loebenstein, G., 1971, Evaluation of Shirofugen as a reliable indicator for rose mosaic virus, *Plant Dis. Rep.* **55**:431.

Free, J. B., 1960, The pollination of fruit trees, *Bee World* **41**:141.

Free, J. B., 1970, *Insect Pollination of Crops*, Academic Press, New York.

Fridlund, P. R., 1976, IR-2, the interregional deciduous tree fruit repository, in: *Virus Diseases and Noninfectious Disorders of Stone Fruits in North America*, pp. 16–22, USDA, Agric. Handbook No. 435, Washington, D.C.

Fulton, R. W., 1964, Transmission of plant viruses by grafting, dodder, seed and mechanical inoculation, in: *Plant Virology* (M. K. Corbett and H. D. Sisler, eds.), pp. 39–67, University of Florida Press, Gainesville.

Fulton, R. W., 1967, Purification and serology of rose mosaic virus, *Phytopathology* **57**:1197.

Fulton, R. W., 1968, Serology of viruses causing cherry necrotic ringspot, plum line pattern, rose mosaic and apple mosaic, *Phytopathology* **58**:635.

Fulton, R. W., 1970a, Prunus necrotic ringspot virus, *CMI/AAB Descriptions of Plant Viruses* No. 5.

Fulton, R. W., 1970b, A disease of rose caused by tobacco streak virus, *Plant Dis. Rep.* **54**:949.

Fulton, R. W., 1971, Tobacco streak virus, *CMI/AAB Descriptions of Plant Viruses* No. 44.

Fulton, R. W., 1972, Apple mosaic virus, *CMI/AAB Descriptions of Plant Viruses* No. 83.

George, J. A., and Davidson, T. R., 1964, Further evidence of pollen transmission of necrotic ring spot and sour cherry yellows viruses in sour cherry, *Can. J. Plant Sci.* **44**:383.

Gerginova, T., 1980, Incidence of prunus necrotic ringspot virus and prune dwarf virus in cherry orchards, *Acta Phytopathol. Acad. Sci. Hung.* **151**:223.

Ghanekar, A. M., and Schwenk, F. W., 1974, Seed transmission and distribution of tobacco streak virus in six cultures of soybeans, *Phytopathology* **64**:112.

Gibbs, A. J., ed., 1977, *Viruses and Invertebrates*, North-Holland, Amsterdam.

Gibbs, A. J., 1980, A plant virus that partially protects its wild legume host against herbivores, *Intervirology* **13**:42.

Gilmer, R. M., 1955, Host range and variable pathogenesis of the necrotic ringspot virus in the genus *Prunus*, *Plant Dis. Rep.* **39**:194.

Gilmer, R. M., 1956, Probable coidentity of Shiro line pattern virus and apple mosaic virus, *Phytopathology* **46**:127.

Gilmer, R. M., and Brase, K. D., 1963, Non-uniform distribution of prune dwarf virus in sweet and sour cherry trees, *Phytopathology* **53**:819.

Gilmer, R. M., and Way, R. D., 1960, Pollen transmission of necrotic ringspot and prune dwarf virus in sour cherry, *Phytopathology* **50**:624.

Halbert, S. E., and Irwin, M. E., 1981, Effect of soybean canopy closure on landing rates of aphids with implications for restricting spread of soybean mosaic virus, *Ann. Appl. Biol.* **98**:15.

Hampton, R. O., 1963, Rate and pattern of prune dwarf virus movement within inoculated cherry trees, *Phytopathology* **53**:998.

Hampton, R. O., 1966, Probabilities of failing to detect prune dwarf virus in cherry trees by bud indexing, *Phytopathology* **56**:650.

Hansen, A. J., and Green, L., 1983, Potential of ribavirin for tree fruit virus inhibition, *Acta Hortic.* **130**:183.

Hardy, A. C., and Milne, P. S., 1937, Insect drift over the North Seas, 1936, *Nature (London)* **139**:510.

Harris, K. F., and Maramorosch, K., eds., 1977, *Aphids as Virus Vectors*, Academic Press, New York.

Havranek, P., and Laska, P., 1972, The course of the migration of cucumber mosaic virus in a vegetable locality, *Sci. Agric. Bohemoslov.* **4**:213.

Hemmati, K., and McLean, D. L., 1977, Gamete-seed transmission of alfalfa mosaic virus and its effect on seed germination and yield of alfalfa plants, *Phytopathology* **67**:576.

Hollings, M., and Stone, O. M., 1964, Investigations of carnation viruses. I. Carnation mottle, *Ann. Appl. Biol.* **53**:103.

Hootman, H. D., and Cale, G. H., 1930, A practical consideration of fruit pollination, *Am. Bee J. Bull.* 579.

Horváth, J., 1979, New artificial hosts and non-hosts of plant viruses and their role in the identification and separation of viruses. X. Cucumovirus group: Cucumber mosaic virus, *Acta Phytopathol. Acad. Sci. Hung.* **14**:285.

Horváth, J., 1980, Viruses of lettuce. II. Host ranges of lettuce mosaic virus and cucumber mosaic virus, *Acta Agron. Acad. Sci. Hung.* **29**:333.

Horváth, J., 1981, New artificial hosts and non hosts of plant viruses and their role in the identification and separation of viruses. XV. Monotypic (almovirus) group: Alfalfa mosaic virus, *Acta Phytopathol. Acad. Sci. Hung.* **16**:315.

Hunter, J. A., Chamberlain, E. E., and Atkinson, J. D., 1958, Note on transmission of apple mosaic by natural root grafting, *N.Z. J. Agric. Res.* **1**:80.

Ignash, Y., 1977, Ogranichenie rasprostraneniya virusa aspermii tomatov v Latviiskoi SSR, *Tr. Latv. Skh. Akad.* **153**:20.

Ikin, R., and Frost, R. R., 1974, Virus diseases of roses, *Phytopathol. Z.* **79**:160.

Johnson, B., 1966, Wing polymorphism in aphids. III. The influence of host plant, *Entomol. Exp. Appl.* **9**:212.

Johnson, C. G., 1969, *Migration and Dispersal of Insects by Flight*, Methuen, London.

Johnson, C. G., and Taylor, L. R., 1957, Periodism and energy summation with special reference to flight rhythms in aphids, *J. Exp. Biol.* **34**:209.

Kaiser, W. J., Wyatt, S. D., and Pesho, G. R., 1982, Natural hosts and vectors of tobacco streak virus in eastern Washington, *Phytopathology* **72**:1508.

Karchi, Z., Cohen, S., and Govers, A., 1975, Inheritance of resistance to cucumber mosaic virus in melons, *Phytopathology* **65**:479.

Kassanis, B., and Posnette, A. F., 1961, Thermotherapy of virus infected plants, *Recent Adv. Bot.* **1**:557.

Kazda, V., and Hervert, V., 1977, Epidemiology of cucumber mosaic virus of glasshouse cucumber, *Ochr. Rostlin* **13**:169.

Kelley, R. D., and Cameron, H. R., 1983, Location of prune dwarf and prunus necrotic ringspot viruses in sweet cherry pollen and fruit, *Phytopathology* **75**:791.

Kennedy, J. S., and Ludlow, A. R., 1974, Co-ordination of two kinds of flight activity in an aphid, *J. Exp. Biol.* **61**:173.

Kennedy, J. S., and Thomas, A. A. G., 1974, Behaviour of some low-flying aphids in wind, *Ann. Appl. Biol.* **76**:143.

Kidd, N. A. C., 1977, The influence of population density on the flight behaviour of the lime aphid, *Eucallipterus tiliae*, *Entomol. Exp. Appl.* **22**:251.

Klos, E. K., and Parker, K. G., 1960, Yields of sour cherry affected by ringspot and yellows viruses, *Phytopathology* **50**:412.

Langridge, D. F., 1969, Effects of temperature, humidity and caging on the concentration of fruit pollen in the air, *Aust. J. Exp. Agric. Anim. Husb.* **9**:549.

Lazar, A. C., and Fridlund, P. R., 1967, The incidence of latent viruses in Washington peach orchards, *Plant Dis. Rep.* **51**:1063.

Lecoq, H., Labonne, G., and Pitrat, M., 1980, Specificity of resistance to virus transmission by aphids in *Cucumis melo*, *Ann. Phytopathol.* **12**:139.

Lees, A. D., 1964, The location of the photoperiodic receptors in the aphid *Megoura viciae* Buckton, *J. Exp. Biol.* **41**:119.

Lindberg, G. D., Hall, D. H., and Walker, J. C., 1956, A study of melon and squash mosaic viruses, *Phytopathology* **46**:489.

Lockhart, B. E. L., and Fischer, H. V., 1976, Cucumber mosaic virus infection of pepper in Morocco, *Plant Dis. Rep.* **60**:262.

Loebenstein, G., Alper, M., and Deutsch, M., 1964, Preventing aphid-spread cucumber mosaic virus with oil, *Phytopathology* **54**:960.

Loebenstein, G., Alper, M., and Levy, S., 1970, Field tests with oil sprays for the prevention of aphid-spread viruses in peppers, *Phytopathology* **60**:212.

Luckwill, L. C., 1954, Virus diseases of fruit trees. IV. Further observations on rubbery wood, chat-fruit and mosaic in apples, *Agric. Hortic. Res. Stn. Annu. Rep.* 1953, pp. 40–46.

MacKay, P. A., and Lamb, R. J., 1979, Migratory tendency in aging populations of the pea aphid, *Acyrthosiphon pisum*, *Oecologia (Berlin)* **39**:301.

MacKay, P. A., and Wellington, W. G., 1977, Maternal age as a source of variation in the ability of an aphid to produce dispersing forms, *Res. Popul. Ecol.* **18**:195.

Maison, P., and Massonié, G., 1982, Premières observations sur la spécificité de la resistance du pêcher à la transmission aphidienne du virus de la Sharka, *Agronomie* **2**:681.

Mali, V. R., and Rajegore, S. B., 1980, A cucumber mosaic virus disease of banana in India, *Phytopathol. Z.* **98**:127.

Marénaud, C., and Saunier, R., 1974, Action des viruses du type Ilar sur le pollen de l'espèce *Prunus persica*, *Ann. Amelior. Plant.* **24**:169.

Marrou, J., Quiot, J. B., Duteil, M., Labonne, G., Leclant, F., and Renoust, M., 1979a, Ecologie et épidémiologie du virus de la mosaïque du concombre dans le sud-est de la France. III - Intérêt de l'exposition de plantes - appâts pour l'étude de la dissémination du virus de la mosaïque du concombre, *Ann. Phytopathol.* **11**:291.

Marrou, J., Quiot, J. B., Duteil, M., Labonne, G., Leclant, F., and Renoust, M., 1979b, Ecologie et épidémiologie du virus de la mosaïque de concombre dans le sud-est de la France. VIII—Influence des brise-vent et de la végétation environnante sur la dissémination du virus de la mosaïque du concombre, *Ann. Phytopathol.* **11**:375.

Matisova, J., 1971a, Alfalfa mosaic virus in lucerne plants and its transmission by aphids in the course of the vegetation period, *Acta Virol.* **15**:411.

Matisova, J., 1971b, Efficiency of alfalfa mosaic virus transmission by aphids in relation to the level of virus in plants, *Acta Virol.* **15**:425.

Mellor, F. C., and Stace-Amith, R., 1977, Virus-free potatoes by tissue culture, in: *Plant Cell Tissue and Organ Culture* (J. Reinert and Y. P. Bajaj, eds.), pp. 616–635, Springer-Verlag, Berlin.

Milbrath, G. M., and Tolin, S. A., 1977, Identification, host range, and serology of peanut stunt virus isolated from soybean, *Plant Dis. Rep.* **61**:637.

Milbrath, J. A., 1961, The relationship of stone fruit ringspot virus to sour cherry yellows, prune dwarf and peach stunt, *Tidsskr. Planteavl* **65**:125.

Mink, G. I., 1980, Identification of rugose mosaic diseased cherry trees by enzyme-linked immunosorbent assay, *Plant Dis. Rep.* **64**:691.

Mink, G. I., 1983, The current status of cherry virus research in Washington, Proceedings of the Washington State Horticultural Association, 1982, 78th Annual Meeting, Yakima, pp. 184–186.

Mittler, T. E., and Dadd, R. H., 1966, Food and wing determination in *Myzus persicae* (Homoptera: Aphididae), *Ann. Entomol. Soc. Am.* **59**:1162.

Moericke, V., 1950, Über das Farbsehen der Pfirsichblattlaus (*Myzodes persicae* Sulz), *Z. Tierpsychol.* **7**:265.

Moericke, V., 1957, Der flug von insekten über pflanzenfreien und pflanzenbewachsenen Flachen, *Z. Pflanzenkr. Pflanzenschutz* **64**:507.

Moran, J. R., Garrett, R. G., and Fairweather, J. V., 1983, A sampling strategy for detecting low levels of virus in crops and its application to surveys of the Victorian Certified Seed Potato Scheme to detect potato viruses X and S, *Plant Dis.* **67**:1325.

Mori, K., 1977, Localisation of viruses in apical meristems and production of virus-free plants by means of meristems and tissue culture, *Acta Hortic.* **78**:389.

Müller, H. J., and Unger, K., 1951, Über die Ursachen der unterschiedlichen resistenz von *Vicia faba* L. gegenüber der bohnenblattlaus Doralis fabae scop. II. Über die Fluggewohnheiten besonders das sommerliche schwärmen, von Doralis fabae und ihre abhangigkeit vom Tagesgang der witterungsfaktoren, *Zuechter* **21**:76.

Nawrocka, B. Z., Eckenrode, C. J., Vyemoto, J. K., and Young, D. H., 1975, Reflective mulches and foliar sprays for suppression of aphid-borne viruses in lettuce, *J. Econ. Entomol.* **68**:694.

Nevkryta, A. N., 1957, Distribution of apiaries for pollinating cherries, *Pchelovodstvo* **34**:34.

Nyéki, J. and Vértesy, J., 1974, Effect of different ringspot viruses on the physiological and morphological properties of Montmorency sour cherry pollen II, *Acta Phytopath. Acad. Sci. Hung.* **9**:23.

Nyland, G., 1960, Heat inactivation of stone fruit ringspot virus, *Phytopathology* **50**:380.

Nyland, G., and Goheen, A. C., 1969, Heat therapy of virus diseases of perennial plants, *Ann. Rev. Phytopath.* **7**:331.

Nyland, G., Gilmer, R. M., and Moore, J. D., 1976, Prunus ringspot group, in *Virus Diseases and Noninfectious Disorders of Stone Fruits in North America*, pp. 104. U.S. Dep. Agric., Agric. Handbook No. 437, Washington, D.C.

Panarin, I. V., 1977, Cereal flea beetles as vectors of Hungarian brome mosaic virus, *Sb. Nauchn. Tr. Krashodar NII Skh.* **13**:158.

Pitrat, M., and Lecoq, H., 1980, Inheritance of resistance to cucumber mosaic virus transmission by *Aphis gossypii* in *Cucumis melo*, *Phytopathology* **70**:958.

Pitrat, M., and Lecoq, H., 1982, Relations génétiques entre les résistances par non-acceptation et par antibiose du melon a *Aphis gossypii*: Recherche de liaisons avec d'autres gènes, *Agronomie* **2**:503.

Posnette, A. F., and Cropley, R., 1956, Apple mosaic virus: Host reactions and strain interference, *J. Hortic. Sci.* **31**:119.

Proeseler, G., 1978, Transmission of brome mosaic virus by cereal beetles (Coleoptera, Chrysomelidae), *Arch. Phytopathol. Pflanzenschutz* **14**:267.

Provvidenti, R., 1975, Natural infection of *Celosia argentea* by cucumber mosaic virus in a commercial greenhouse, *Plant Dis. Rep.* **59**:166.

Provvidenti, R., 1976, Reaction of *Phaseolus* and *Macroptilium* species to a strain of cucumber mosaic virus, *Plant Dis. Rep.* **60**:289.

Provvidenti, R., Robinson, R. W., and Shail, J. W., 1980, A source of resistance to a strain of cucumber mosaic virus in *Lactuca saligna* L., *Hortic. Sci.* **15**:528.

Purivirojkul, W., Sittiyos, P., Hsu, C. H., Poehlman, J. M., and Sehgal, O. P., 1978, Natural infection of mungbean (*Vigna radiata*) with cucumber mosaic virus, *Plant Dis. Rep.* **62**:530.

Quiot, J. B., Douine, L., Marchoux, G., and Devergne, J. C., 1976, Ecologie du virus de la mosaïque du concombre dans le sud-est de la France. I. - Recherche des plantes spontanées hôtes du CMV et caractérisation des populations virales, *Agric. Conspec. Sci.* **39**:533.

Quiot, J. B., Marrou, J., Labonne, G., and Verbrugghe, M., 1979a, Ecologie et épidémiologie du virus de la mosaïque du concombre dans le sud-est de la France: Description du dispositif expérimental, *Ann. Phytopathol.* **11**:265.

Quiot, J. B., Verbrugghe, M., Labonne, G., Leclant, F., and Marrou, J., 1979b, Ecologie et épidémiologie du virus de la mosaïque du concombre dans le sud-est de la France. IV — Influence des brise-vent sur la répartition des contaminations virales dans une culture protégée, *Ann. Phytopathol.* **11**:307.

Quiot, J. B., Marchoux, G., Douine, L., and Vigouroux, A., 1979c, Ecologie et épidémiologie du virus de la mosaïque du concombre dans le sud-est de la France. V. - Rôle des espèces spontanées dans la conservation du virus, *Ann. Phytopathol.* **11**:325.

Quiot, J. B., Devergne, J. C., Marchoux, G., Cardin, L., and Douine, L., 1979d, Ecologie et épidémiologie du virus de la mosaïque du concombre dans le sud-est de la France. VI. - Conservation de deux types de populations virales dans les plantes sauvages, *Ann. Phytopathol.* **11**:349.

Quiot, J. B., Devergne, J. C., Cardin, L., Verbrugghe, M., Marchoux, G., and Labonne, G., 1979e, Ecologie et épidémiologie du virus de la mosaïque du concombre dans le sud-est de la France. VII - Répartition de deux types de populations virales dans des cultures sensibles, *Ann. Phytopathol.* **11**:359.

Quiot, J. B., Leroux, J. P., Labonne, G., and Renoust, M., 1979f, Epidémiologie de la maladie filiforme et de la nécrose de la tomate provoquées par le virus de la mosaïque du concombre dans le sud-est de la France, *Ann. Phytopathol.* **11**:393.

Schmelzer, K., 1969, The elm mottle virus, *Phytopathol. Z.* **64**:39.

Shepherd, R. J., 1972, Transmission of virus through seed and pollen, in: *Principles and Techniques in Plant Pathology* (C. I. Kado and H. A. Agarwal, eds.), pp. 267–292, Van Nostrand–Reinhold, Princeton, N.J.

Smith, P. R., 1983, The Australian Fruit Variety Foundation and its role in supplying virus-tested planting material to the fruit industry, *Acta Hortic.* **130**:263.

Smith, P. R., and Stubbs, L. L., 1976, Transmission of prune dwarf virus by peach pollen and latent infection in peach trees, *Aust. J. Agric. Res.* **27**:839.

Smith, P. R., Stubbs, L. L., and Challen, D. I., 1977a, Studies on the epidemiology of peach rosette and decline disease in Victoria, *Aust. J. Agric. Res.* **28**:103.

Smith, P. R., Stubbs, L. L., and Challen, D. I., 1977b, The detection of prune dwarf virus in peach trees affected with peach rosette and decline with Golden Queen peach as an indicator and the distribution of the virus in affected trees, *Aust. J. Agric. Res.* **28**:115.

Stubbs, L. L., and Smith, P. R., 1971, The association of prunus necrotic ringspot, prune dwarf and dark green sunken mottle viruses in the rosetting and decline disease of peach, *Aust. J. Agric. Res.* **22**:771.

Sweet, J. B., 1980, Fruit tree virus infection of ornamental Rosaceous trees and shrubs, *J. Hortic. Sci.* **55**:103.

Sweet, J. B., and Barbara, D. J., 1979, A yellow mosaic disease of horse chestnut (*Aesculus* spp.) caused by apple mosaic virus, *Ann. Appl. Biol.* **92**:335.

Taylor, L. R., 1963, Analysis of the effect of temperature on insects in flight, *J. Anim. Ecol.* **32**:99.

Thomas, W. D., and Graham, R. W., 1951, Seed transmission of red node in pinto beans, *Phytopathology* **41**:959.

Thomsen, A., 1961, Termoterapeutiske behandlinger af nellike, *Horticultura* **15**:136.

Tolin, S. A., and Miller, J. D., 1975, Peanut stunt virus in crown vetch, *Phytopathology* **65**:321.

Tomlinson, J. A., and Walker, V. M., 1973, Further studies on seed transmission in the ecology of some aphid-transmitted viruses, *Ann. Appl. Biol.* **73**:293.

Van der Meer, F. A., and Huttinga, H., 1979, Lilac ring mottle virus, *CMI/AAB Descriptions of Plant Viruses* No. 201.

Van Emden, H. F., ed., 1972, *Aphid Technology*, Academic Press, New York.

Vértesy, J., 1976, Embryological studies of Ilar-virus infected cherry seeds, *Acta Hortic.* **67**:245.

Watson, M. A., and Roberts, F. M., 1939, A comparative study of the transmission of Hyoscyanus virus 3, potato virus Y and cucumber virus 1 by the vectors *Myzus persicae* (Sulz.), *M. circumflexus* (Buckton) and *Macrosiphon gei* (Koch), *Proc. R. Soc. London Ser. B* **127**:543.

Welsh, M. F., 1976, Control of stone fruit virus diseases, in: *Virus Diseases and Noninfectious Disorders of Stone Fruits in North America*, pp. 10, USDA, Agric. Handbook No. 437, Washington, D.C.

Willison, R. S., Berkeley, G. H., and Chamberlain, G. C., 1948, Yellows and necrotic ringspot of sour cherries in Ontario—distribution and spread, *Phytopathology* **38**:776.

Wood, G. A., Chamberlain, E. E., Atkinson, J. D., and Hunter, J. A., 1975, Field studies with apple mosaic virus, *N.Z. J. Agric. Res.* **18**:399.

Wyatt, S. D., and Kuhn, C. W., 1980, Derivation of a new strain of cowpea chlorotic mottle virus from resistant cowpeas, *J. Gen. Virol.* **49**:289.

Yarwood, C. E., 1955, Mechanical transmission of an apple mosaic virus, *Hilgardia* **23**:613.

Zimmerman-Gries, S., 1979, Reducing the spread of potato leaf roll virus, alfalfa mosaic virus and potato virus Y in seed potatoes by trapping aphids on sticky yellow polyethylene sheets, *Potato Res.* **22**:123.

Zitter, T. A., 1980, Management of viruses by alteration of vector efficiency and by cultural practices, *Annu. Rev. Phytopathol.* **18**:289.

Index